ATLAS

of the

BRITISH FLORA

Edited by

F. H. PERRING and S. M. WALTERS

Published by the

BOTANICAL SOCIETY OF THE BRITISH ISLES

1990

First edition originally published by Thomas Nelson & Sons Ltd
Second impression 1963
Second edition 1976
Second impression 1977
Third impression 1979
Third edition 1982
Reprinted 1990 1992
©

BOTANICAL SOCIETY OF THE BRITISH ISLES
1962, 1976, 1982, 1990

ISBN 0 90115 819 4

Printed in Great Britain by Redwood Press Limited,
Melksham, Wiltshire

Contents

Preface

THIS REPRINT OF THE third edition of the *Atlas* is designed to satisfy a continuing demand from a wide range of professional and amateur biologists whose work involves the wild flora of the British Isles. The stimulus to undertake the reprint arose from a joint Conference of the British Ecological Society and the Botanical Society of the British Isles held in Exeter in April 1989, under the title 'Species Mapping and the Biology of Plant Distribution', at which S.M.W. gave the opening address on the subject of the *Atlas of the British Flora*. Two strong impressions remained from this occasion. Firstly, the *Atlas* continues to be greatly valued as a basic document in a very wide range of studies, not least in those increasingly professional areas concerning nature conservation and the care of the environment. Secondly, and ironically, most people who know the value of the *Atlas* and continually refer to its distribution maps do not themselves possess a copy. Preliminary discussion at the Conference encouraged some of us to explore the possibility of a cheap facsimile reprint at a reduced page-size to satisfy this clear demand until such time as a new *Atlas* is available, which could not be before 1995. Encouraged by this development, we submitted the project to the Botanical Society, and were pleased to have it accepted before the end of 1989.

It is a source of satisfaction to the editors that the *Atlas* has been used to monitor changes in the British flora. One in nine of the 10-kilometre squares surveyed in the 1950's were re-surveyed in 1987-8 under the auspices of the B.S.B.I. Monitoring Scheme. We await the publication of the results of this scheme with great interest.

The work here reprinted is the third and most recent edition of the *Atlas* published in 1982. We have included (as in that third edition) the original foreword provided for the first edition by the then doyen of studies of the geographical distribution of British plants, Professor J.R. Matthews, and also the detailed introduction to the first edition in which the history of mapping British plants and the origin and aims of the BSBI Distribution Maps Scheme are explained. The very short extra prefaces to both the second edition (1976) and the third edition (1982) are omitted.

Apart from our new preface, the only new material is the *Index and Bibliography* specially prepared by Chris Preston to enable users of the Atlas to check whether for a particular genus or species there is a more up-to-date map published elsewhere. The six pages of overlay maps which were supplied with the first and third editions[1] have been reproduced as three ordinary pages, so that users who wish may make their own overlays by copying these pages on to transparent paper or film, whilst everyone can buy the reprint at a reasonable price.

No attempt has been made to change the contents either in terms of nomenclature or taxonomy (eg. inclusion of 'new' species), and the statement on p. xvi of the original introduction still stands, namely: 'this volume . . . contains maps of all generally accepted native British species (excluding critical segregates) and most well-established introductions'. Some naturalised species that have become widespread since 1960 are, of course, not included. Nomenclature remains as in J.E. Dandy's

---

[1] The second edition, on grounds of cost, was provided with only one overlay page of two maps.

'*List of British Vascular Plants*' (1958), with synonyms where Clapham, Tutin and Warburg's *Flora* (ed. 1, 1952) differed from Dandy's *List*. Both in the second and the third editions, some maps were up-dated from the first edition, and in this reprint such maps are marked with the symbols + or *. An explanation of these symbols is given below. It can be assumed that any unmarked map therefore shows the distribution as we knew it at the end of the Distribution Maps Scheme in 1960, and much has happened to the British countryside to change plant distributions in the last 30 years. It is not, however, correct to assume that an updated map necessarily contains much new information: in many cases, the updating involved only the deletion of a few erroneous records. Users of the *Atlas* who need up-to-date information on the distribution of particular species are encouraged to write to Mr C.D. Preston, Biological Records Centre, Monks Wood Experimental Station, Abbots Ripton, Huntingdon, PE17 2LS, if no updated map is cited in the Bibliography.

In conclusion we would like to express our thanks to the following people whose critical interest in the project has made it possible: A.O. Chater, Mrs M.E. Briggs and M. Walpole, all Officers of the Botanical Society, and in particular Chris Preston, who not only supplied the material for the new *Bibliography*, but made valuable suggestions about the use of the third edition. Finally, we record our pleasure and satisfaction with the work of Geoff Green and Martin Walters, who have professionally undertaken the printing and production, including the cover design.

F.H. PERRING
S.M. WALTERS

January 1990

EXPLANATION OF THE SYMBOLS WHICH MAY APPEAR
BY THE SPECIES NUMBER ON THE MAPS

+   map updated for the second edition
*   map updated for the third edition
NB   This updating may involve only minimal correction of the original map in the first edition.

Foreword

A FEATURE OF MANY FLORAS, including some which have long been in general use in Britain, is a reference under the description of each species to its geographical distribution. Such references can only be in general terms, since it is not the purpose of an ordinary descriptive Flora to provide detailed distributional data about every species. Nevertheless, the need for factual information about the distribution of plants in Britain was recognised long ago by H. C. Watson, who was the first to undertake a scientific study of the subject—a study which occupied his attention for more than forty years, and culminated in the production of two major works, *Cybele Britannica* and *Topographical Botany*, a second edition of which appeared in 1883, two years after the author's death. The vice-comital system which Watson finally adopted as a method for recording the distribution of species in this country is familiar to British botanists and has long been in general use.

Watson did not deal with Ireland, but an essentially similar scheme of vice-county divisions for that country was developed by Praeger, who gives in *Irish Topographical Botany* a carefully prepared and reliable record of the distribution of plants in Ireland.

These topographical statistics both for Britain and Ireland are conveniently summarised in *The Comital Flora of the British Isles* by Druce, and allowing for corrections and additions, they have frequently served as a basis for the construction of maps to illustrate species distribution as well as for more general studies in the phytogeography of the British Isles.

Mapping on a vice-county basis, however, may and frequently does obscure important distributional features which become apparent only when localised records are used; and at a conference on the distribution of British plants held by the Botanical Society of the British Isles in 1950, it became evident that an advance on the Watsonian vice-county system was not only desirable but long over-due. To ensure a greater measure of detail and greater scientific accuracy in the production of maps of British plants, it was suggested by Professor Clapham that they should be based on the 10 km. square of the national grid as the unit area, a unit very much smaller than the average vice-county.

This proposal was welcomed and it was decided to refer the matter to a special "Maps Committee" which would consider ways and means. After making detailed investigations requiring several years to complete, an agreed scheme was officially launched in April 1954, and it requires no words of mine to emphasise the magnitude of the task to be undertaken—a task now successfully accomplished. Within the space of five full summers records have been obtained of the plants occurring in approximately 3,500 squares covering the British Isles. Tribute must be paid to the officials for their sustained efforts and enthusiasm in carrying out the plan and to the large body of workers in the field for their active co-operation. For the first time we have before us a comprehensive series of maps illustrating in detail the distribution of plants in Great Britain and Ireland—maps which I, personally, would have greatly welcomed when engaged in my own studies in British phytogeography.

The Atlas of the British Flora takes its place alongside Hulten's *Atlas över Kärlväxterna i Norden* published in 1950, and will be invaluable to botanists, professional and amateur alike, who are in any way interested in the study of our native flora. It will make a special appeal to the ecologist as well as to the taxonomist and plant-geographer, and in view of the possible significance of polyploidy in phytogeography, the cytogeneticist may turn over its pages not without interest. Examination of the maps reveals a surprising number of distribution patterns, some of which recall Watson's "Types of Distribution", but the interpretation of the range of any particular species equires an investigation of all the factors which play a part in determining its range. Marked discontinuities in distribution are exhibited by some species, the explanation of which may be found by taking into account what is known of their history in late-glacial and post-glacial times, as detailed by Godwin in *The History of the British Flora*. Present and past distributions can thus be compared, but whatever the interpretation of different distribution patterns, it is hoped that the maps themselves, whether of common or rare species, may serve as an impetus to further autecological studies of our native plants. Be that as it may, the publication of the Atlas is an important landmark in the scientific progress of British Botany.

Dinnet, Aberdeenshire
March 1961

J.R. MATTHEWS

Introduction

HISTORY OF THE MAPPING OF PLANT DISTRIBUTION

EARLY HISTORY

As long ago as 1836, Hewett Cottrell Watson, an English amateur botanist who pioneered the study of the distribution of British vascular plants, had seen the possibility of preparing a series of plant distribution maps. In an article on the construction of maps for illustrating the distribution of plants he wrote as follows:

To represent the distribution of individual forms or species let us first imagine a common geographical map, in outline, of such dimensions as would render it possible to mark every locality for any given species, by some sign, or spot of colour, covering a corresponding space on the map. This would give an exact picture of topographical distribution; but, as it would require to be made on the scale of at least a yard to the mile, it is obviously out of the question. With less precision . . . we might greatly reduce the scale by indicating all localities within certain distances of each other as single ones.

From this passage it is clear that Watson would have been a wholehearted supporter of the work which has gone on during the last eight years to produce this Atlas, and that he might have attempted such a work himself had he lived long enough and had the circumstances been more favourable. As it was, during his lifetime he gradually divided Great Britain into smaller and smaller recording units: from the six of *Outlines of the Distribution of British Plants* (1832) to the 112 vice-counties of *Topographical Botany* (1873). In 1843, in the third edition of *The Geographical Distribution of British Plants* we find what might be called "provincial distribution maps" for thirty-nine native British species, together with the data on which the maps are based. At this period Watson had divided Britain into eighteen provinces and for each species he has a diagram in which "by omitting the figures from those squares which correspond to districts within which the species has not been ascertained to grow, a tolerably exact notion of its topographical area may be instantly conveyed to the eye of the reader". He goes on to say that botanists who wished might colour the maps following the data given in the text, and that had no expense ruled it out he would have produced a printed map for each of the 1,200 species listed.

Finally, to appreciate how completely Watson had grasped the essential points to be borne in mind in recording and mapping plant distributions, one should turn to the introduction to *Cybele Britannica* (1847). Here he distinguishes two "circumstances of distribution—first the extent of geographical surface over which the plant is spread", for which he suggests the term *Area* should be used, "and secondly the greater or less frequency of the species within that space", which he suggests might be called *Census*. He sees clearly that, by making the recording units sufficiently small, some sort of measure of the second attribute of distribution can be obtained. Thus:

Tormentilla officinalis [i.e. *Potentilla erecta*] and *Hypericum pulchrum* are found in every province; and it is not improbable that they would equally be found in every county, if looked for. But if we could subdivide all the counties into sections of a square mile each, the *Tormentilla* would assuredly be found in many more of these square mile sections than would the *Hypericum*.

Now, well over 100 years later, we can test the validity of Watson's hypothesis, if for "county" we read "vice-county", and for "square mile" we read "a 10-kilometre square", by comparing the maps of the *Potentilla* and the *Hypericum* on pages 122 and 59 respectively. He might equally well have referred to the two species of nettle, *Urtica dioica* and *U. urens*; both are recorded from every county in the British Isles and so have similar "Areas", but a glance at page 182 will show that their "Census" is very different.

Before he died in 1881 Watson had therefore prepared the way for the study of the distribution of British plants on a unit area basis. Unfortunately, he had no successor with the same interest and ability, and as late as the early 1950's British botanists were still using the vice-county system, as for example in the accounts in the "Biological Flora" series published in the *Journal of Ecology*.

Development of techniques of mapping plant distribution in the late nineteenth and the first half of the twentieth century nearly all took place on the continent of Europe. One of the most outstanding contributions was made by Hermann Hoffmann, who between 1860 and 1880 published a series of papers which contain not only much the earliest published examples of the use of dot distribution maps, both on a local and on a European scale, but also an example of the use of the grid square method about half a century earlier than any other that has been traced. How far Hoffmann was influenced by Watson's earlier works we may never know: there is no indication in his writings that he had ever heard of Watson. However, Watson's work was widely known to Continental plant-geographers by the middle of the nineteenth century and there is nothing improbable in Hoffmann having read, for example, the introduction to *Cybele* or *Remarks on the Distribution of British Plants* which was translated into German in 1837. The first of Hoffmann's published works of interest was a paper in 1860 in which he deals with two calcicolous species *Prunella grandiflora* and *Dianthus carthusianorum*. Maps of these two species which accompany this paper have a good claim to be the earliest published examples of dot distribution maps. Two further papers (1867 and 1869) provide many examples of Hoffmann's local mapping, and also, quite unexpectedly, what must be the first European dot-map, that of *Gentiana verna*, based on data from De Candolle's *Prodromus*. Finally, Hoffmann invented a simple grid method of recording and used it throughout his *Flora des Mittelrheingebietes* (1879).

Hoffmann's example fell on more fertile ground than had

Watson's, and from his time dot-mapping in Europe became a more or less continuously expanding activity. Already, by 1879, Ihne had published a dot-map using several symbols to distinguish records made in different years and therefore to indicate change of range. In Germany others began to follow the established methods and the influence gradually spread to Scandinavia and Finland—but not to the British Isles. There appear to be no published distribution maps of any British plant in the nineteenth century after Watson's originals, except the classic of Miller Christy on the oxlip *(Primula elatior)*, first published in 1884. It was not until the beginning of this century that interest revived, when we find that the "pioneer" was Praeger (1902 and 1909) for the Irish flora only. It was, in fact, 1917 before a distribution map for the British Isles as a whole appeared, when several were published by Moss in his *Cambridge British Flora*. Further maps were published by Stapf (1917), by Sir Edward Salisbury in his deservedly famous study of the East Anglian flora (1932), and by Professor J. R. Matthews in his important paper on the *Geographical relationships of the British flora* (1937), though none of these used dot-maps.

THE PRODUCTION OF DOT-MAPS IN BRITAIN

The credit for the publication of the first dot-map of a British plant must go to Good who included one in his study of the Lizard Orchid, *Himantoglossum hircinum* (1936). But it is not until after the war of 1939–45 and the advent of the "Biological Flora" and the "New Naturalist" series that dot-distribution maps may be said to have become a standard requirement of British ecologists and taxonomists.

Could this need be met from the data available in 1950? For the rare species the answer was probably "yes". At a Conference of the Botanical Society of the British Isles that year on *Aims and Methods in the Study of the Distribution of British Plants* (published in 1951) a number of speakers produced evidence to show that data for rare species were available and how much more informative were dot-maps than those based on vice-comital distributions. It was also made clear that for common species the answer would have to be "No"; for many parts of Scotland, Ireland and Wales no other data were available than that a species does or does not occur in a particular vice-county. During this conference the difficulties and possibilities of mapping the British flora were fully discussed. Notice was taken of recent work on the Continent, notably that of Hultén, whose magnificent atlas of maps showing the distribution of vascular plants in north-west Europe, *Atlas över Kärlväxterna i Norden* (1950), was published shortly after the conference finished. The work of the Instituut voor det Vegetatie-Onderzoek in Nederland (I.V.O.N.), which was carrying out a survey of the distribution of vascular plants in Holland using a basic square of approximately 5 × 4 kilometres, was also carefully considered.

The final paper of the conference was read by Professor A. R. Clapham. In the course of it he suggested that the Botanical Society of the British Isles should take steps to ensure that before long we had a set of distribution maps of British species; that the maps, when produced, should be comprehensive and accurate; that they should also be of small scale (1 in 10 million) so that they could be printed four on each page of a single volume; that the unit area should be the 10-kilometre grid square; that overlays should show ground over 1,000 and 2,000 feet and also basic substrata; and that certain further information about infraspecific units and certain historical data should also be supplied if possible. After these suggestions had been discussed the following proposal was made:

That the conference expresses its wish that the Council of the B.S.B.I. should discuss the possibility of preparing and producing a series of maps of the British Flora and that if they consider it practicable they should enlist the co-operation of such other bodies as may be desirable. . . .

This proposal was carried with acclamation.

THE LAUNCHING OF THE MAPS SCHEME

The Council of the B.S.B.I. met in May 1950 and discussed the proposal. They appointed a committee to consider the part the Society might play in the project. This Maps Committee consisted of:

> Professor A. R. Clapham
> Mr J. E. Lousley
> Mr E. Milne-Redhead
> Professor T. G. Tutin
> Mr E. C. Wallace
> Dr E. F. Warburg

and at its first meeting Mr Lousley was elected Chairman and Professor Clapham, Secretary. After much careful investigation of the practical difficulties the Committee felt, early in 1953, that, given adequate financial support, the project to prepare and publish an atlas of British plants was a practical one, and accordingly an approach was made to the Nuffield Foundation for a grant for a five years' project. This approach was successful and the offer of a grant of £10,000 for the scheme was gratefully accepted by the Council in December 1953. Valuable financial assistance was also given by the Nature Conservancy who made a grant of £1,000 per annum for four years (commencing in April 1955) and also agreed to meet the cost of the Powers-Samas punched-card recording system adopted by the Committee for the incorporation of the data and the mechanised map production, on the understanding that, at the end of the Scheme, the Conservancy would take over the machinery and the punched cards as the basis of a permanent recording system. Through the kind co-operation of the Professor of Botany, Professor G. E. Briggs, and the General Board of the University, accommodation was made available to the organisation in the Botany School, Cambridge. The following societies have made grants to the Scheme since it commenced: The Botanical Society of the British Isles, The British Ecological Society, and the Royal Irish Academy.

Now that the future of the project was assured the Maps Committee was reformed and enlarged, with powers to co-opt representatives of other bodies, and the following also served on the committee or its sub-committees during the course of the scheme: R. W. David, Dr J. G. Dony, J. C. Gardiner, Professor

H. Godwin, R. A. Graham, Professor J. Heslop Harrison, R. D. Meikle, Dr F. H. Perring, Dr C. D. Pigott, Professor M. E. D. Poore, J. G. Skellam and other representatives of the Nature Conservancy, Dr W. T. Stearn, E. L. Swann, Dr S. M. Walters, Professor D. A. Webb. To run the scheme the following appointments were made:

★Director:	S. M. Walters from 6 April 1954
Senior Worker:	F. H. Perring from 1 October 1954
Secretary:	Miss A. Matthews from 6 April 1954
Punched-card Operator:	Mrs S. Fincham from 1 October 1954

★ On 31 March 1959 Dr Walters' term of office came to an end and the scheme has been directed by Dr Perring since that date.

The "Maps Office" thus came into being just before the scheme was officially launched at the Conference of the Botanical Society on 9 April 1954, exactly four years after its creation had been first suggested.

In addition to the central organisation in Cambridge regional offices were set up in Scotland and Ireland, and part-time paid assistants organised the enrolment of volunteers and the collection and refereeing of data before passing the information on to headquarters. Mrs M. E. D. Poore served as the Scottish Regional officer from 1954–6, and was followed by Miss E. Beattie who carried on until the office was closed in 1959. Accommodation for the office was kindly made available by the Scottish Nature Conservancy. In Ireland, Professor D. A. Webb set up a small regional office to assemble the data for Eire, whilst Professor J. Heslop Harrison co-ordinated workers in Northern Ireland, with the able assistance of Miss M. P. H. Kertland. The National Museum of Wales in Cardiff has provided advice and assistance with Welsh records throughout the Scheme.

OUTLINE OF THE METHOD USED IN PREPARING THE MAPS

THE RECORDING UNIT

The basis of the scheme is to indicate by means of a symbol the presence of each species of vascular plant in every 10-kilometre square of the Ordnance Survey National Grid in which it occurs, thus producing a distribution map of each. The National Grid covers Great Britain and the eastern half of Ireland, and is represented on all recent British Ordnance Survey Maps of a scale between 1 : 1,000,000 and 1 : 25,000. The maps in the 1 : 25,000 or 2½ inches to 1 mile scale are issued in sheets which exactly correspond to the 10-kilometre square, and these were recommended for field workers, being available for almost the whole of Great Britain. In Ireland not only was the National Grid missing from half the country, but no large-scale maps were available even for the gridded half. The problem of producing gridded maps for Ireland was presented to Professor Webb, who solved it elegantly and published the skeleton of his agony in a short paper in 1955. He prepared originals on the ½ inch to 1 mile Irish Ordnance Survey Maps and copies were made for use in Cambridge and Northern Ireland. Volunteer recorders in Ireland have had their own maps marked with the grid for the area they have been visiting, or have been requested to limit their recording of one list of plants to within a mile radius of any one locality, the list being conventionally gridded when sent in. The grid system has been strictly adhered to with a few exceptions:

(i) Some coastal squares contain such a small area of land that a round dot on the final map would appear to fall in the sea. In these cases the records for the square concerned have been incorporated with those of an adjacent square.

(ii) The records for the Orkneys and Shetlands have been conventionally gridded so that they are placed in an inset. This is more convenient for publication and also avoids the difficulty of duplicated grid references: N30, 31, 40, 41 and 42 become 57, 58, 67, 68 and 69.

(iii) The records for the Channel Isles have also been conventionally gridded so that they appear in an inset: 90/15 Guernsey, 90/25 Herm, 90/35 Sark, 90/42 Jersey, 90/48

Alderney. All records for those islands are placed in these squares.

N.B. Records for N.E. Caithness (v.-c. 109), S. Orkneys (v.-c. 111) and Alderney in the Channel Isles appear twice on the map, once in the true position and once in the conventional one.

THE RECORD CARDS

Regional

The first task of the scheme was to produce record cards which would enable data to be collected quickly and accurately in the field, whilst at the same time enabling the processing of the data to be carried out efficiently at headquarters. To this end seven Regional Record Cards were designed, each an 8 × 5-inch card printed with six columns (five in Ireland) on each side, carrying up to 900 species which were perhaps the most likely to be recorded in each region (Fig. 1). The scientific names of the species are conventionally abbreviated to five letters of the genus and three of the species (with few exceptions where this would be ambiguous). Each species has a code number—the species number—used to identify the species in the punched card system. In addition space is provided at the head of the card for grid reference, locality, habitat, date, altitude, vice-county and the recorder's personal code number. At the bottom of the reverse side space is left for "Other Species". Such cards had been pioneered by I.V.O.N. in Holland, and in England by the Cambridge Natural History Society. The seven regions were:

(1) South-west (v.-cs 1–14 and 22).
(2) South-east (v.-cs 15–21, 24–32).
(3) Midlands (v.-cs 23, 33, 34, 36–40, 53–65).
(4) North (v.-cs 66–71 and Scotland south of the Clyde-Forth Line).
(5) Scotland (north of the Clyde-Forth line).
(6) Wales.
(7) Ireland.

The nomenclature used was that of the *Flora of the British Isles* by Clapham, Tutin and Warburg (1952), though the Irish card differs in some respects. In addition a card was produced based on the nomenclature used by Bentham and Hooker (ed. vii, 1924) for the benefit of those who had not adopted the new names. The grasses and sedges were largely excluded from this card. A ninth card was called the "Common Species Card" and carried the English and scientific names of 120 widespread species, which were selected as being readily recognisable and having no taxonomic complications, so that it might be used by school children and other beginners. When the scheme started we were uncertain of the response which would come from volunteers and we felt these two additional cards might be important. In practice, however, their use has been very limited in the field, and because their design differed from the regional cards they proved to be very inconvenient in the office when data had to be transferred from one type of card to another.

Individual

In addition field recorders were supplied with individual record cards (Fig. 2). These are $4\frac{3}{4} \times 2$ inches with spaces reserved for the following information: species, locality, habitat, grid reference, date, vice-county, altitude, status, collector's name, determiner's name (if other than the collector), and source (whether field record, or from literature or herbarium). There were three colours of card to be used for the three main sources:

(i) Buff—field records
(ii) Yellow—herbarium records
(iii) Pink—literature records.

The use of these cards is described on page xviii. The buff cards are the same as those used to incorporate the field records; they are dual-purpose cards written on and punched as individual records, punched only as field records.

COLLECTION OF THE FIELD RECORDS

One of the essential requirements of a scheme of this kind is that it should cover the country as evenly as possible. Our main objective therefore has been to try to acquire lists of species from each of the approximately 3,500 10-kilometre squares which cover the British Isles. To this end appeals for help were made in 1954 to all members of the Botanical Society, the British Ecological Society, local Natural History Societies, Field Study Centres, University Botany Departments, and to the public in general through articles and letters to national and local newspapers and scientific periodicals. As far as possible particular squares were allocated to individuals or groups in an attempt to avoid duplication. Cards and instructions were sent to all who volunteered with a request for them to send in their data at the end of each season. Further stimulus was given from time to time by publishing, or circulating to selected centres of activity, situation maps which showed squares from which records had already been received or were expected. Gradually the cover of records became more complete. Botanical Society field meetings were organised mainly to visit underworked areas, and in this work the Council for the Study of the Scottish Flora joined with enthusiasm. During the last two

or three years of the scheme, when some information had been received from almost every square, volunteers were asked to make a copy of our master card (see page xviii) for a square, and send only additions to us at the end of each year. New people were asked to concentrate on underworked squares whereas they had been directed to unworked squares in the past.

Enormous credit is due to British botanists for their achievement in the five effective full seasons which were available for them to complete the task. Over 3,000 volunteered, though only about 1,500 actually sent in returns and the great bulk of the records were contributed by about 250. Nevertheless during that period they visited all but seven of the 3,500 squares, and failed to record more than 150 species in only 137 squares. The total number of records made was about $1\frac{1}{2}$ million—an average of about 400 records per square (Fig. 3). The whole effort was almost entirely voluntary; people gave freely of their time and went to considerable expense to pay their own travelling and hotel bills, and without this service the Atlas could never have been produced.

When the scheme started some data were already available in the form of published lists and others in private notebooks. It was our intention to concentrate on lists made from 1950 onwards and to distinguish between these and earlier records by the use of symbols on the published maps. However, as the scheme developed it became clear that 1930 should be used as the basic date-line, for a number of reasons:

(i) The botanical survey of Dorset made by Professor Good was started in 1931. As Professor Good's survey had been intensive and thorough, there seemed no point in repeating his work when he had generously agreed to make his, largely unpublished, basic data available to us.

(ii) The study of the flora of the Inner and Outer Hebrides by Professor J. W. Heslop Harrison *et al.* had begun about 1935 and had continued ever since. The work of A. J. Wilmott *et al.* in the Outer Hebrides also began in the 1930s. There seemed no justification for repeating this work in one of the most inaccessible parts of Britain, and one where the flora changes least, when so many other areas had never been visited.

(iii) The second supplement to *Topographical Botany* was published in 1929–1930 and the *Comital Flora* by G. C. Druce in 1932. These combined to provide an effective summary of knowledge up to about 1930, and by extracting the reports of the *Botanical Exchange Club* and the *Botanical Society of the British Isles* from 1930 onwards we hoped to be able to locate most vice-county records.

It was also fortunate that at the time the scheme commenced a number of botanists were engaged in writing Floras of particular counties, and all generously agreed to make their invaluable data available to us—in most cases producing cards for each square in the county themselves and in others lending us the data in a readily assimilable form. Others have subsequently begun to work on their Floras and have contributed likewise. We wish to acknowledge with sincere thanks the contribution of all those who have supplied more or less complete county lists (see Appendix III, page 419).

In many cases we know that the county organisers did not work

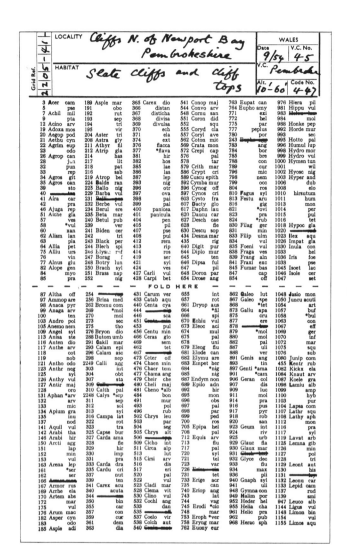

Fig. 1. REGIONAL RECORD CARD

Fig. 2. INDIVIDUAL RECORD CARD

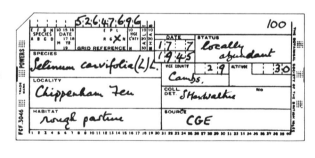

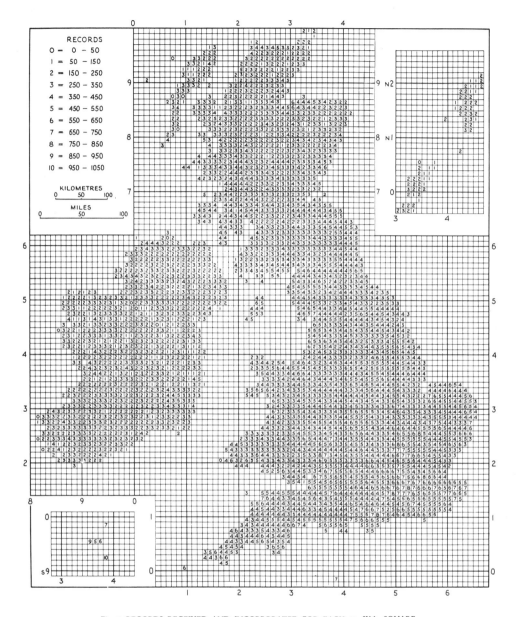

Fig. 3. RECORDS RECEIVED AND INCORPORATED FOR EACH 10 KM. SQUARE, INCLUDING PRE-1930 RECORDS BUT EXCLUDING INDIVIDUAL RECORDS

alone. That we never corresponded with their assistants directly may mean we are not aware of their help and cannot acknowledge their contribution personally. We must apologise if their names do not appear in the list of acknowledgements but we are not unmindful of the importance of their work. Many others have made contributions at least equal in quantity to those listed, but because their loyalties were not so restricted to particular counties it is less easy to acknowledge their work. We hope they will forgive this omission; it does not mean we are any the less grateful.

Frequency data

The list of species from a 10-kilometre square has been used only qualitatively in the production of the maps. At the beginning of the scheme it was realised that with a great many recorders over a short period it would be senseless to ask for frequency data; nevertheless we thought that it might be possible to get an estimate of rarity on the following basis:

A species known or reasonably suspected to be restricted to one particular locality in the 10-kilometre square, the "locality" not exceeding 1-kilometre square, should be marked as rare.

This request was only complied with by those submitting complete county lists or by those working intensively particular squares; it was found to be impractical over large areas of the country where squares were only visited once by a party for a day or less, so this distinction between the records has not been used in compiling the maps.

THE EVENNESS OF THE SURVEY

Despite the vigorous attempts made, the survey has not been as even as we would have liked: some parts of the British Isles have been less well recorded than others. An impression of this uneven coverage can be obtained by studying the map (Fig. 3) showing by hundreds the records received from each square. Thus, when analysing a distribution map, apparent areas of rarity or absence for which no plant geographical reason appears likely should be regarded with caution, and the possibility of a recording omission should be considered. This is particularly aggravating for the common species, because local floras often give no localities for such species. Other inadequately recorded areas which are not clear from the map (Fig. 3) are listed below. Some visual impression of these areas is given by overlay 6a (see page xxv).

Great Britain
(1) V.-cs 24 and 32, N. Buckinghamshire, S. Northamptonshire. Squares 42/62–5, 72–5, 82, 83 were under-recorded, particularly for late-flowering species. The area was mainly covered in May 1956, by a B.S.B.I. meeting, but has not been re-visited later in the year.
(2) V.-c. 35, Monmouth, near Newport. Squares 31/28, 29, 38 have not been visited recently. Records used date from *circa* 1928 and were made by A. E. Wade. The absence of post-1930 records from them is not significant.

Ireland
(1) Because of the scarcity of botanists living in the country and the difficulty of persuading great numbers of volunteers to make a visit, much greater use has had to be made of old (pre-1930) records in Ireland than elsewhere. That only an old record is shown for any particular square is *not* significant. However, these old records are reasonably evenly distributed and the absence of any recent (post-1930) records within an area showing old records may be significant, particularly if the change in range suggested corresponds with a similar trend in Great Britain.
(2) An additional complication in Ireland is that, in the south-west particularly, the flowering season starts earlier than elsewhere in the British Isles. Moreover, the south-west is so far removed from the main centres of botanical activity that visitors tend to arrive rather late in the season. Therefore some species, e.g. *Anemone nemorosa* (page 19) and *Ranunculus ficaria* (page 24) may appear to be rarer than they really are in south-west Ireland because they had died down by the time the observers arrived.

Lists of old (pre-1930) records

Besides the post-1950 records collected in the field since the scheme began (and post-1930 in the exceptions cited), a considerable amount of data has been assembled from work done before 1930 which has either been published in Floras and journals, or preserved in manuscripts and notebooks. An above-average concentration of old records may be expected for the following vice-counties:

7 & 8 Wiltshire	70 Cumberland
21 Middlesex	71 Isle of Man
30 Bedfordshire	S Sarnia
35 Monmouthshire	78 Peeblesshire
36 Herefordshire	80 Roxburghshire
40 Shropshire	81 Berwickshire
41 Glamorgan	109 Caithness
42 Breconshire	H. 20 Wicklow
60 West Lancashire	H. 34 & 35 Donegal

In addition to the "county" lists there are two other fairly large sources of data:

(1) Scotland. G. West, *Flora of Scottish Lakes* (1904, 1910).
(2) Ireland. Lists of data collected by Praeger and others when preparing *Irish Topographical Botany* (1901) and the *Botanist in Ireland* (1934).

Some other lists were acquired, but only those west of the major grid line 4 were incorporated. There has not been time to deal with old records east of this line—but as this is the most heavily populated part of the British Isles and the most heavily botanised the omission will, if anything, redress the balance rather than disturb it.

Additional records from herbaria and literature

Besides the 1½ million or so field records at least 150,000 individual record cards have been accumulated. Many of the buff cards were duplicates sent in to amplify records submitted on regional cards—this information was asked for if the record was of special interest for reasons of taxonomy, status, rarity, etc. Most were collected deliberately and represent an important additional source of grid square records, particularly for the rare species.

Information has been extracted from the following herbaria:

Royal Botanic Gardens, Kew
British Museum (Natural
   History)
Royal Botanic Garden,
   Edinburgh
National Museum of Wales
National Museum of Ireland
Cambridge University
Oxford University
Trinity College, Dublin
Queen's University, Belfast
Belfast Museum
Stratford Museum, London
   E.11

Somerset County Museum,
   Taunton
Warwick County Museum
Horsham School, Horsham,
   Sussex
Sexey's School, Bruton,
   Somerset
T. J. and E. Foggitt,
   Huddersfield, W. Yorks
R. A. Graham
J. E. Lousley
N. D. Simpson
E. C. Wallace

The following publications have been widely used:

Botanical Exchange Club Reports (1930–47).
Watsonia and *Proceedings of the Botanical Society* (1947–).
County Floras, etc. of which an almost complete set was made
   available on loan to the scheme through the generosity of
   the University Library, the Botany School Library and the
   University Botanic Garden Library, Cambridge, and many
   members of the Botanical Society of the British Isles who
   gave or lent copies of Floras or papers.

INCORPORATION OF THE RECORDS

Field records

From the field record cards for a square, of which many may have been received from different recorders, a single master card is compiled; thus we have collected 3,500 master cards, one for each square. It is from these cards that the data are incorporated into the "system" by punching a single Powers-Samas forty-column card for each species recorded in each square. All the information can be coded into numerical form: each species has a number with suffixes for sub-species, etc., and the date (usually 1950, which was used conventionally for all records made during the period 1950–60), grid reference of the 10-kilometre square and the vice-county,* are already numerical. Using an Automatic Key Punch (A.K.P.) it is only necessary to punch separately the code number of each species which has been recorded from the square. The other information which is common for all the species is punched into the pack of cards thus produced by setting up the A.K.P. in the "repeat" position and passing the pack through a second time.

The punched cards are checked by feeding them into a Powers 3 Tabulator which prints a list of the data punched into the cards. First the "common" information is checked, then the variable information; the species numbers are compared with the master

* Sarnia was numbered 113 but for all other numbers Watson and Praeger were followed.

card by two members of the office staff calling over the numbers and checking the tabulation visually.

After checking, the cards are sorted according to the species number, so as to bring together all the cards for each species, which are then filed in numerical order to await preparation for mapping.

Individual records

The individual record cards are edited to ensure that all the essential information is complete and has been altered into a suitable numerical code for punching, and then passed to the punched-card operator who punches each card separately on the A.K.P. This machine is ideal for this purpose as the face of the card is visible whilst the keys are being set up. The whole card is then punched in one action. These are then sorted and stored in numerical order. Every individual card is punched in a certain position so that all the individual cards for a species can be retrieved from the pack for the whole species in order that the extra information they give can be available for scrutiny.

The information on these cards (including date to the nearest month, an eight-figure grid reference, status, source, vice-county, altitude and habitat) can also be produced in numerical form on the tabulator.

CONTENTS OF THE ATLAS

The decision as to which species were to be included in the Atlas was left to a sub-committee of the Maps Committee. The resulting list has been used as a basis for the contents of the Atlas, but at a late stage it was realised that it was not practical to include all the micro-species listed in the *List of British Vascular Plants* by Dandy (1958), and that though data on some groups were adequate it would be better to leave all microspecies and hybrids until a later volume for two reasons:

(1) Future users of the Atlas would find it more convenient for reference if there were two distinct volumes with a clear difference of contents.

(2) The preparation of maps of some microspecies for the main volume would hold up its publication and leave so few maps for a second volume that it would perhaps be too small to produce at an economic price.

This volume therefore contains maps of all generally accepted native British species (excluding critical segregates) and most well-established introductions. The choice of the latter group has necessarily been arbitrary, and opinions will differ as to the wisdom of the final selection. In certain cases it has only been possible to publish aggregate maps; the reasons for these are various but are outlined on page xx.

PREPARATION OF THE MAPS

The list of species approved by the sub-committee was divided into three classes depending upon how much editing a species would probably require before an adequate map could be produced. These classes were:

A. Rare species—recorded from not more than twenty vice-counties in the *Comital Flora*.

B. Medium species—recorded from 21–100 vice-counties.

C. Common species—recorded from more than 100 vice-counties.

In the Atlas each species is labelled with its category, indicating the amount of editing which it has received. The vice-county distribution has only been used as a guide, and some species have received more attention than their vice-comital distribution alone demanded. Special attention was given to species formerly more common whose distribution has become much more restricted during the last few centuries. For example, *Lycopodium selago* (page 1) is recorded in the *Comital Flora* from 129 vice-counties, but to have obscured its almost complete extinction in southern England would clearly have been misleading, so it has been treated as a *B* instead of a *C* species. The maps were prepared in taxonomic order and final revision and despatch to the printers began in December 1960. Records have continued to come in and many have been incorporated; thus the later maps should be more complete than the earlier ones. However, as it is the grasses and sedges which have benefited from this, and as they always tend to be less well recorded, this fact may have preserved the balance between the beginning and end of the Atlas.

A. Rare species

The distribution of these species has been investigated as fully as possible. In each case the following procedure was adopted. To the field records were added data from the herbarium of Cambridge University, the relevant county Floras, papers and monographs and individual records from other sources. After editing, all the cards were punched and a number of tabulations prepared on to which the localities were typed. The tabulations were circulated to the following:

J. E. Lousley
A. E. Wade, National Museum of Wales
E. C. Wallace
Dr E. F. Warburg, The Herbarium, University of Oxford
Mrs B. Welch, British Museum (Natural History)
P. S. Green and (later) Dr J. Milne, The Herbarium, Royal Botanic Garden, Edinburgh (only if species occurred in Scotland)
Professor D. A. Webb, Trinity College, Dublin (only if species occurred in Ireland)

These authorities checked their herbaria for additional records and added considerably from their own knowledge of the species in the field or of overlooked sources in the literature. They also made recommendations about taxonomy, status and validity. Further individual cards were prepared and others altered before the data were finally assembled for map-making. On these maps the old and recent records are distinguished by symbols except in a few cases where taxonomic difficulty has resulted in under-recording, so that the distinction would be meaningless, or where the species is now so rare that to disclose its present distribution might be dangerous.

Another device has also been used occasionally to protect very rare species. The dot for a locality, instead of being placed in the correct grid square, has been placed in one of the eight adjacent grid squares.

B. Medium species

The distribution of these species has been checked at the vice-county level and records have been added from the readily available literature for all those vice-counties from which no *recent* (post-1930) record has been fed into the system.

In most cases records from about 10 per cent of the counties listed in the *Comital Flora* still remained untraced at this stage. Thanks to the unselfish co-operation of a number of people, who worked upon lists of desiderata sent to them from time to time, it was possible to reduce the number of vice-county records untraced. Credit for this considerable contribution is due to the following:

A. E. Wade, Wales
Miss J. Gibbons, Lincolnshire
Mr and Mrs R. C. L. Howitt, Nottinghamshire
Miss M. Scannell, Ireland

and to many others whom we have bothered less persistently but who have always replied so readily.

On these maps the old and the new records are distinguished by symbols. It must be remembered that a *complete* search for old records has not been made. They are only included for those counties where no recent records are available and for a few others which have been fed into the system. The maps are not as complete as they could be, but we have tried to include the most important data. What we aim to show is the outline of the present distribution and an indication of those areas where the species is very rare or entirely absent.

C. Common species

The distribution of these species has been checked at the vice-county level and records have been added from the literature from all those vice-counties from which no records have been received. Otherwise the procedure has been identical to that carried out for *B* species. However, on these maps no distinction is made between old and new records.

Despite the efforts made, records from many vice-counties have remained untraced, particularly in Scotland where there are few county Floras and where many species reach the limit of their range and are probably rare or casual only. These missing vice-counties are listed in Appendix II (page 409). Search of H. C. Watson's herbarium at Kew, the Herbarium of the Royal Botanic Garden, Edinburgh, or the *Annals of the Scottish Natural History Society* would probably produce records from a great many of these counties, but in very few cases are these omissions so great as to be misleading. There is, however, one group of plants, the arable weeds, for which some of the maps are unsatisfactory. This is particularly true of those species which have decreased considerably in abundance and frequency since the improvements in the purity of seed at the beginning of this century. *Agrostemma githago* (page 65) for example was so widespread in eastern Scotland in the nineteenth century that it is recorded as "Common" without localities in all the Floras. It is now apparently absent, but very few old localities have been traced and placed on the map, so that the dramatic change which has taken place is under-emphasised unless reference is also made to the list of "missing vice-counties".

Special help

Though most of the editing has been carried out at headquarters with the guidance of the various people mentioned above, we have also been fortunate in the number of those who have provided data for particular species in which they were interested. We wish to acknowledge this help, which means that the maps of many species are much more complete than would otherwise have been possible. See Appendix III (p. 419).

Special mention must also be made of the work of A. Slack and A. McG. Stirling who unearthed many additional records from literature and herbaria for the less rare Scottish mountain plants.

VERIFICATION OF RECORDS

The records received by the scheme have been checked in a number of ways and elimination of challenged records has taken place at various stages.

County referees

The field cards, grouped into counties or larger areas, were sent to the referees who had agreed to check the cards and who had a special knowledge of the area concerned. Upon the basis of their observations records were challenged or withdrawn.

We wish to acknowledge the assistance of all those who undertook this work. A list of their names appears in Appendix III (page 420).

Preparation of Master Cards

All the field records were transferred to Master Cards, and this process gave the staff an opportunity of considering every record. Possible errors overlooked by the county referees could often be reconsidered at this stage.

Preparation of data for mapping

When the cards for a species were arranged in vice-county and grid reference order, apparent new county records or outlying squares could be noted and rejected unless confirmation was found in Floras published since the *Comital Flora*, or in the revision of the Irish Flora by Webb (1952), or in the form of reliable individual record cards. There are also certain counties, e.g. E. Ross and E. Sutherland (v.-cs 106 and 107), which were very much underrecorded in the past, particularly in the lowlands, and from which a great many new county records have been made in recent years, often by several independent recorders; it became clear that these should be accepted. There are also a number of species about which no taxonomic confusion seems likely (e.g. *Cymbalaria muralis*), which have been spreading continuously over the last thirty years and for which the vice-county distribution in the *Comital Flora* is now out of date.

Inspection of the map

The map itself when produced may show certain isolated records and the origin and validity of these can be checked. A final inspection of all the maps was made by the editors with the assistance of C. D. Pigott.

By these methods we hope to have eliminated any flagrant

errors, so that the general distribution pattern is not misleading. Nevertheless errors within the distribution range may still remain. These are unfortunate but not serious.

Despite the use of these various devices the maps of some species still suggest difficulties in identification reflected in a high percentage of apparent new county records or a great disjunction of distribution between one recorder's area and another's. Various methods have been used to reduce the seriousness of these difficulties.

Reliance on experts

For two difficult genera, *Fumaria* and *Potamogeton*, we have been fortunate in having the assistance of acknowledged experts who have generously made available to us the results of their work over many years. In these cases we have used symbols to distinguish between those records which have been confirmed by the experts and those which have not. For most species of these genera we have not included any records from a county from which the experts have seen no material, except in Ireland where the amount of herbarium material was limited and to follow this rule would have been more misleading than to ignore it. In one or two species of a very critical nature only the records passed by the experts have been accepted.

Mapping of aggregates

This device has been used in a number of cases where two or more closely related species have been confused or seriously underrecorded. In some of these cases the specific distinctions are not clear; in other cases it has been apparent that some recorders have found more difficulty than others.

Such aggregate maps are clearly labelled and a complete list will be found in Appendix I (page 407).

Aggregate maps have also been included for a number of familiar critical genera. The segregates will in most cases be included in the critical supplement, but it was thought worth while to include maps of the following in this volume: *Rubus fruticosus, Alchemilla vulgaris, Sorbus aria, Euphrasia officinalis*.

Provisional maps

Where treatment as an aggregate would have concealed some valuable information, and yet doubt still remained about the accuracy or completeness of the records, it has been decided to include maps in the Atlas but label them provisional. These maps are clearly marked with a "P" in the top right-hand corner of the legend. A list of provisional maps with explanation is given in Appendix I (page 407).

STATUS OF RECORDS

There is no difficulty with two classes of species when maps are being prepared. One class consists of species certainly native throughout their range, the other of alien species entirely introduced. These aliens are marked as such with an asterisk before the species name in the legend. In a few cases aliens have persisted for many years in some stations whilst being only casual in others; here a symbol to indicate this has been used (e.g. *Crocus purpureus*, page 330).

Some indication of the history of the spread of the alien species has been given in five cases where adequate information in the form of dated records existed. These species are: *Buddleja davidii*,

Veronica filiformis, *Galinsoga parviflora*, *Galinsoga ciliata* and *Senecio squalidus*. In addition the first record has been indicated for eleven species where the spread has been rapid.

The maps of a few conspicuous aliens, particularly trees, e.g. *Laburnum anagyroides* and *Aesculus hippocastanum*, are inadequate because some recorders ignored them.

Other species vary widely in the relative proportion of native and introduced records and the ease with which their status in different areas can be ascertained. Each species has been treated on its merits, but three main types of treatment have been used.

(i) Rare natives, widely introduced.

Native, or possibly native records alone, have been mapped (e.g. *Pinus sylvestris*, *Tilia platyphyllos*, *Buxus sempervirens*).

(ii) Fairly frequent natives, rarely introduced.

This is the largest group. An attempt has been made based on statements in local Floras, the *Comital Flora* and field observations to distinguish between native and introduced localities (e.g. *Helleborus* spp., *Tilia cordata*, *Viburnum lantana*).

(iii) Common natives, commonly introduced.

This group consists mainly of our common trees which have been so widely planted as to make determination of native status impossible. In these cases we have made a statement on the possible native range based on the judgement of experts (e.g. *Carpinus betulus*, *Fagus sylvatica*, *Salix alba*.)

Recorders were asked to note on their field cards whether or not a species was in their opinion only an introduction in the square. Though the request was not always complied with, the distribution of these observations is often a useful guide. These records have been distinguished by using the formula "recorded introduction" (one symbol) and "all other records" (another symbol). The maps of *Ilex aquifolium*, *Fraxinus excelsior* and *Sambucus nigra*, for example, have been prepared in this way, and the information on introduction, though incomplete, seems helpful.

MAP PRODUCTION

At a late stage in the planning of the Maps Scheme, when it had already been agreed that the use of punched-card machinery was essential for the preparation of the data for mapping, the idea arose that it might be possible to use the punched cards themselves to produce the maps mechanically. The problem was presented to L. R. Smith of Powers-Samas (now part of International Computers and Tabulators) and within days he had solved it and was able to demonstrate the technique when the scheme was launched in April 1954. So thorough was Mr Smith's first analysis of the problem that the technique has been altered little since it was first developed.

The forty-column cards are prepared for mapping by sorting all those for one species on the first figure of the grid reference. The cards are divided into four groups, each for a vertical strip of the British Isles. The four groups proceeding from west to east across the map have the first figure of the grid reference as follows:

8 and 9   0 and 1   2 and 3   4, 5 and 6

Each of these groups is twenty 10-kilometre squares wide except the last which is twenty-five—the maximum number of print-units available on the tabulator used.

Within each strip the cards are arranged in vertical order: 100 master cards are interspersed between them, one for each of the 100 possible vertical positions. (By insetting the Channel Islands, Orkneys and Shetlands, the British Isles can be shown on a map as less than 1,000 kilometres from north to south.)

The cards are transferred to a standard tabulator modified only so that the vertical and horizontal throws are equal. An outline map is set in a fixed position in the tabulator, and the 100 master cards operate the mechanism so that the map turns through 100 positions, one for each 10-kilometre square, during the course of a run. When a species card punched with a grid reference is fed into the tabulator the map remains stationary, the card is sensed and the print-unit is operated to bring up a symbol in the correct position.

By this method a map of 3,500 dots can be produced in less than an hour, and an average map of about 1,000 dots in 20 minutes.

During the scheme the tabulator has been used to place the solid dots on the maps. Circles and other symbols have been added by hand.

We would like to take this opportunity of expressing our gratitude to Mr Smith for his help with this particular problem, and also for the many other ways in which he ensured that the mechanisation equipment gave such reliable service. To his name we would like to add that of W. Wadeson whose administration on behalf of Powers-Samas meant that our association with that firm was always a happy one.

INTERPRETATION OF THE MAPS

It was never intended that the Atlas should include a phytogeographical essay based upon the distribution maps. It is a factual document which, we hope, details the distribution of the British flora with sufficient accuracy to make it a valuable tool for biologists and all others whom it may happen to interest.

So far the introduction has explained exactly how the maps have been prepared, a necessary preliminary to any interpretation of the published maps. What follows is an attempt to interpret to the reader the maps in relation to the work done in preparing them. This we feel is of greater importance than a detailed analysis of types of distribution. Such an analysis may well be considered worth while in the future, but if so it can be published elsewhere. Nevertheless the reader may wish to consider for himself the distribution patterns in relation to possible limiting factors, and as an aid to this interpretation we have included in a pocket six transparent overlays each consisting of a pair of maps on the same scale as those in the text. The six pairs are:

1*a* Vice-counties   *b* Rivers
2*a* and *b* Altitude
3*a* Geology   *b* Temperature
4*a* and *b* Temperature
5*a* Rainfall   *b* Humidity
6*a* Underworked squares   *b* National Grid

VICE-COUNTIES

1a This shows the sub-divisions of the British Isles laid down for Britain by H. C. Watson and for Ireland by R. Ll Praeger. The map should be useful in at least two ways. It can be used in conjunction with the data on missing vice-county records to extend, in the mind's eye, the limits of distribution; it may also be useful to local botanists who wish to know which species occur in their county. We are grateful to J. E. Dandy for his advice in the preparation of this map. The lists of vice-counties and of vice-county records omitted from the maps are given in Appendix II (page 409).

RIVERS

1b The main rivers have been included both as a topographical guide and for their occasional value in interpreting distribution patterns.

ALTITUDE

2a *For mountain plants.* This overlay shows the distribution of high ground by means of three symbols. The circle indicates those 10-kilometre squares which have some land over 1,000 feet but not exceeding 2,500 feet. The black dot indicates squares with some land between 2,500 feet and 3,500 feet, and the dot surrounded by a circle shows squares where the altitude exceeds 3,500 feet.

The main value of this map will be found in studying the tolerances of mountain plants. It is possible to build up a series of increasing tolerance to low altitudes from those like *Gnaphalium norvegicum* (page 278) confined to the highest mountains, through intermediate species like *Saxifraga aizoides* (page 139) which indicates so clearly the "Highland Line" in Scotland, to widespread mountain plants like *Saxifraga stellaris* (page 136) and *Oxyria digyna* (page 178) which are found as low down as 1,000 feet or so throughout their range. Some species occur at even lower altitudes in the north and west where many of our mountain plants, apparently intolerant of high summer temperatures, come down to sea-level. Examples of this type are *Carex capillaris* (page 356) and *Sedum rosea* (page 133).

2b *For lowland plants.* This overlay shows by dots all those squares which contain no land below 500 feet. This is intended to be useful in interpreting the distribution of some of our strictly lowland species, particularly the arable weeds which are scarce or absent from most of north and west Britain. Here again a series can be produced from the more exacting species like *Papaver rhoeas* (page 28), through more tolerant species like *Potentilla reptans* (page 123), to species which only fail to reach the highest areas like *Stellaria media* (page 70). The three common species of *Sonchus* (pages 297–8) also form an interesting and informative series.

GEOLOGY

3a It would have been impossible on this small scale to produce a map showing all the main geological divisions for an area as complex as the British Isles. Instead we have included a simple map which shows by dots those 10-kilometre squares in which there is an outcrop of limestone which exceeds 5 per cent of the area of the square. We would have liked also to include calcareous

shell sands, but adequate data were not available for the whole area. These are found on the west and north coasts of Britain, and are absent from the east and south coasts from Durham to Cornwall, where calcareous species only occur on the coast if there is a limestone outcrop. The value of this map in distinguishing between calcicole and calcifuge species can perhaps be best appreciated by comparing a widespread calcicole, *Anthyllis vulneraria* (page 109) with a widespread calcifuge, *Erica cinerea* (page 195). These illustrate the basic patterns of the two types. Less widespread species reflect their preferences over a part of the British Isles, but are limited by climate. Thus we have southern calcicoles like *Phyteuma tenerum* (page 258) and *Thesium humifusum* (page 152), intermediate species like *Actaea spicata* (page 19) and *Sesleria caerulea* (page 384), and northern ones like *Veronica fruticans* (page 230).

This overlay is of help in interpreting local distribution patterns in many of the maps. The only other significant group of species for which another geological overlay might have been valuable is the plants of the heavy clays. *Pimpinella major* (page 161) and *Sanguisorba officinalis* (page 126) are good examples of this type of distribution, and may be used as reference maps in discovering other species with similar preferences.

CLIMATE

There is an infinite number of ways in which the climate, the average weather of a period of years, can be presented in graphic or cartographic form. The effect of climate upon different species varies widely, and it may take years of research to determine what is the critical factor which has limited a particular species to its present range. Moreover, the limiting factor is not the same in different parts of the range and in many cases it will be a combination of factors which will be of importance. Thus it must be clear that the selection of climatic maps which we have chosen to include is a subjective one, and is only to be used as a general guide to the type of factors which may be of prime importance for a particular species. However, our experience suggests that the maps we have chosen are helpful with a wide range of species.

3b *Average mean of daily minimum temperature, February* (1901–30). The lowest mean temperatures occur in February. The most obvious features are that the minimum values occur in eastern Scotland and England (34°F., 1.1°C.), and the lowest values in Ireland (36°F., 2.2°C.) are found in the north. In contrast high values (over 40°F., 4.4°C.) are confined to south and west Ireland and to south-west England.

The distribution of two groups of species seems to be closely correlated with winter temperature. Firstly there are those which are mainly northern and eastern: *Linnaea borealis* (page 265) occurs only in eastern Scotland from Sutherland southwards, though it was once known from Northumberland (v.-c. 67) and north-east Yorkshire* (v.-c. 62), and appears to be limited by a February mean minimum temperature of over 34°F. A more widespread species with a similar distribution pattern but which probably tolerates slightly higher winter temperatures is *Trientalis europaea* (page 204). This almost reaches the west coast of Scotland, but it is still essentially an eastern species, reaching its southern limit in Britain near the east coast in Suffolk (v.-c. 25). Both these species

* There is a possibility that *Linnaea* was an introduction with pine trees here.

occur widely in the coniferous forests of northern Europe. The climate of north-east Scotland has affinities with that of northern continental Europe. It is possible that the Scots Pine, *Pinus sylvestris* (page 16), is truly native only in this area.

A more widespread species which may nevertheless belong to this group is *Cardamine amara* (page 44). Though it occurs from Inverness to the south coast of England it reaches the west coast only between Lancashire (v.-c. 60) and Argyll (v.-c. 98) and it may be limited by the 36°F. isotherm. It is therefore significant to find that this species also occurs in Ireland, but only in the north. *Geranium pratense* and *Geum rivale* are rather common species which might be included in this group. The former (page 90) is also confined in Ireland to the north, and it becomes much rarer near the south coast of England where it is often only of garden origin. The latter (p. 124) does occur more widely in Ireland, but it is more frequent in the north and more or less confined to the mountains in the south.

It is possible that in some of these cases at least a low winter temperature is necessary for the production of viable seed. Matthews (1942) found evidence to suggest that low temperatures increase the germination of *Trientalis europaea*, and Kinzel (in Skene, 1924, page 425) showed that the seeds of *Menyanthes trifoliata*, *Gentiana nivalis* and *Adoxa moschatellina* require freezing before germination takes place in nature.

There is thus a fairly large group of species with a mainly eastern tendency in Britain which only occur in the north of Ireland, where they are often restricted to a single locality (e.g. *Helianthemum chamaecistus*, page 60, and *Adoxa moschatellina*, page 266). These isolated occurrences are not concentrated in any one area; they suggest a relict distribution from a period when winter temperatures were lower in the north of Ireland than they are today.

A second group of species whose distribution appears to be controlled in part at least by winter temperatures are those which are intolerant of low values. The most exacting species are those which occur in south and west Ireland and in south-west England. Mean February minimum temperatures have almost the same values in these two areas, 41°F. (5.0°C.) being the highest temperature on the mainland. The species in this group are clearly relics of a time when a warmer winter prevailed, and their exact localities now show no significant grouping into particular areas. Thus it is possible to arrange a series of species which are mainly to be found in the Mediterranean or south-west Europe, which occur in different parts of the total area with mean February minima above 38°F. *Erica mediterranea* (page 195) is concentrated in West Mayo (v.-c. H.27). *Daboecia cantabrica* (page 193) is confined to a small area centred on West Galway (v.-c. H.16). *Pinguicula grandiflora* (page 239) is almost restricted to Cork and Kerry (v.-cs H.1-5) but has become naturalised when introduced outside the main area in Clare, Wicklow, Devon and Cornwall. *Erica vagans* (page 195) has generally been regarded as native only on the Lizard Peninsula of West Cornwall (v.-c. 1), but there are old records from Tramore, Waterford (v.-c. H.6) and a colony has recently been discovered some distance from human habitation in Fermanagh (v.-c. H.33); both these localities are within the total range of this group of south-western species. A last example in the series could be *Erica ciliaris* (page 194) which has not been found in Ireland but is limited to two main areas with temperatures between about 38 and 40°F.

Other species showing similar tendencies but which are more widespread include *Cicendia filiformis* (page 207), *Carex punctata* (page 354), and *Ranunculus lenormandii* (page 23). It is interesting to note that the last two occur in south-west Scotland where there are relatively high values along the northern edge of the Irish Sea. Likewise all three species have been recorded from East Anglia mainly on or near the coast, and here again the maritime influence maintains higher winter temperatures near the coast than inland. The end-point of this series might be seen in species like *Phyllitis scolopendrium* (page 7) and *Epipactis helleborine* (page 332) which occur throughout Great Britain except in north-east Scotland, and are thus almost complete opposites of the northern-continental *Linnaea* and *Trientalis*.

It is probable that this intolerance of low winter temperatures involves at least two different factors limiting plant distribution. Firstly low temperatures affect those species which are truly frost-sensitive. Amongst these one would include *Arbutus unedo*, the Strawberry Tree. Many of our aliens which have found normally suitable growing conditions in the south and east nevertheless suffer occasional catastrophes if caught by a rare frost, when, for example, whole populations of *Carpobrotus edulis* (page 80), and *Eucalyptus* spp. are destroyed. The map of *Fuchsia magellanica* (page 149), which indicates the areas where it is naturalised, is probably one of the best to indicate the effectiveness of frost. It would be interesting to study this in relation to a map showing the mean of the lowest annual temperature recorded for the last thirty years.

The other effect of low winter temperatures is to limit those species which use the winter and spring for growth and become dormant during the heat of high summer. These species are not necessarily frost-sensitive but probably fail to compete successfully after a series of cold winters. Some of the most widespread examples include *Umbilicus rupestris* (page 135), and *Ceterach officinarum* (page 9).

4a Average mean of daily mean temperature, January (1901–30). This map may perhaps give a better correlation with the group of species mentioned in the last paragraph, where it is the cumulative effect of winter temperatures above a certain figure which is important for full development.

4b Average mean of daily mean temperature, July (1901–30). The main feature of this map is that the isotherms run almost horizontally across our islands, with a slight dip in the west. Highest values occur in the Channel Islands and the Shetlands, with a difference between the two of almost 9°F. (5°C.). The highest values in Ireland occur in the south-east.

A series can be made of species showing a decreasing demand for high summer temperatures. The northern limits of these species are almost horizontal. There are about forty species recorded from the Channel Islands which do not occur as natives in the rest of the British Isles, where the summer temperatures are presumably not high enough. There is a fairly significant group of species which are confined to England south of the Thames, e.g. *Phyteuma tenerum* (page 258), which appear to be limited by the 62°F. isotherm. Other species, less exacting, reach their northern limit in the south Midlands, e.g. *Polygala calcarea* (page 56), *Cerastium pumilum* (page 68), and *Lathyrus aphaca* as a native (page 116). None of these species occurs in Ireland. However, *Campanula trachelium* (page 256) is more widespread in England reaching north to Lincolnshire (v.-c. 54) and south Yorkshire (v.-c. 63), a region which lies between the 60° F. and 61° F. isotherm. It is perhaps significant that this species *does* occur in Ireland, and in the south-east, the only part of that country with mean July temperatures over 60° F. Another species of very similar type is *Euphorbia amygdaloides* (page 173), very local in southern Ireland though not in the same area as the *Campanula*. As with the apparently relict species of the north of Ireland, which have a scattered, almost random, distribution, there is a similar group of species of southern distribution in Britain which occur somewhere in south-east Ireland, but are not concentrated in one particular area.

The series can be extended northwards with species of increasing tolerance. *Blackstonia perfoliata* (page 208) has a northern limit in Berwickshire, whilst *Rosa arvensis* (page 127) occurs throughout Ireland and just reaches the south of Scotland as a native.

The converse of the northern limit is the southern limit. This is most clearly seen in coastal plants where the distribution is not confused by altitude. Here again a series can be recognised. *Blysmus rufus* (page 351) reaches as far south as Kerry (v.-c. H.1) and North Lincolnshire (v.-c. 54) and is probably limited by temperatures over 60° F. (15.6° C.); *Mertensia maritima* (page 216) was once reported as far south as Kerry and West Norfolk (v.-c. 28), but now has its southern limit in Down (v.-c. H.38) and West Lancashire (v.-c. 60), and appears to be limited by slightly lower temperatures than *Blysmus*. *Ligusticum scoticum* (page 166) is restricted even further—to the coasts of Scotland and the north of Ireland. These species which appear to be intolerant of high summer temperatures are mainly found in northern Europe, some even having a circumpolar distribution, whereas the species with horizontal northern limits are found in continental and southern Europe.

5a Rainfall. This map may be useful in understanding the limits of hydrophytes on the one hand and xerophytes on the other, but it is unsatisfactory in that it does not include the factor of evaporation, which varies considerably throughout the British Isles and is of equal importance with rainfall in determining the water regime of the soil. However, what this map shows clearly is the very low rainfall experienced on the east side of Britain as far north as Caithness (v.-c. 109). This may account for the rarity or absence from such areas of species like *Chrysosplenium oppositifolium* (page 139) and *Nardus stricta* (page 405), and the presence of many of our arable weeds and plants of waste places in a narrow coastal strip, particularly concentrated at the head of the Moray Firth in v.-cs 95, 96, 106 and 107.

5b Humidity. Precipitation/Saturation deficit ratio. This map is based upon the method described by Meyer (1926). The Precipitation/Saturation deficit ratio (P/SD) is calculated by dividing the annual precipitation in millimetres by the absolute saturation deficit of the air expressed in millimetres of mercury. This ratio can be determined for any meteorological station recording precipitation, temperature and vapour pressure. In the British Isles these data are available for forty-five stations, and by interpolating from a map of mean annual vapour pressure (Meteorological Office, 1938), data from another sixty stations can be used. This is rather too few for detailed work, but this map should give a general indication of the trend of humidity distribution in these islands.

The most outstanding features are the very high values experienced all along the west coast of Ireland and Scotland; the low values in the counties of Essex and Cambridgeshire (v.-cs 18, 19 and 29); and the strong influence of the coast in maintaining high values, there being a very rapid gradient between the south coast of England and the inland areas to the north.

Here again we can recognise two main groups of species, those which are confined to the driest area and those which are absent from it. Of those which occur within it, the main concentration is in the "Breckland" on the borders of West Norfolk, West Suffolk and Cambridgeshire (v.-cs 26, 28 and 29). This is an area of very porous sands which, combined with the low humidity of the climate, make it probably the driest ecologically in the British Isles. Examples of this group include *Silene otites* (page 63), *Veronica praecox* and *Veronica verna* (pages 230 and 231). Other species with different edaphic requirements are also confined to central-eastern England; *Trifolium ochroleucon* (page 105) in hedgerow banks on the clay; *Primula elatior* (page 201) mainly in boulder-clay woods, and *Melampyrum cristatum* (page 234) on their margins. *Calamintha nepeta* (page 244) is locally abundant in Essex and Suffolk. Many of these species of eastern England are truly continental and some have their centre of distribution as far east as the Russian steppes.

A species of wider tolerance is *Pulsatilla vulgaris* (page 19) which occurs almost entirely in areas with P/SD values below 150. The absence of this species from the chalk of the North and South Downs may be for purely climatic reasons. It does not occur immediately on the other side of the Channel in France, but appears again near Rouen where the P/SD value is *c.* 125, about the same as Cambridge.

Other species have more widespread but parallel types of distribution, and a remarkably large group have a northern and western limit along a line joining the Humber to the Severn Estuary. This line is probably the resultant of a number of factors; nearly all our lowland limestones lie to the south and east of it, and the higher ground to the north and west begins to affect the species most sensitive to altitude; nevertheless humidity must also be of great importance. Good examples are *Chenopodium hybridum* (page 82), *Lactuca serriola* (page 296) and *Medicago polymorpha* (page 103) as a native. It is noteworthy that in a number of these cases the species become rare near the coast, particularly in Sussex (v.-cs 13 and 14) and in north-east Norfolk (v.-c. 27). More tolerant species occur widely over south, east and central England but become limited to the coast in the west, and may be almost entirely coastal in Ireland, e.g. *Cerastium semidecandrum* (page 69), or only occur inland in the south-east corner of that country, e.g. *Erodium cicutarium* (page 95).

Humidity alone can hardly be sufficient explanation for all species having distribution patterns of this type; some of them, for example, occur in seasonally wet soils, e.g. *Pulicaria vulgaris* (page 276) and *Lythrum hyssopifolia* (page 143). These species have declined with improved agricultural drainage. The probable explanation of their occurrence in "continental" Britain is that in this area alone is there a combination of a high winter water-table followed by a dry spring and early summer during which they can develop and reach maturity in the absence of competition.

In contrast to the species mainly found in the drier parts of the British Isles are those which are confined to or are most abundant in the north and west. Although the uneven distribution of acid and basic rocks, and the high winter temperatures of the west compared with the east, make interpretation difficult, it is hard to believe that in a group like the ferns humidity is not one of the most significant limiting factors. This is emphasised by the fact that the area which lacks many of these apparently high-humidity demanding species is almost the same as that in which the drought-resistant species are centred, i.e. central East Anglia. In Cambridgeshire (v.-c. 29) ferns are generally rare and only thirteen of the forty-seven species in the British flora occur at the present day, whereas Sussex (v.-cs 13 and 14) on the same longitude, despite the fact that it is 100 miles further south, has twenty-six species. This must surely be due to the coastal position and higher humidity values of the latter county.

UNDERWORKED SQUARES

6a Certain variations in the evenness with which the British Isles have been surveyed during the scheme have already been referred to (page xvii). A map (fig.3) gives an account of the actual numbers of species recorded from each 10-kilometre square. However, this does not readily give a visual impression of the areas which have been somewhat neglected. This overlay is therefore provided which indicates by dots those squares which we believe have been underworked. It attempts to take account of the total flora likely to be found in an area before deciding whether the list we have received is adequate or not. "Adequate" might be taken as meaning that over 60 per cent of the possible flora has been recorded.

NATIONAL GRID

6b This overlay makes it possible to find the four-figure co-ordinates of any dot in the Atlas. Those interested in having further details on any of the 10-kilometre square records in this work can obtain them by writing to the headquarters of the scheme and quoting species number and grid reference.

Besides those to whom we have paid tribute in the text, and those many others whose help in the collection of the records we have acknowledged in Appendix III, we would like to thank most sincerely the office staff who have been responsible for dealing with all the correspondence and the incorporation of the records during the eight years it has taken to produce this Atlas. Most particularly we would like to mention Mrs S. Fincham, the punched-card operator, who has been with us throughout, and whose cheerful enthusiasm has never wavered, and Miss A. Matthews, who was our secretary from 1954–9, and whose able direction of the office administration enabled us to deal smoothly with some almost overwhelming situations. We would also like to thank Miss J. Nicholson, who was secretary from July 1959 to July 1961, and Miss A. Hughes, who has been helping us for the last few months. We are very grateful to Mrs R. Marriott and Miss J. Daltry, who gave general assistance in the office between 1957 and 1959.

This list of acknowledgements would not be complete without special reference to those members of the Society who have served on the Maps Committee and have given freely of their time to advise and encourage. We are particularly grateful to J. E. Lousley and A. R. Clapham, respectively the permanent chairman and secretary of that Committee, and to the two treasurers of the Society, E. L. Swann and J. C. Gardiner, who served during the course of the scheme and who undertook the not inconsiderable additional business without demur.

We also wish to record our sincere thanks to the Nuffield Foundation for their grant for the first five years of the scheme, and to the Nature Conservancy for their continued financial assistance which has enabled us to produce a more complete work than was at first thought to be possible. To the General Board of the University of Cambridge we are grateful for making accommodation available throughout the scheme and, for the last few years, providing it without charge; and to our hosts, the Professor of Botany (1954–5) and the Director of the University Botanic Garden (since 1955) for tolerating our presence in their departments and making us feel so much at home.

Finally, we wish to record our appreciation of the help given us by W. T. Stearn in preparing this Introduction, by C. D. Pigott in editing the maps, by Mrs M. Pigott in producing the base-map, and by Miss C. A. Lambert in drawing the overlays.

University Botanic Garden,
Cambridge, March 1962

F. H. PERRING
S. M. WALTERS

The Distribution Maps

EXPLANATION OF THE LEGEND

Amount of editing

A Search made for all records: probably almost complete. Old and recent records generally distinguished.

B Search made for records from vice-counties from which no recent (post-1930) records had been received. Old and recent records distinguished.

C As for B but old and recent records not distinguished on the map.

For a fuller account of these categories see the Introduction (page xix).

Provisional maps

P These species are discussed in the Introduction (page xx), and a complete list with explanation appears in Appendix 1 (page 407).

The species number

These are taken from Dandy (1958). The maps follow the order of that work except in cases where it has been convenient to place two species on one map. This has generally only been done with species of the same genus and, with large genera, only with species of the same section.

Nomenclature

Dandy (1958) has been followed with one or two minor exceptions. Synonyms have been given where the nomenclature used by Clapham, Tutin and Warburg (1952) differs from that used by Dandy.

Status

* An asterisk before a species name indicates that the species is almost certainly not native anywhere in the British Isles including the Channel Islands.

Symbols

The meaning of the symbols is given on each map so that care should be taken, when comparing maps with identical symbols, to ensure that they are being used in the same way. In preparing the maps primary consideration has always been given to making the symbols easily visible. No symbol has any quantitative significance: each one indicates only that one or more specimens of a species have been recorded in a 10-kilometre square.

Dates

onwards

A date followed by "onwards" refers to records made in 10-kilometre squares in that or subsequent years (recent records).

Before

A date preceded by "Before" refers to records made in 10-kilometre squares from which no more recent record has been traced (old records). Recent records which are now known or thought to be extinct are also mapped as though they were old. The *spread* of a few interesting aliens has been shown in a special way by indicating the first record and giving its date or by showing all the 10-kilometre squares from which a species was known before a certain date. In all other cases the old records only indicate the possible *decline* of a species.

Scale

The maps in this Atlas are reproduced at a scale of approximately 1 to 8,000,000.

The maps published in this Atlas disclose the distribution of species to the nearest 10-kilometre square, and this may give rise to misgivings that the publication of this information will endanger the survival of some of our rarest species. To avoid possible risks we have, in certain cases, conventionally disguised the 10-kilometre squares where the species are known to occur today (see Introduction, page xix): in all cases the more exact localities for rare species which are held in our files are treated as confidential and are revealed, if at all, only to bona fide students and research workers.

The Atlas also shows how local many quite common species have become in certain parts of the British Isles in recent times, and stresses the importance of conserving locally not only the classical rarities but some of the more widespread British species. This publication should guide naturalists in two ways to a more effective conservation of our flora; firstly by acting as a warning about areas where species should on no account be collected, and secondly by pointing to areas where active conservation is necessary and giving factual evidence to support arguments against the destruction of threatened sites.

For both rare and local species we believe that education and information will serve the ends of conservation better than secrecy and ignorance. We are confident that in the hands of British botanists, without whose enthusiasm this volume could not have been produced, the information published will never be misused.

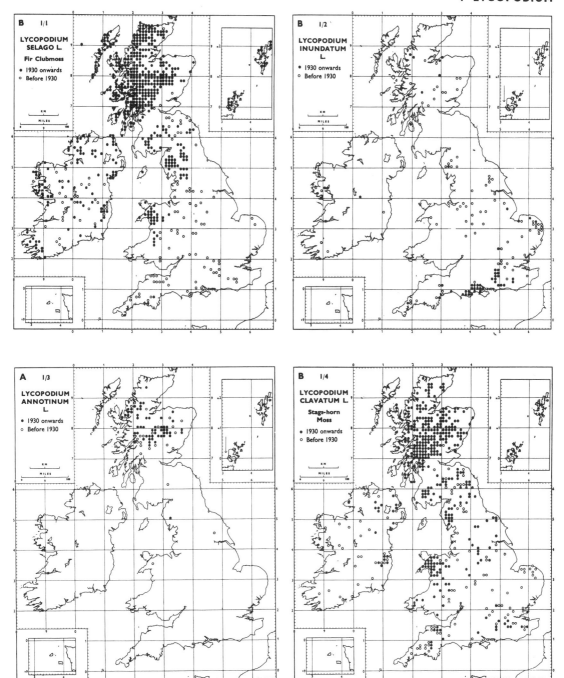

B 1/1
LYCOPODIUM
SELAGO L.
Fir Clubmoss
• 1930 onwards
○ Before 1930

B 1/2
LYCOPODIUM
INUNDATUM
L.
• 1930 onwards
○ Before 1930

A 1/3
LYCOPODIUM
ANNOTINUM
L.
• 1930 onwards
○ Before 1930

B 1/4
LYCOPODIUM
CLAVATUM L.
Stags-horn
Moss
• 1930 onwards
○ Before 1930

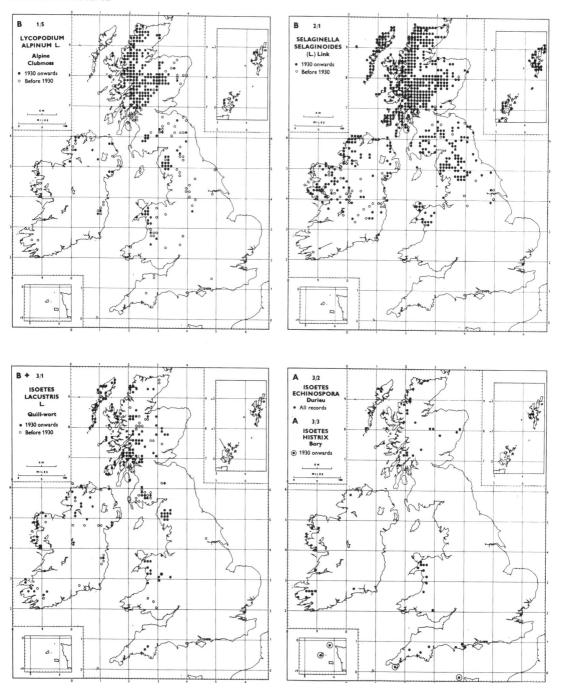

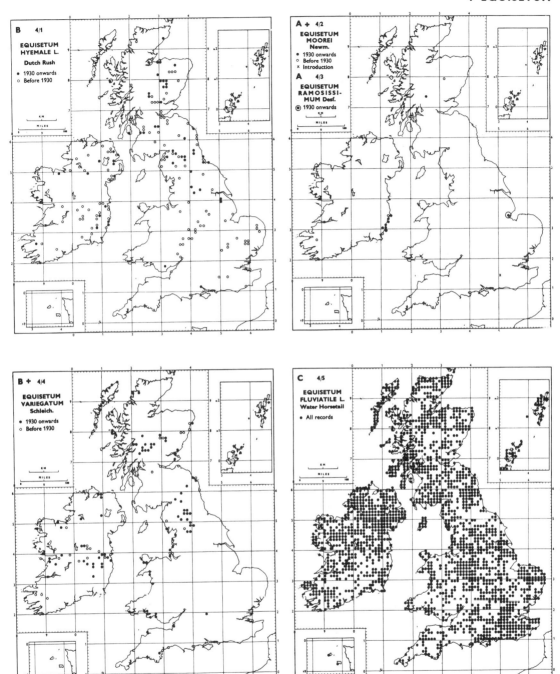

B 4/1

EQUISETUM HYEMALE L.
Dutch Rush

- 1930 onwards
○ Before 1930

A + 4/2

EQUISETUM MOOREI Newm.

- 1930 onwards
○ Before 1930
× Introduction

A 4/3

EQUISETUM RAMOSISSIMUM Desf.

⊚ 1930 onwards

B + 4/4

EQUISETUM VARIEGATUM Schleich.

- 1930 onwards
○ Before 1930

C 4/5

EQUISETUM FLUVIATILE L.
Water Horsetail

- All records

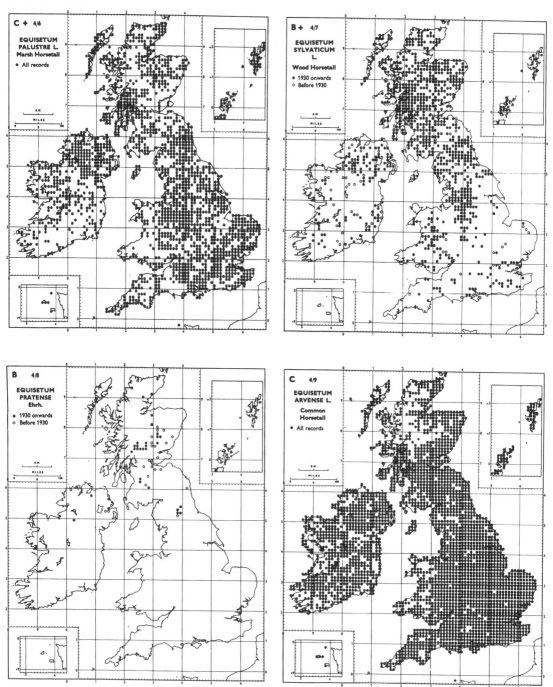

C + 4/6

EQUISETUM PALUSTRE L.
Marsh Horsetail
• All records

B + 4/7

EQUISETUM SYLVATICUM L.
Wood Horsetail
• 1930 onwards
○ Before 1930

B 4/8

EQUISETUM PRATENSE Ehrh.
• 1930 onwards
○ Before 1930

C 4/9

EQUISETUM ARVENSE L.
Common Horsetail
• All records

4

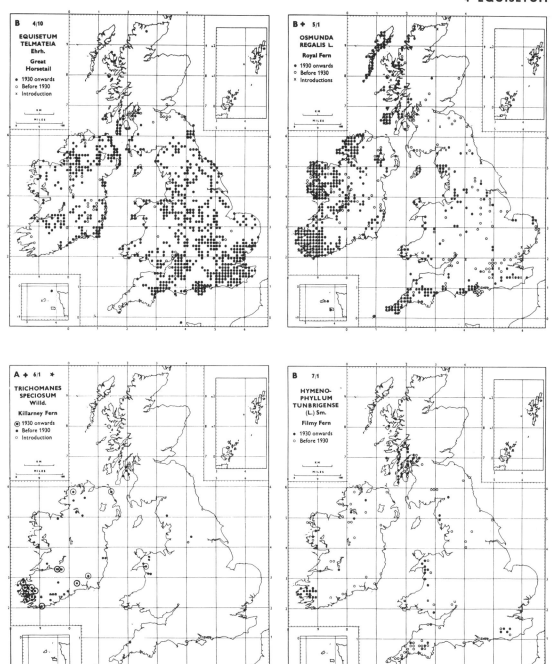

B 4/10

EQUISETUM
TELMATEIA
Ehrh.

Great
Horsetail

• 1930 onwards
○ Before 1930
× Introduction

KM
MILES

B + 5/1

OSMUNDA
REGALIS L.

Royal Fern

• 1930 onwards
○ Before 1930
× Introductions

KM
MILES

A + 6/1 ★

TRICHOMANES
SPECIOSUM
Willd.

Killarney Fern

⊙ 1930 onwards
• Before 1930
○ Introduction

KM
MILES

B 7/1

HYMENO-
PHYLLUM
TUNBRIGENSE
(L.) Sm.

Filmy Fern

• 1930 onwards
○ Before 1930

KM
MILES

HYMENOPHYLLACEAE

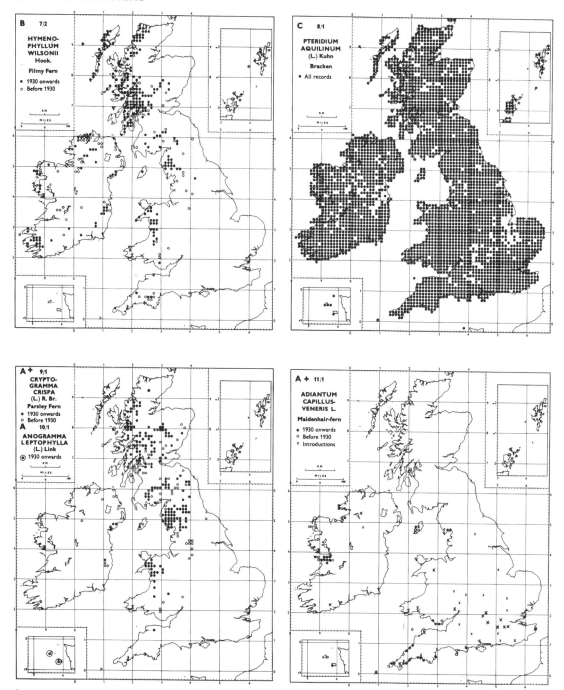

B 7/2

HYMENO-
PHYLLUM
WILSONII
Hook.

Filmy Fern

● 1930 onwards
○ Before 1930

C 8/1

PTERIDIUM
AQUILINUM
(L.) Kuhn

Bracken

● All records

A + 9/1

CRYPTO-
GRAMMA
CRISPA
(L.) R. Br.

Parsley Fern

● 1930 onwards
○ Before 1930

A 10/1

ANOGRAMMA
LEPTOPHYLLA
(L.) Link

⊚ 1930 onwards

A + 11/1

ADIANTUM
CAPILLUS-
VENERIS L.

Maidenhair-fern

● 1930 onwards
○ Before 1930
× Introductions

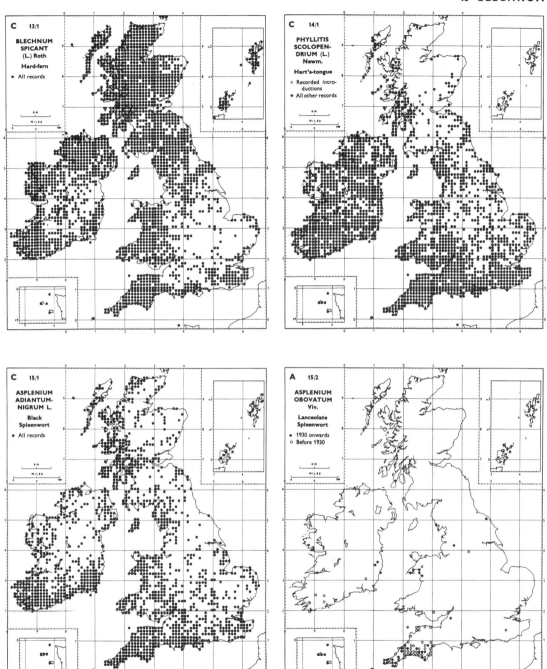

ASPLENIACEAE

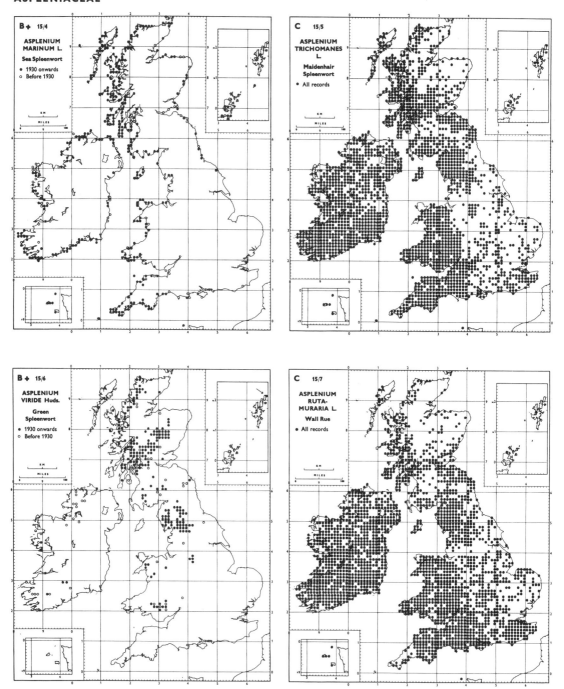

B✦ 15/4

ASPLENIUM
MARINUM L.

Sea Spleenwort

● 1930 onwards
○ Before 1930

KM
MILES

C 15/5

ASPLENIUM
TRICHOMANES
L.

Maidenhair
Spleenwort

● All records

KM
MILES

B✦ 15/6

ASPLENIUM
VIRIDE Huds.

Green
Spleenwort

● 1930 onwards
○ Before 1930

KM
MILES

C 15/7

ASPLENIUM
RUTA-
MURARIA L.

Wall Rue

● All records

KM
MILES

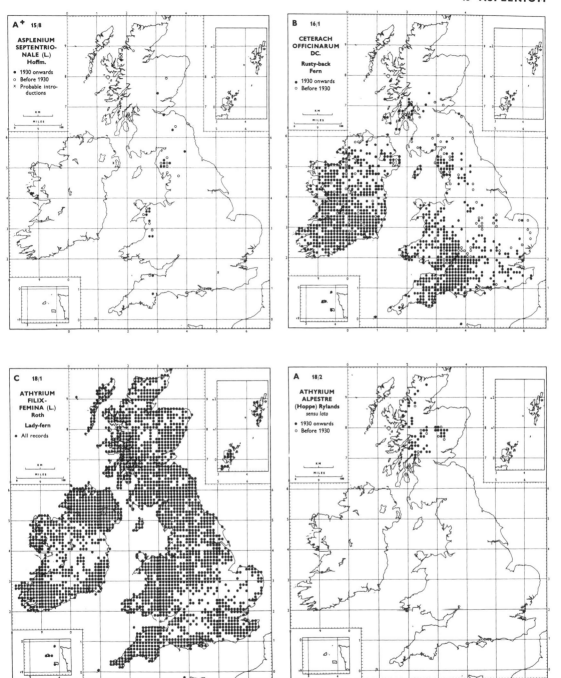

A⁺ 15/8

ASPLENIUM
SEPTENTRIO-
NALE (L.)
Hoffm.

● 1930 onwards
○ Before 1930
✕ Probable intro-
   ductions

KM
MILES

B 16/1

CETERACH
OFFICINARUM
DC.

Rusty-back
Fern

● 1930 onwards
○ Before 1930

KM
MILES

C 18/1

ATHYRIUM
FILIX-
FEMINA (L.)
Roth

Lady-fern

● All records

KM
MILES

A 18/2

ATHYRIUM
ALPESTRE
(Hoppe) Rylands
sensu lato

● 1930 onwards
○ Before 1930

KM
MILES

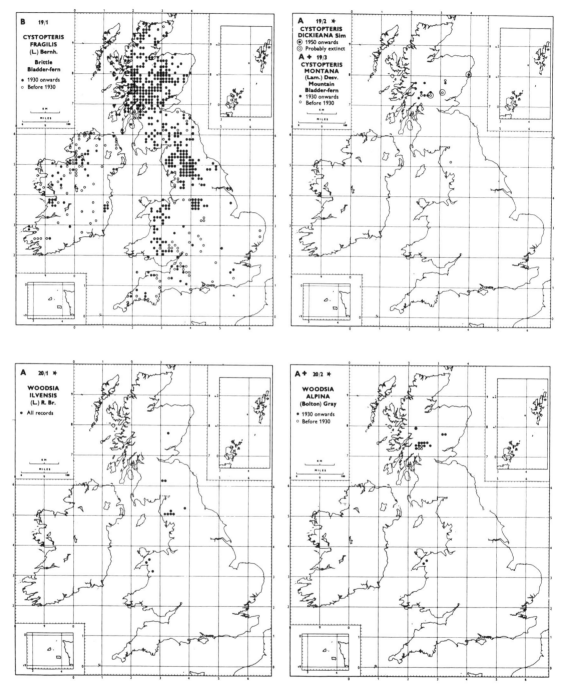

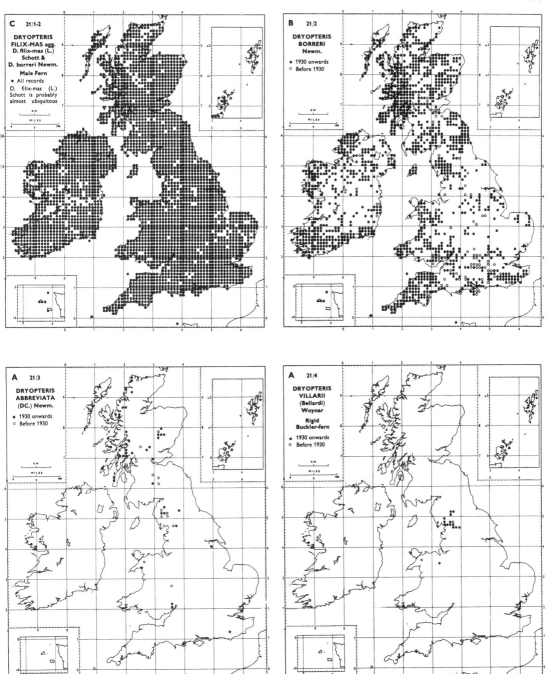

C 21/1-2

DRYOPTERIS
FILIX-MAS agg.
D. filix-mas (L.)
Schott &
D. borreri Newm.
Male Fern
● All records
D. filix-mas (L.)
Schott is probably
almost ubiquitous

B 21/2

DRYOPTERIS
BORRERI
Newm.
● 1930 onwards
○ Before 1930

A 21/3

DRYOPTERIS
ABBREVIATA
(DC.) Newm.
● 1930 onwards
○ Before 1930

A 21/4

DRYOPTERIS
VILLARII
(Bellardi)
Woynar
Rigid
Buckler-fern
● 1930 onwards
○ Before 1930

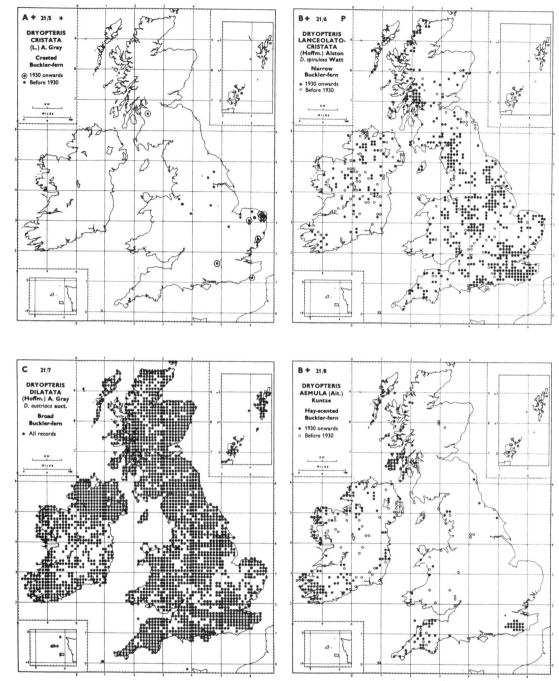

A ✦ 21/5 ✶

DRYOPTERIS
CRISTATA
(L.) A. Gray

Crested
Buckler-fern

⊙ 1930 onwards
• Before 1930

B ✦ 21/6 P

DRYOPTERIS
LANCEOLATO-
CRISTATA
(Hoffm.) Alston
D. spinulosa Watt

Narrow
Buckler-fern

• 1930 onwards
○ Before 1930

C 21/7

DRYOPTERIS
DILATATA
(Hoffm.) A. Gray
D. austriaca auct.

Broad
Buckler-fern

• All records

B ✦ 21/8

DRYOPTERIS
AEMULA (Ait.)
Kuntze

Hay-scented
Buckler-fern

• 1930 onwards
○ Before 1930

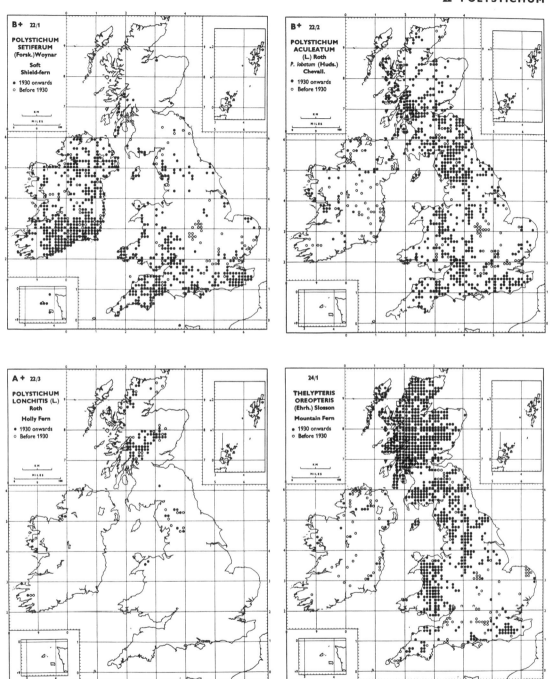

B+ 22/1

POLYSTICHUM SETIFERUM
(Forsk.)Woynar

Soft
Shield-fern

• 1930 onwards
○ Before 1930

B+ 22/2

POLYSTICHUM ACULEATUM
(L.) Roth
P. lobatum (Huds.)
Chevall.

• 1930 onwards
○ Before 1930

A+ 22/3

POLYSTICHUM LONCHITIS (L.)
Roth

Holly Fern

• 1930 onwards
○ Before 1930

24/1

THELYPTERIS OREOPTERIS
(Ehrh.) Slosson

Mountain Fern

• 1930 onwards
○ Before 1930

THELYPTERIDACEAE

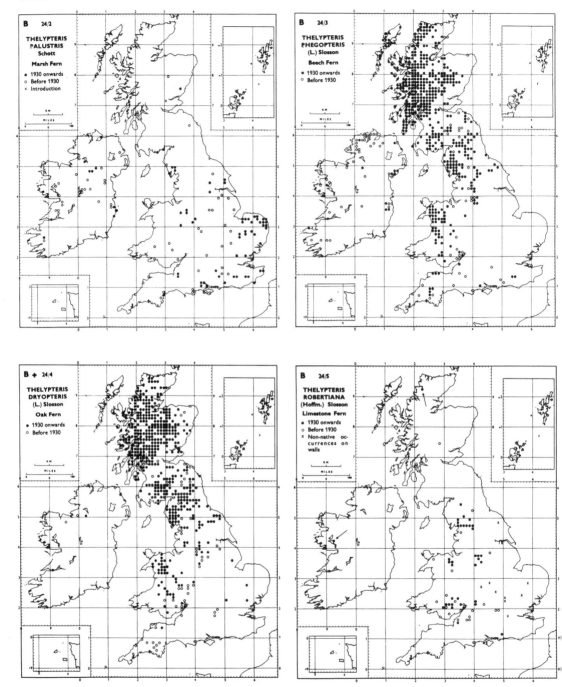

B 24/2

THELYPTERIS
PALUSTRIS
Schott

Marsh Fern

• 1930 onwards
○ Before 1930
× Introduction

B 24/3

THELYPTERIS
PHEGOPTERIS
(L.) Slosson

Beech Fern

• 1930 onwards
○ Before 1930

B ✛ 24/4

THELYPTERIS
DRYOPTERIS
(L.) Slosson

Oak Fern

• 1930 onwards
○ Before 1930

B 24/5

THELYPTERIS
ROBERTIANA
(Hoffm.) Slosson

Limestone Fern

• 1930 onwards
○ Before 1930
× Non-native oc-
  currences on
  walls

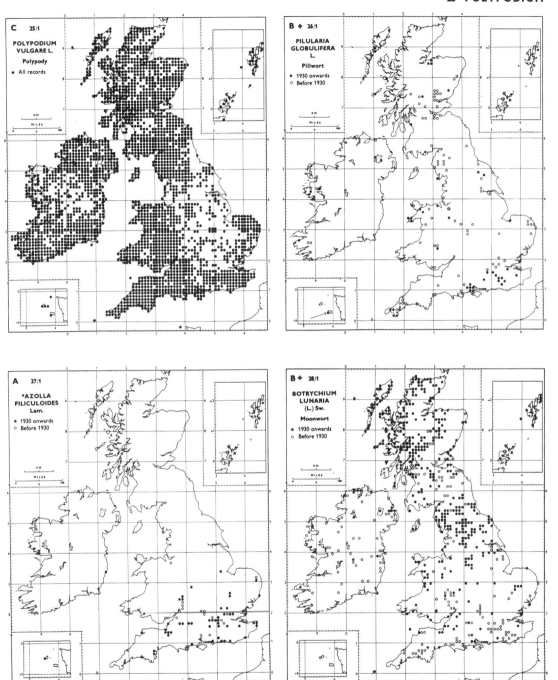

C 25/1
POLYPODIUM
VULGARE L.
Polypody
● All records

B + 26/1
PILULARIA
GLOBULIFERA
L.
Pillwort
● 1930 onwards
○ Before 1930

A 27/1
*AZOLLA
FILICULOIDES
Lam.
● 1930 onwards
○ Before 1930

B + 28/1
BOTRYCHIUM
LUNARIA
(L.) Sw.
Moonwort
● 1930 onwards
○ Before 1930

OPHIOGLOSSACEAE

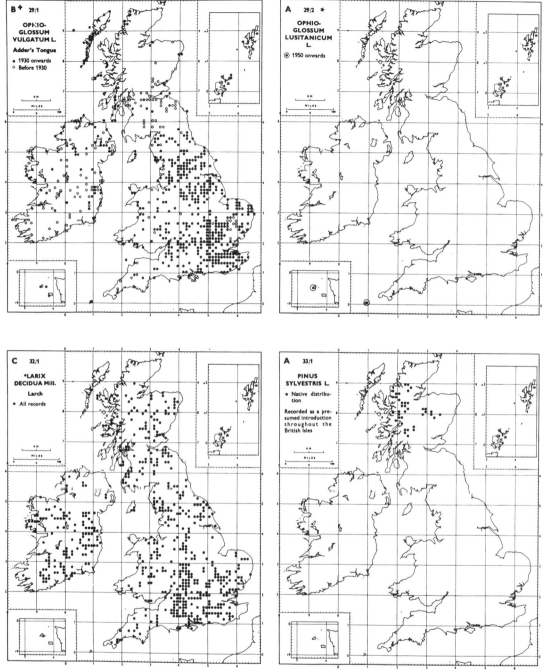

B ✛ 29/1

OPHIO-
GLOSSUM
VULGATUM L.

Adder's Tongue

● 1930 onwards
○ Before 1930

A 29/2 ✳

OPHIO-
GLOSSUM
LUSITANICUM
L.

⊙ 1950 onwards

C 32/1

*LARIX
DECIDUA Mill.

Larch

● All records

A 33/1

PINUS
SYLVESTRIS L.

● Native distribu-
tion

Recorded as a pre-
sumed introduction
throughout the
British Isles

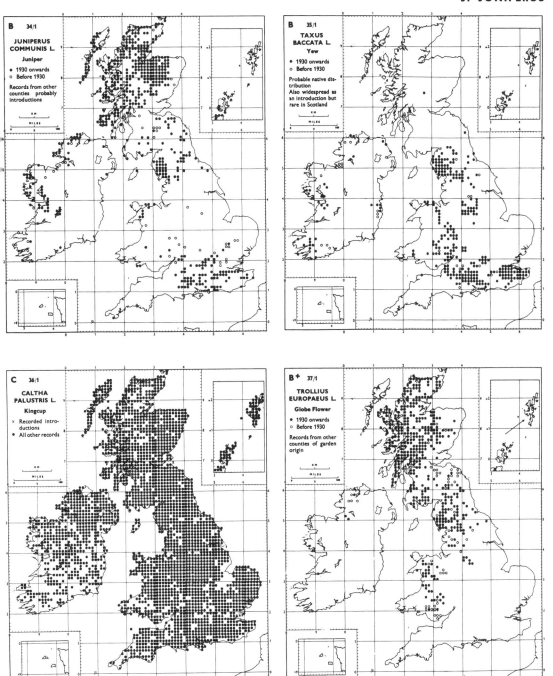

B 34/1

JUNIPERUS
COMMUNIS L.
Juniper

• 1930 onwards
○ Before 1930

Records from other
counties probably
introductions

KM
MILES

B 35/1

TAXUS
BACCATA L.
Yew

• 1930 onwards
○ Before 1930

Probable native dis-
tribution
Also widespread as
an introduction but
rare in Scotland

KM
MILES

C 36/1

CALTHA
PALUSTRIS L.
Kingcup

× Recorded intro-
ductions
• All other records

KM
MILES

B+ 37/1

TROLLIUS
EUROPAEUS L.
Globe Flower

• 1930 onwards
○ Before 1930

Records from other
counties of garden
origin

KM
MILES

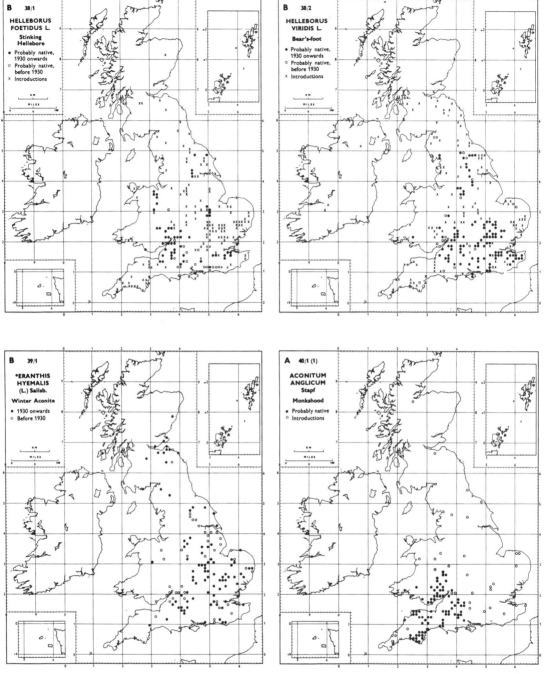

B 38/1

HELLEBORUS
FOETIDUS L.

**Stinking
Hellebore**

- Probably native,
  1930 onwards
○ Probably native,
  before 1930
× Introductions

B 38/2

HELLEBORUS
VIRIDIS L.

Bear's-foot

- Probably native,
  1930 onwards
○ Probably native,
  before 1930
× Introductions

B 39/1

•ERANTHIS
HYEMALIS
(L.) Salisb.

Winter Aconite

- 1930 onwards
○ Before 1930

A 40/1 (1)

ACONITUM
ANGLICUM
Stapf

Monkshood

- Probably native
○ Introductions

18

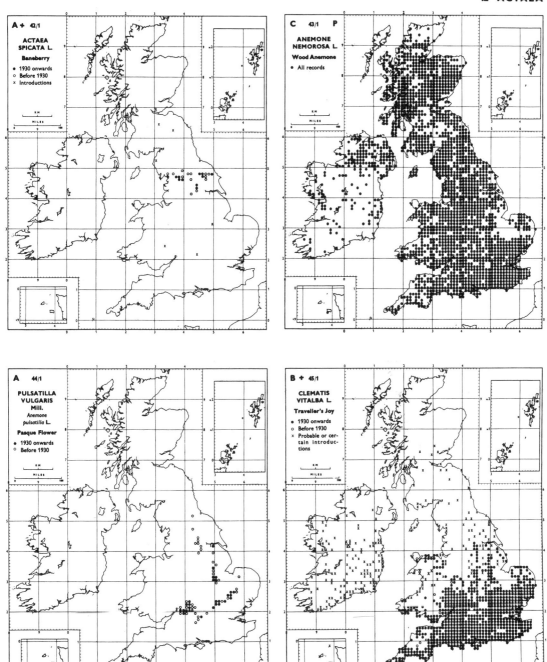

A + 42/1

**ACTAEA
SPICATA L.**

Baneberry

● 1930 onwards
○ Before 1930
× Introductions

C 43/1 **P**

**ANEMONE
NEMOROSA L.**

Wood Anemone

● All records

A 44/1

**PULSATILLA
VULGARIS
Mill.**
*Anemone
pulsatilla* L.

Pasque Flower

● 1930 onwards
○ Before 1930

B + 45/1

**CLEMATIS
VITALBA L.**

Traveller's Joy

● 1930 onwards
○ Before 1930
× Probable or cer-
  tain introduc-
  tions

19

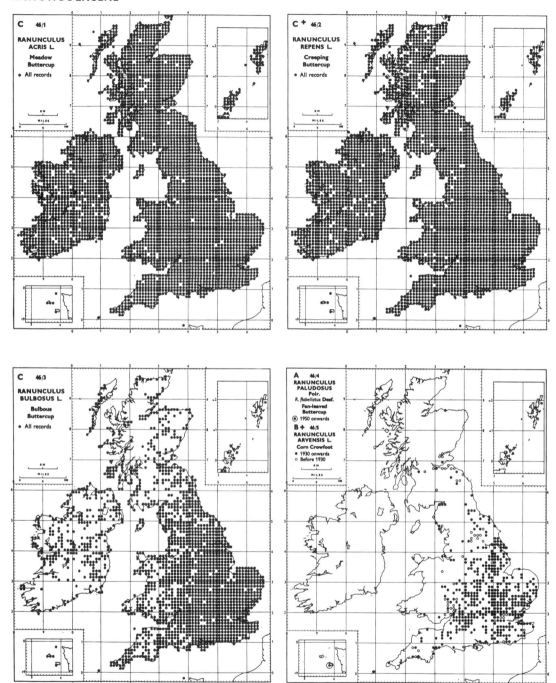

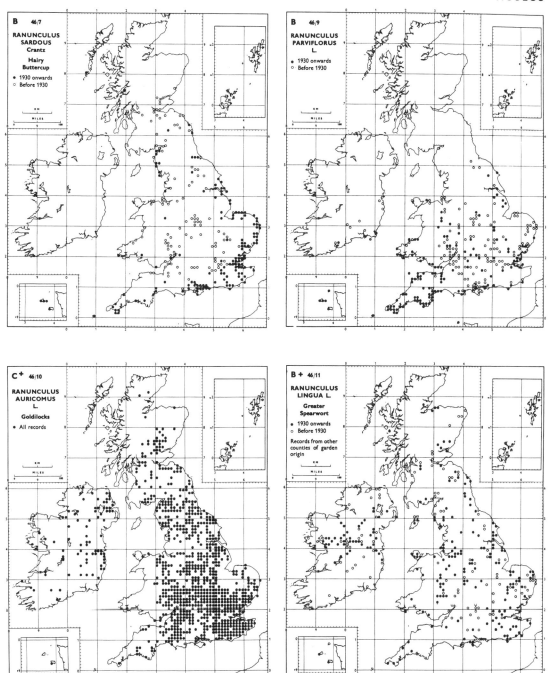

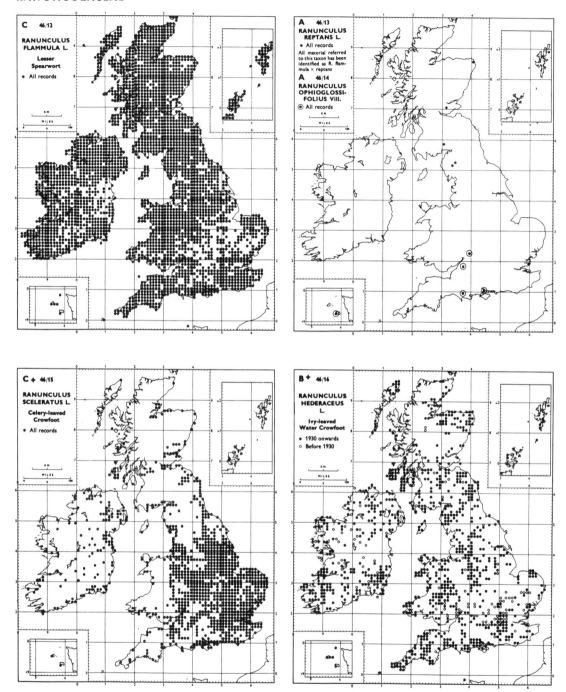

C 46/12
RANUNCULUS
FLAMMULA L.
Lesser
Spearwort
● All records

A 46/13
RANUNCULUS
REPTANS L.
● All records
All material referred
to this taxon has been
identified as R. flam-
mula × reptans
A 46/14
RANUNCULUS
OPHIOGLOSSI-
FOLIUS Vill.
◉ All records

C + 46/15
RANUNCULUS
SCELERATUS L.
Celery-leaved
Crowfoot
● All records

B + 46/16
RANUNCULUS
HEDERACEUS
L.
Ivy-leaved
Water Crowfoot
● 1930 onwards
○ Before 1930

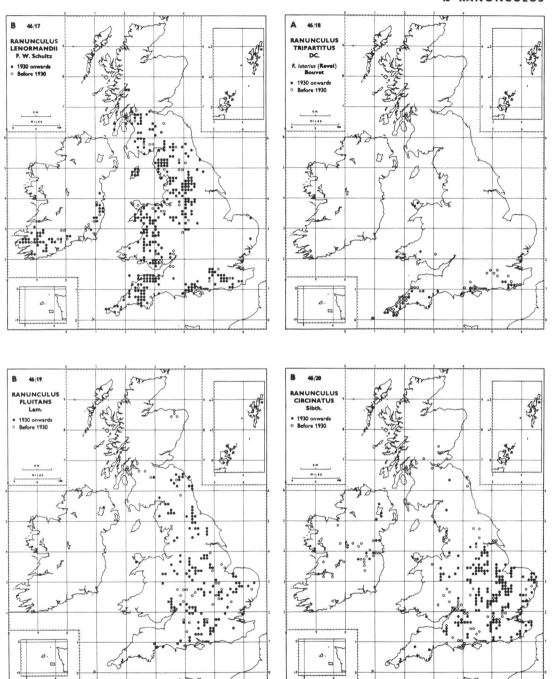

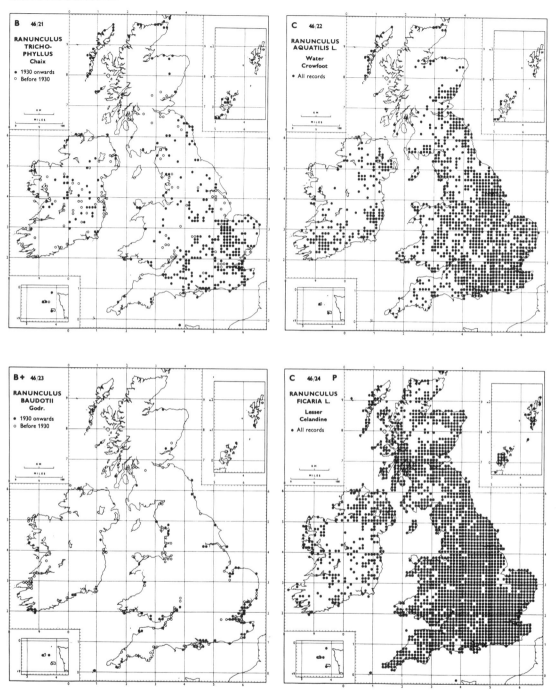

B 46/21

RANUNCULUS
TRICHO-
PHYLLUS
Chaix

● 1930 onwards
○ Before 1930

KM
MILES

C 46/22

RANUNCULUS
AQUATILIS L.

Water
Crowfoot

● All records

KM
MILES

B+ 46/23

RANUNCULUS
BAUDOTII
Godr.

● 1930 onwards
○ Before 1930

KM
MILES

C 46/24 **P**

RANUNCULUS
FICARIA L.

Lesser
Celandine

● All records

KM
MILES

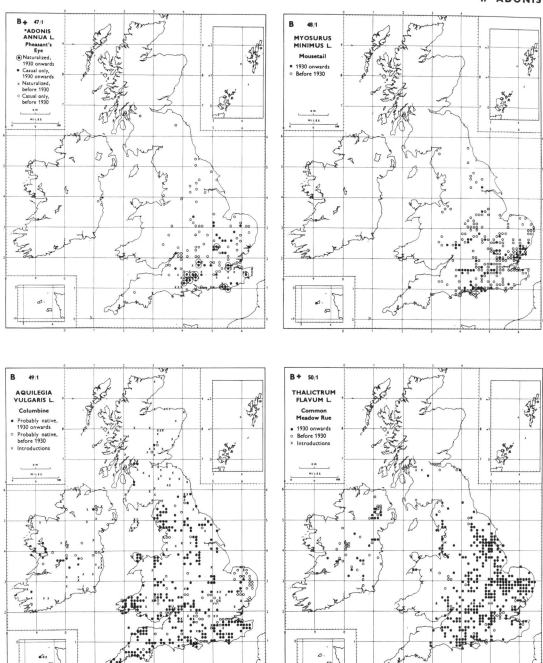

B ✛ 47/1
*ADONIS
ANNUA L.
Pheasant's
Eye
⊙ Naturalized,
   1930 onwards
• Casual only,
   1930 onwards
× Naturalized,
   before 1930
○ Casual only,
   before 1930

B   48/1
MYOSURUS
MINIMUS L.
Mousetail
• 1930 onwards
○ Before 1930

B   49/1
AQUILEGIA
VULGARIS L.
Columbine
• Probably native,
   1930 onwards
○ Probably native,
   before 1930
× Introductions

B ✛ 50/1
THALICTRUM
FLAVUM L.
Common
Meadow Rue
• 1930 onwards
○ Before 1930
× Introductions

RANUNCULACEAE

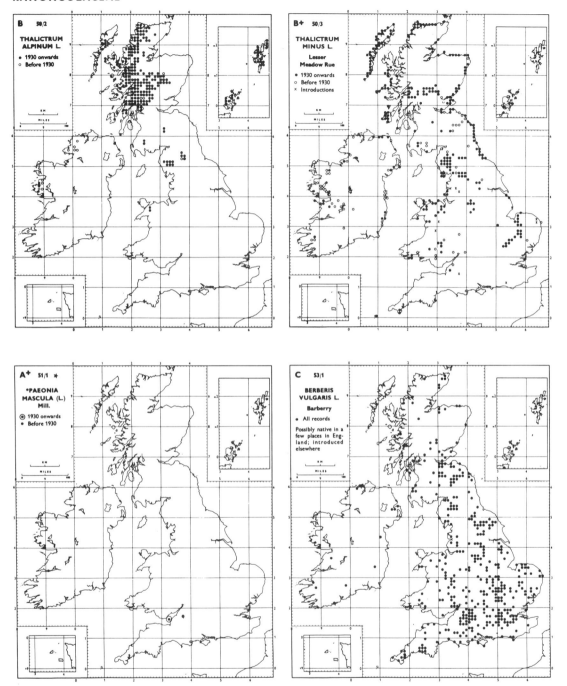

B 50/2

THALICTRUM
ALPINUM L.

- 1930 onwards
○ Before 1930

KM
MILES

B+ 50/3

THALICTRUM
MINUS L.

**Lesser
Meadow Rue**

- 1930 onwards
○ Before 1930
× Introductions

KM
MILES

A+ 51/1 ✳

*PAEONIA
MASCULA (L.)
Mill.

⊙ 1930 onwards
● Before 1930

KM
MILES

C 53/1

BERBERIS
VULGARIS L.

Barberry

- All records

Possibly native in a
few places in Eng-
land; introduced
elsewhere

KM
MILES

26

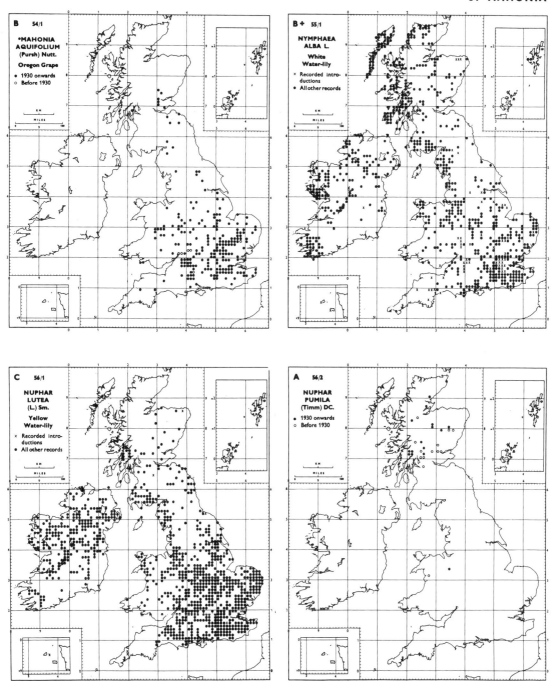

B 54/1

*MAHONIA
AQUIFOLIUM
(Pursh) Nutt.

Oregon Grape

• 1930 onwards
○ Before 1930

B+ 55/1

NYMPHAEA
ALBA L.

White
Water-lily

× Recorded intro-
ductions
• All other records

C 56/1

NUPHAR
LUTEA
(L.) Sm.

Yellow
Water-lily

× Recorded intro-
ductions
• All other records

A 56/2

NUPHAR
PUMILA
(Timm) DC.

• 1930 onwards
○ Before 1930

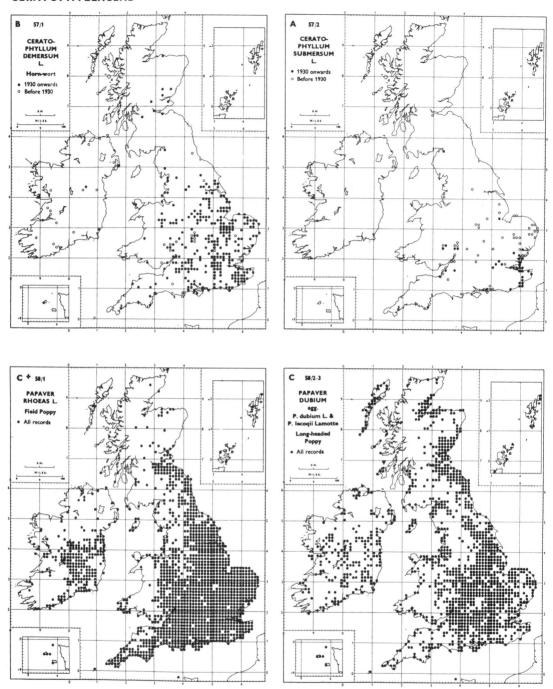

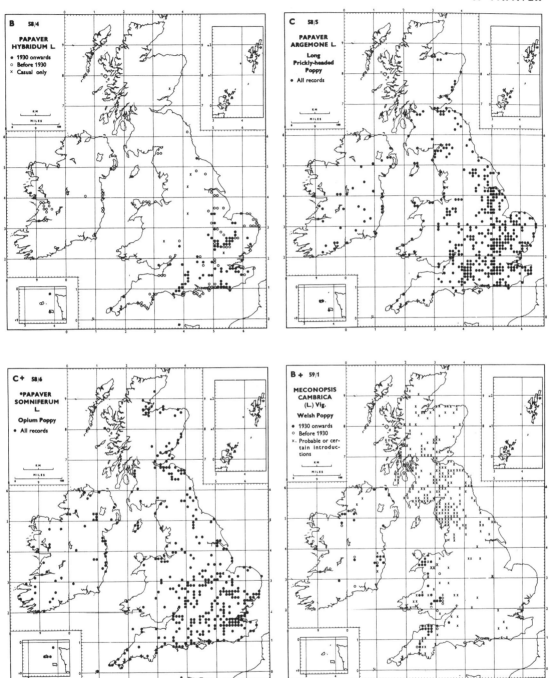

PAPAVERACEAE

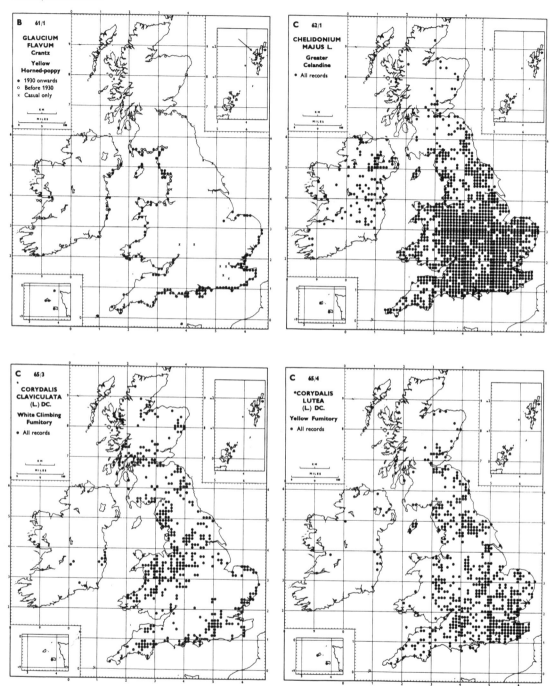

B 61/1

GLAUCIUM
FLAVUM
Crantz

Yellow
Horned-poppy

- 1930 onwards
- Before 1930
- × Casual only

C 62/1

CHELIDONIUM
MAJUS L.

Greater
Celandine

- All records

C 65/3

CORYDALIS
CLAVICULATA
(L.) DC.

White Climbing
Fumitory

- All records

C 65/4

°CORYDALIS
LUTEA
(L.) DC.

Yellow Fumitory

- All records

30

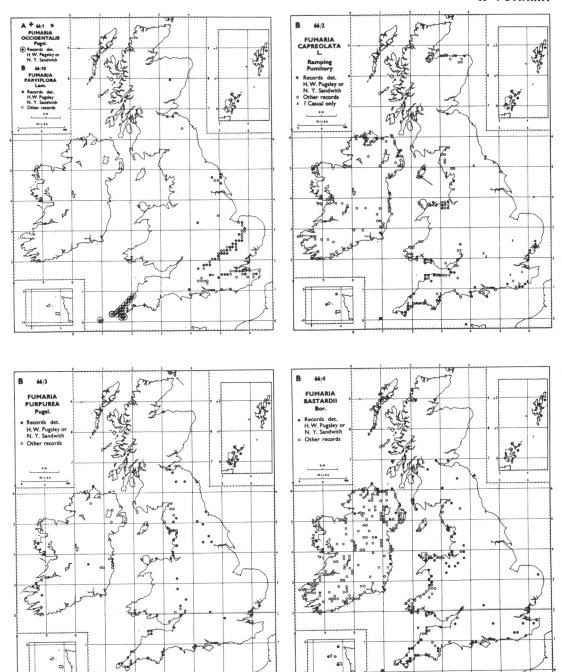

FUMARIACEAE

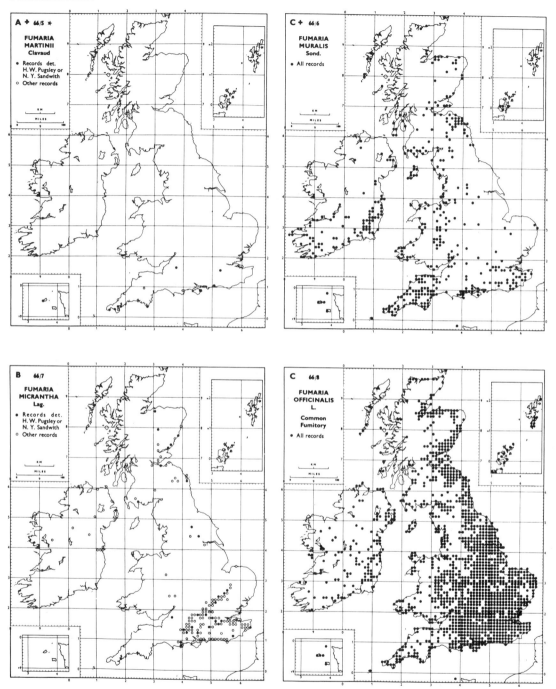

A + 66/5 ⚹

FUMARIA MARTINII
Clavaud

- Records det. H. W. Pugsley or N. Y. Sandwith
- ○ Other records

C + 66/6

FUMARIA MURALIS
Sond.

- All records

B 66/7

FUMARIA MICRANTHA
Lag.

- Records det. H. W. Pugsley or N. Y. Sandwith
- ○ Other records

C 66/8

FUMARIA OFFICINALIS
L.

Common Fumitory

- All records

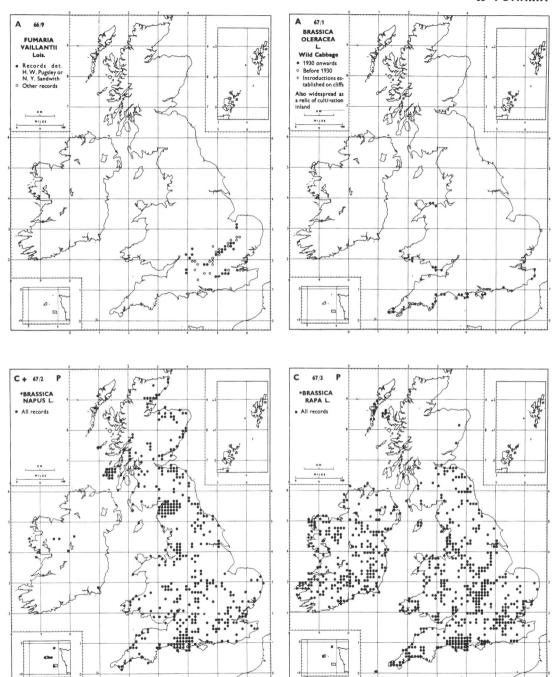

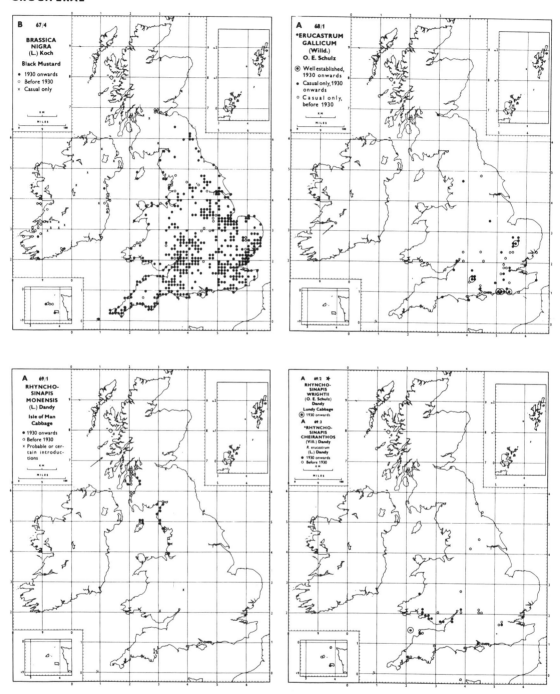

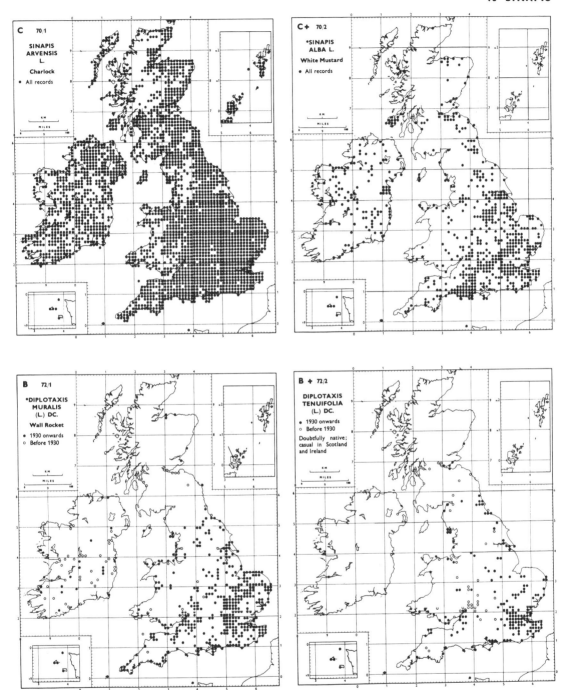

C 70/1

SINAPIS
ARVENSIS
L.

Charlock

● All records

C+ 70/2

*SINAPIS
ALBA L.

White Mustard

● All records

B 72/1

*DIPLOTAXIS
MURALIS
(L.) DC.

Wall Rocket

● 1930 onwards
○ Before 1930

B+ 72/2

DIPLOTAXIS
TENUIFOLIA
(L.) DC.

● 1930 onwards
○ Before 1930

Doubtfully native;
casual in Scotland
and Ireland

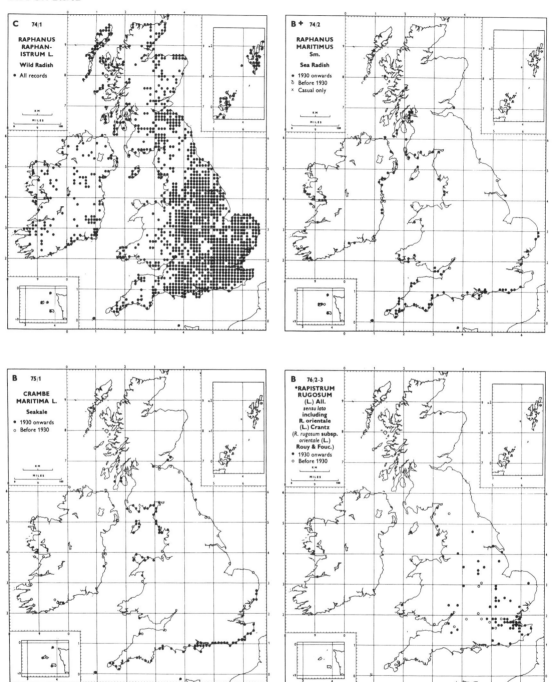

C 74/1

RAPHANUS
RAPHAN-
ISTRUM L.

Wild Radish

• All records

B + 74/2

RAPHANUS
MARITIMUS
Sm.

Sea Radish

• 1930 onwards
ठ Before 1930
× Casual only

B 75/1

CRAMBE
MARITIMA L.

Seakale

• 1930 onwards
o Before 1930

B 76/2-3

*RAPISTRUM
RUGOSUM
(L.) All.
sensu lato
including
R. orientale
(L.) Crantz
(R. rugosum subsp.
orientale (L.)
Rouy & Fouc.)

• 1930 onwards
o Before 1930

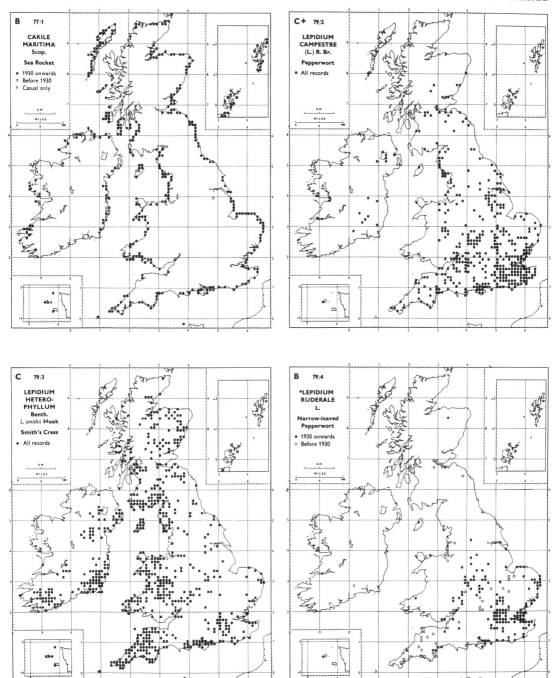

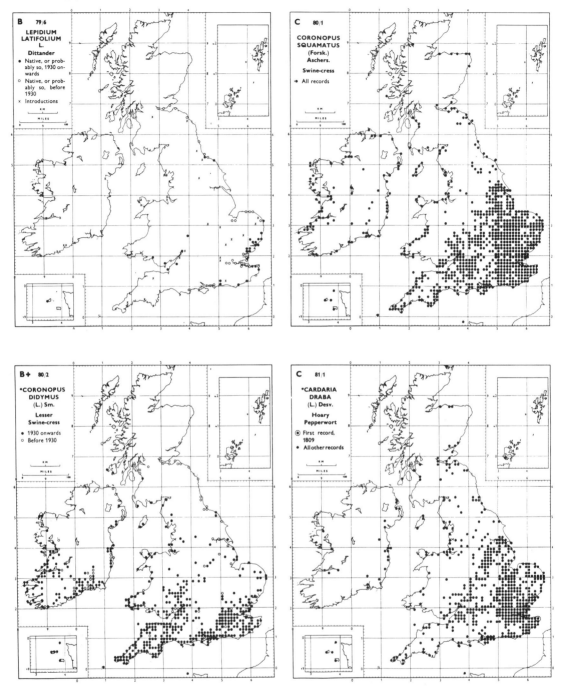

B 79/6
LEPIDIUM
LATIFOLIUM
L.
Dittander
● Native, or prob-
ably so, 1930 on-
wards
○ Native, or prob-
ably so, before
1930
× Introductions

C 80/1
CORONOPUS
SQUAMATUS
(Forsk.)
Aschers.
Swine-cress
•○ All records

B+ 80/2
*CORONOPUS
DIDYMUS
(L.) Sm.
Lesser
Swine-cress
● 1930 onwards
○ Before 1930

C 81/1
*CARDARIA
DRABA
(L.) Desv.
Hoary
Pepperwort
⊙ First record,
1809
● All other records

38

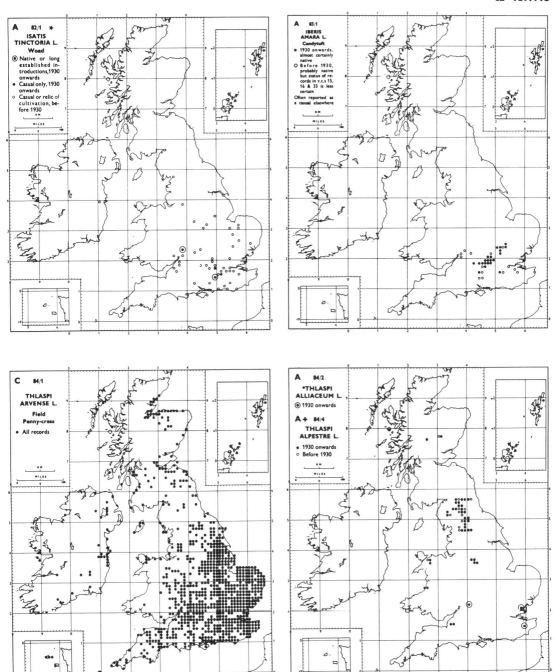

A 82/1 ★
ISATIS
TINCTORIA L.
Woad
⊙ Native or long established introductions,1930 onwards
● Casual only, 1930 onwards
○ Casual or relic of cultivation, before 1930

A 83/1
IBERIS
AMARA L.
Candytuft
● 1930 onwards, almost certainly native
○ Before 1930, probably native but status of records in v.c.s 15, 16 & 33 is less certain
Often reported as a casual elsewhere

C 84/1
THLASPI
ARVENSE L.
Field
Penny-cress
● All records

A 84/2
°THLASPI
ALLIACEUM L.
⊙ 1930 onwards
A+ 84/4
THLASPI
ALPESTRE L.
● 1930 onwards
○ Before 1930

CRUCIFERAE

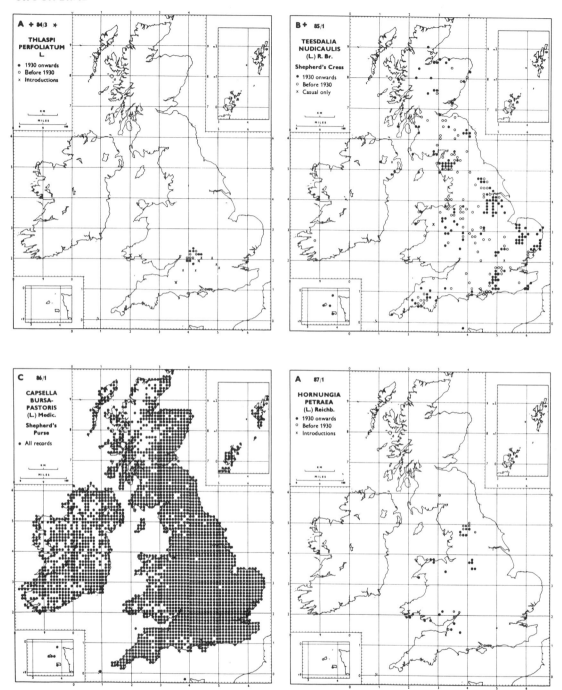

A ✦ 84/3 ★

THLASPI
PERFOLIATUM
L.

● 1930 onwards
○ Before 1930
× Introductions

KM
MILES

B ✦ 85/1

TEESDALIA
NUDICAULIS
(L.) R. Br.

Shepherd's Cress

● 1930 onwards
○ Before 1930
× Casual only

KM
MILES

C 86/1

CAPSELLA
BURSA-
PASTORIS
(L.) Medic.

Shepherd's
Purse

● All records

KM
MILES

A 87/1

HORNUNGIA
PETRAEA
(L.) Reichb.

● 1930 onwards
○ Before 1930
× Introductions

KM
MILES

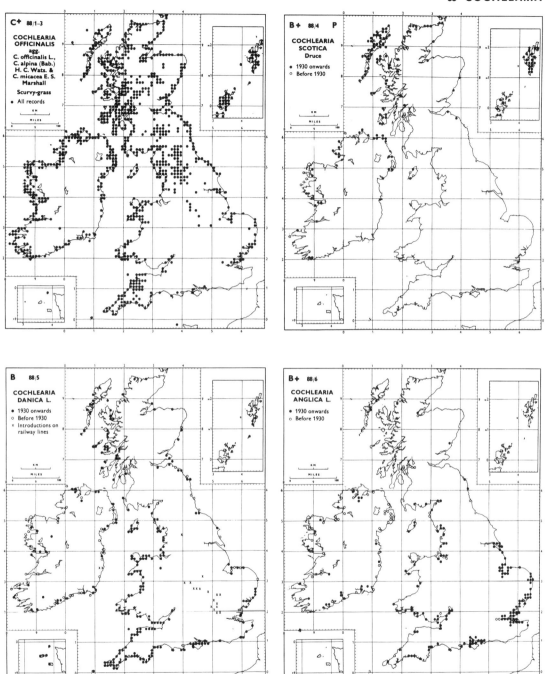

C+ 88/1–3

COCHLEARIA
OFFICINALIS
agg.
C. officinalis L.,
C. alpina (Bab.)
H. C. Wats. &
C. micacea E. S.
Marshall

Scurvy-grass

• All records

B+ 88/4 P

COCHLEARIA
SCOTICA
Druce

• 1930 onwards
○ Before 1930

B 88/5

COCHLEARIA
DANICA L.

• 1930 onwards
○ Before 1930
× Introductions on
   railway lines

B+ 88/6 P

COCHLEARIA
ANGLICA L.

• 1930 onwards
○ Before 1930

CRUCIFERAE

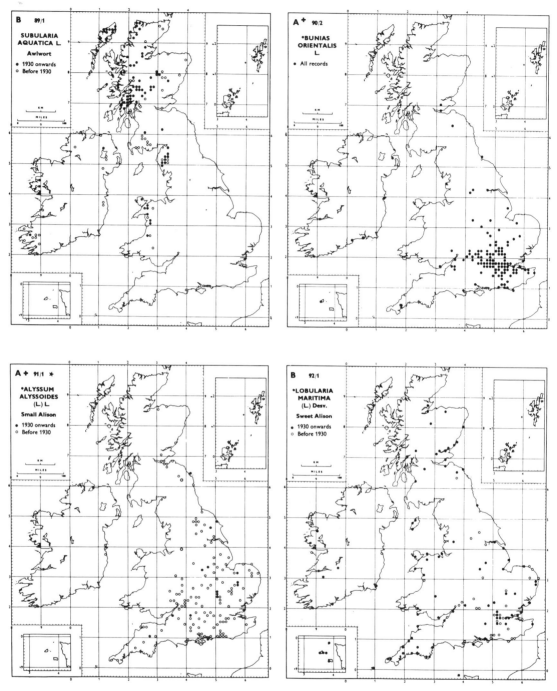

B   89/1

SUBULARIA
AQUATICA L.

Awlwort

● 1930 onwards
○ Before 1930

A+   90/2

*BUNIAS
ORIENTALIS
L.

● All records

A+   91/1 ✳

*ALYSSUM
ALYSSOIDES
(L.) L.

Small Alison

● 1930 onwards
○ Before 1930

B   92/1

*LOBULARIA
MARITIMA
(L.) Desv.

Sweet Alison

● 1930 onwards
○ Before 1930

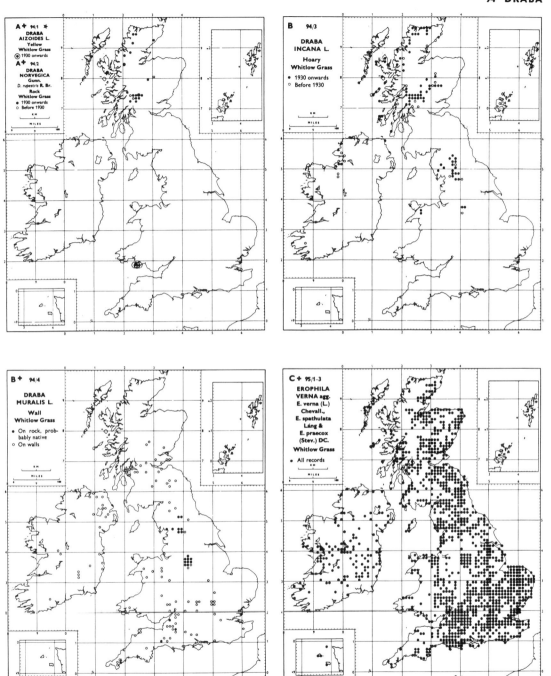

CRUCIFERAE

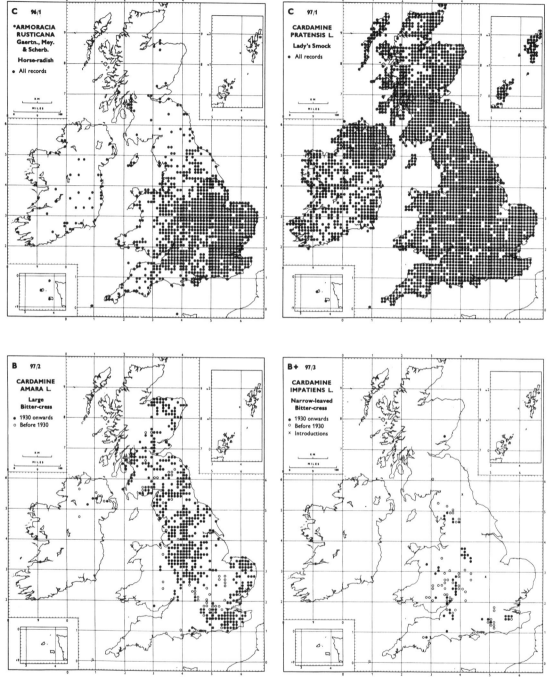

C 96/1

*ARMORACIA
RUSTICANA
Gaertn., Mey.
& Scherb.

Horse-radish

• All records

C 97/1

CARDAMINE
PRATENSIS L.

Lady's Smock

• All records

B 97/2

CARDAMINE
AMARA L.

Large
Bitter-cress

• 1930 onwards
○ Before 1930

B+ 97/3

CARDAMINE
IMPATIENS L.

Narrow-leaved
Bitter-cress

• 1930 onwards
○ Before 1930
× Introductions

44

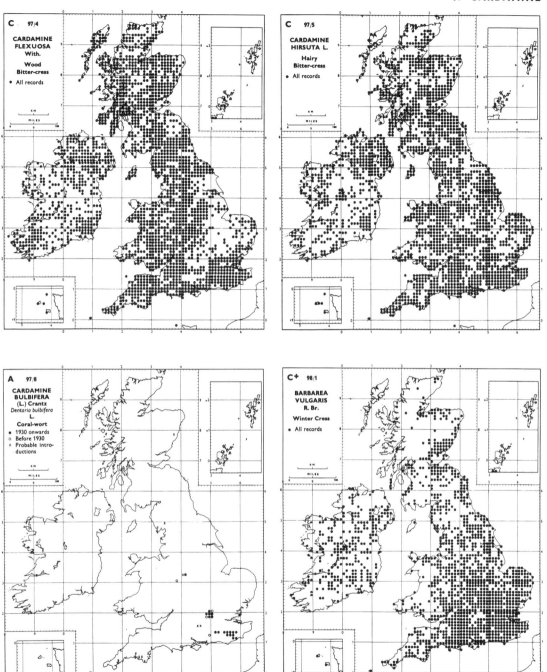

C · 97/4

CARDAMINE
FLEXUOSA
With.

Wood
Bitter-cress

• All records

C 97/5

CARDAMINE
HIRSUTA L.

Hairy
Bitter-cress

• All records

A 97/8

CARDAMINE
BULBIFERA
(L.) Crantz
Dentaria bulbifera
L.
Coral-wort

• 1930 onwards
○ Before 1930
× Probable intro-
  ductions

C⁺ 98/1

BARBAREA
VULGARIS
R. Br.

Winter-Cress

• All records

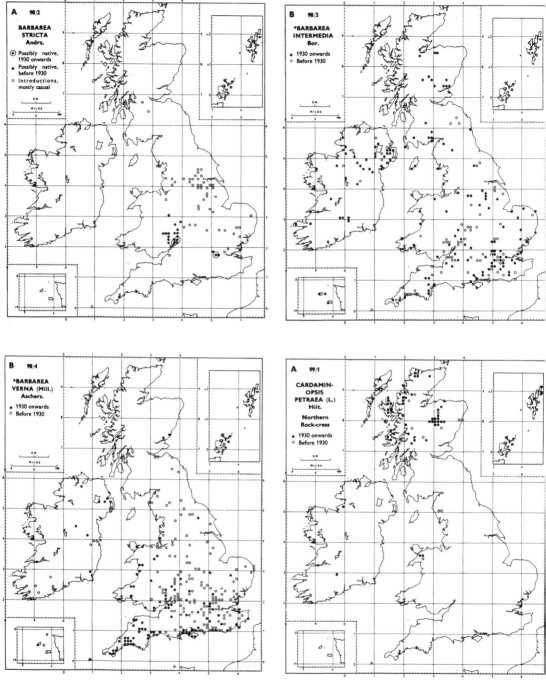

A 98/2

BARBAREA
STRICTA
Andrz.

⊙ Possibly native,
  1930 onwards
● Possibly native,
  before 1930
○ Introductions,
  mostly casual

B 98/3

*BARBAREA
INTERMEDIA
Bor.

● 1930 onwards
○ Before 1930

B 98/4

*BARBAREA
VERNA (Mill.)
Aschers.

● 1930 onwards
○ Before 1930

A 99/1

CARDAMIN-
OPSIS
PETRAEA (L.)
Hiit.

Northern
Rock-cress

● 1930 onwards
○ Before 1930

46

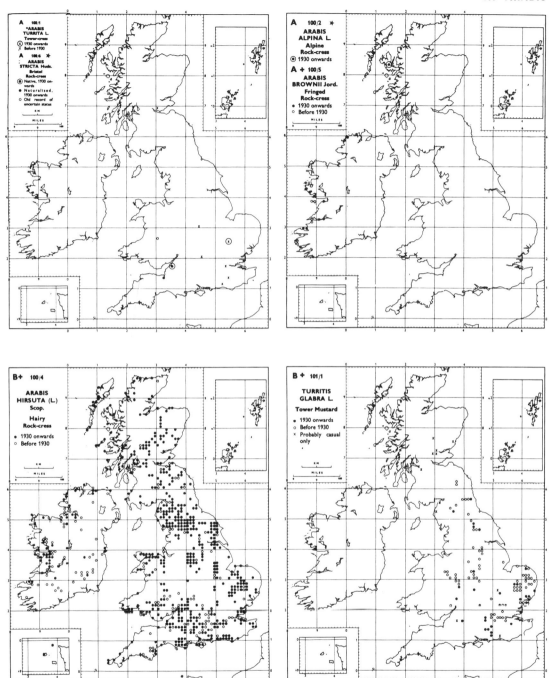

A   100/1
*ARABIS
TURRITA L.
Tower-cress
⊗ 1930 onwards
✕ Before 1930

A   100/6   ✱
ARABIS
STRICTA Huds.
Bristol
Rock-cress
⊙ Native, 1930 on-
wards
● Naturalized,
1930 onwards
○ Old record of
uncertain status

A   100/2   ✱
ARABIS
ALPINA L.
Alpine
Rock-cress
⊙ 1930 onwards

A + 100/5
ARABIS
BROWNII Jord.
Fringed
Rock-cress
● 1930 onwards
○ Before 1930

B + 100/4
ARABIS
HIRSUTA (L.)
Scop.
Hairy
Rock-cress
● 1930 onwards
○ Before 1930

B + 101/1
TURRITIS
GLABRA L.
Tower Mustard
● 1930 onwards
○ Before 1930
✕ Probably casual
only

47

CRUCIFERAE

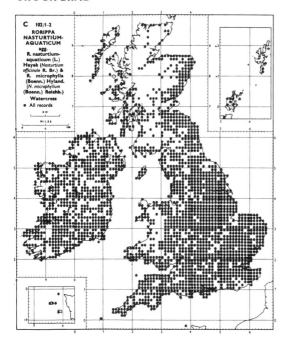

It was intended that
Rorippa nasturtium-aquaticum
and *R. microphylla*
should be mapped separately,
but the data received
proved to be inadequate.

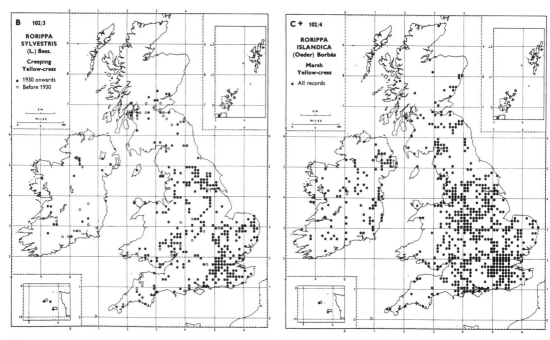

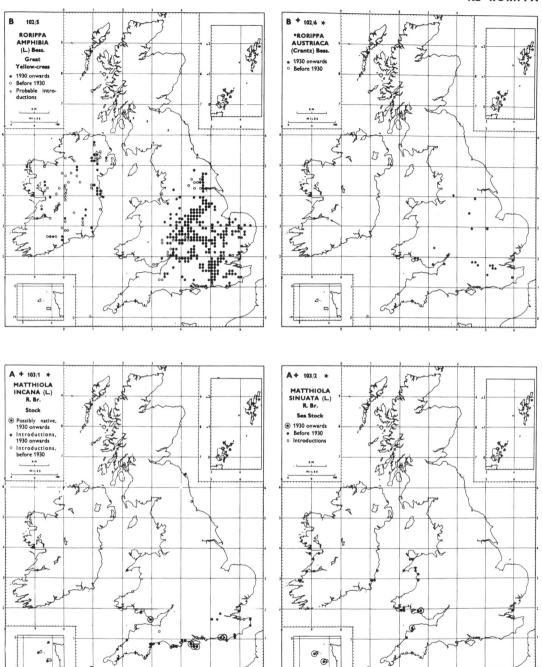

B 102/5

RORIPPA
AMPHIBIA
(L.) Bess.

Great
Yellow-cress

• 1930 onwards
○ Before 1930
× Probable intro-
  ductions

B + 102/6 *

*RORIPPA
AUSTRIACA
(Crantz) Bess.

• 1930 onwards
○ Before 1930

A + 103/1 *

MATTHIOLA
INCANA (L.)
R. Br.

Stock

◉ Possibly native,
  1930 onwards
• Introductions,
  1930 onwards
○ Introductions,
  before 1930

A + 103/2 *

MATTHIOLA
SINUATA (L.)
R. Br.

Sea Stock

◉ 1930 onwards
• Before 1930
○ Introductions

49

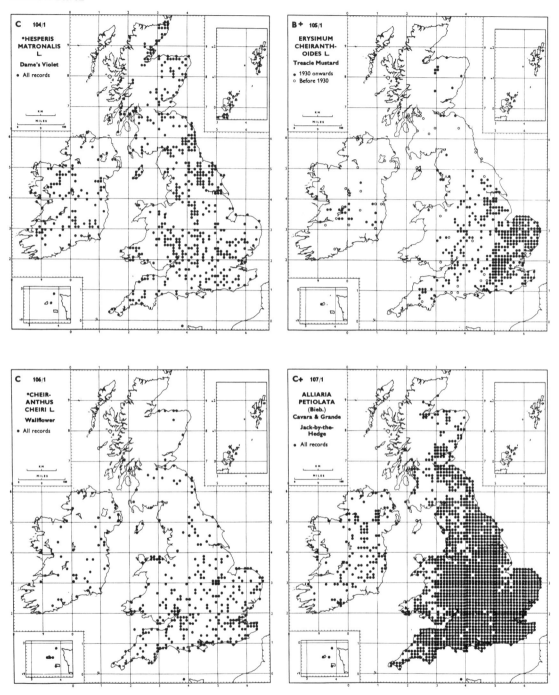

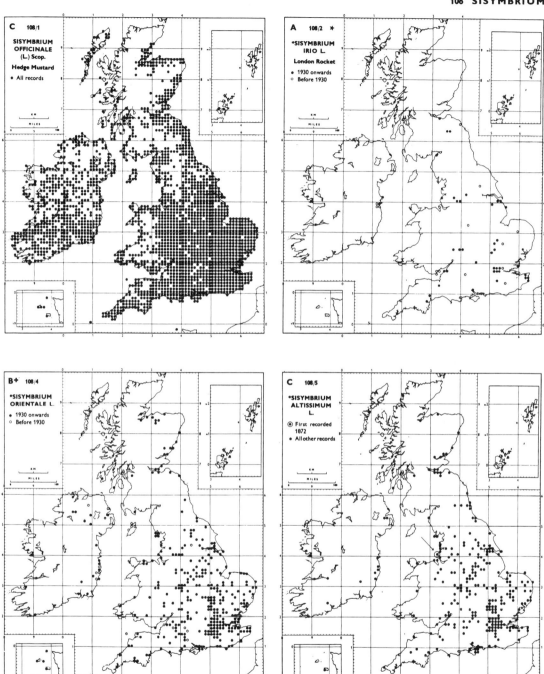

C 108/1

SISYMBRIUM
OFFICINALE
(L.) Scop.

Hedge Mustard

● All records

A 108/2 ✳

*SISYMBRIUM
IRIO L.

London Rocket

● 1930 onwards
○ Before 1930

B+ 108/4

*SISYMBRIUM
ORIENTALE L.

● 1930 onwards
○ Before 1930

C 108/5

*SISYMBRIUM
ALTISSIMUM
L.

◉ First recorded
  1872
● All other records

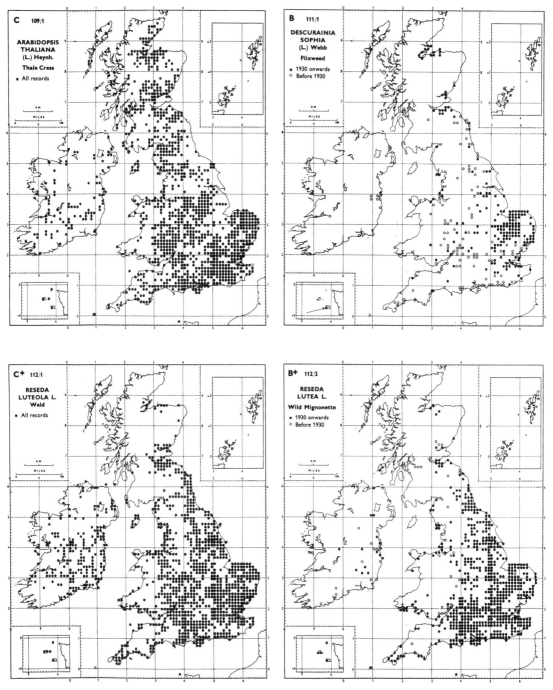

C 109/1

ARABIDOPSIS
THALIANA
(L.) Heynh.

Thale Cress

• All records

B 111/1

DESCURAINIA
SOPHIA
(L.) Webb

Flixweed

• 1930 onwards
○ Before 1930

C+ 112/1

RESEDA
LUTEOLA L.
Weld

• All records

B+ 112/2

RESEDA
LUTEA L.

Wild Mignonette

• 1930 onwards
○ Before 1930

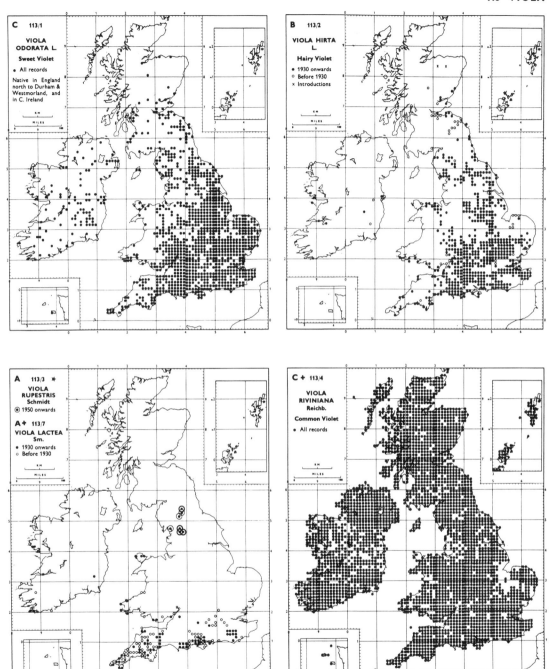

C 113/1

VIOLA
ODORATA L.

Sweet Violet

• All records

Native in England
north to Durham &
Westmorland, and
in C. Ireland

B 113/2

VIOLA HIRTA
L.

Hairy Violet

• 1930 onwards
○ Before 1930
× Introductions

A 113/3 ✱

VIOLA
RUPESTRIS
Schmidt

◉ 1950 onwards

A + 113/7

VIOLA LACTEA
Sm.

• 1930 onwards
○ Before 1930

C + 113/4

VIOLA
RIVINIANA
Reichb.

Common Violet

• All records

VIOLACEAE

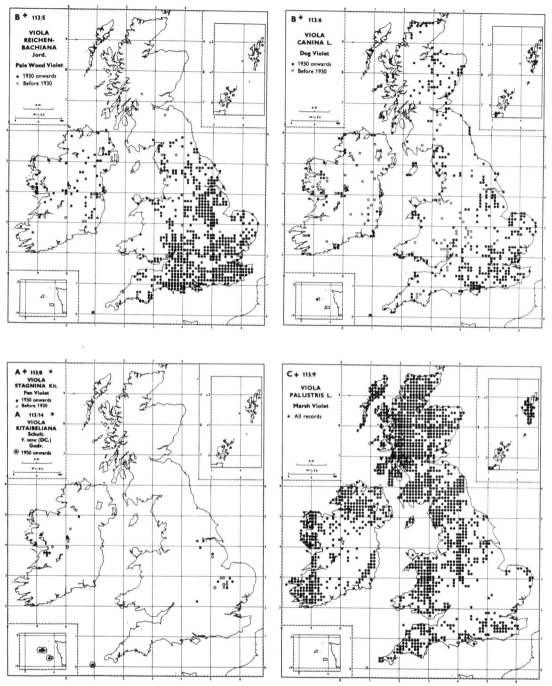

B + 113/5

VIOLA
REICHEN-
BACHIANA
Jord.

Pale Wood Violet

• 1930 onwards
○ Before 1930

B + 113/6

VIOLA
CANINA L.

Dog Violet

• 1930 onwards
○ Before 1930

A + 113/8 *

VIOLA
STAGNINA Kit.
Fen Violet
• 1930 onwards
○ Before 1930

A   113/14 *

VIOLA
KITAIBELIANA
Schult.
V. nana (DC.)
Godr.
◉ 1950 onwards

C + 113/9

VIOLA
PALUSTRIS L.

Marsh Violet

• All records

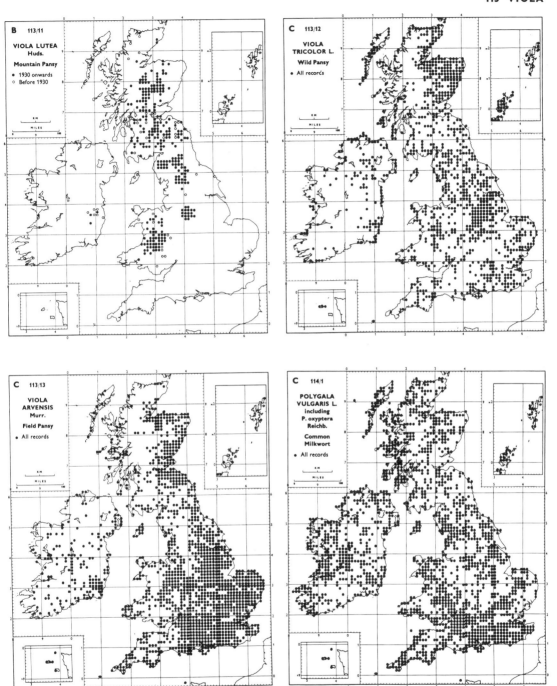

B 113/11

VIOLA LUTEA
Huds.
Mountain Pansy

- 1930 onwards
- Before 1930

C 113/12

VIOLA TRICOLOR L.
Wild Pansy

- All records

C 113/13

VIOLA ARVENSIS
Murr.
Field Pansy

- All records

C 114/1

POLYGALA VULGARIS L.
including
P. oxyptera
Reichb.

Common Milkwort

- All records

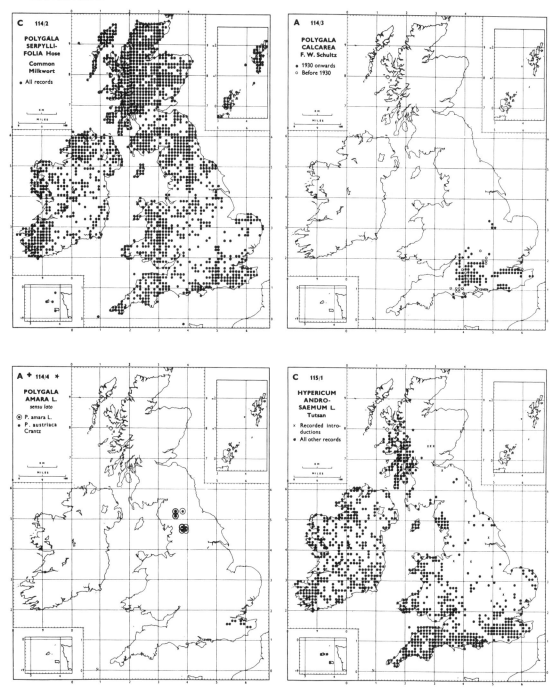

C 114/2

POLYGALA
SERPYLLI-
FOLIA Hose

Common
Milkwort

• All records

KM
MILES

A 114/3

POLYGALA
CALCAREA
F. W. Schultz

• 1930 onwards
○ Before 1930

KM
MILES

A + 114/4 *

POLYGALA
AMARA L.
sensu lato

⊙ P. amara L.
• P. austriaca
Crantz

KM
MILES

C 115/1

HYPERICUM
ANDRO-
SAEMUM L.
Tutsan

× Recorded intro-
ductions
• All other records

KM
MILES

56

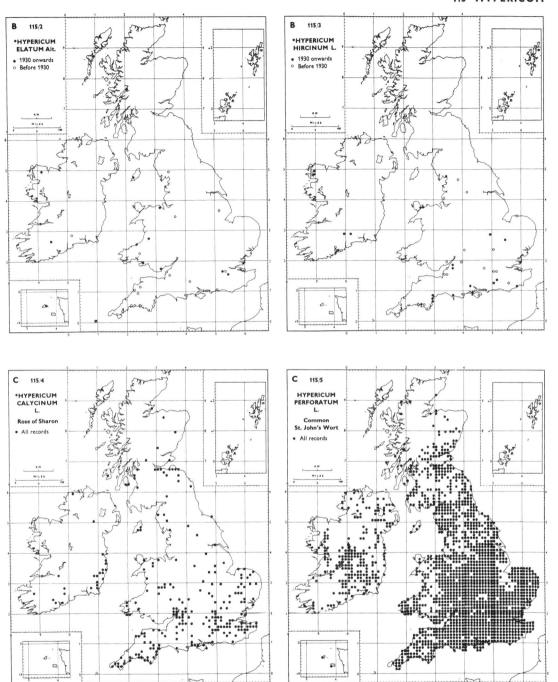

GUTTIFERAE

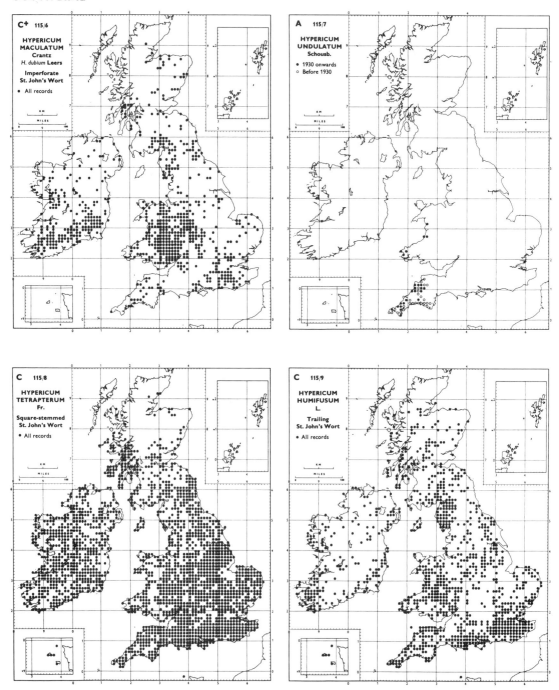

C+ 115/6

HYPERICUM
MACULATUM
Crantz
H. dubium Leers

Imperforate
St. John's Wort

● All records

KM
MILES

A 115/7

HYPERICUM
UNDULATUM
Schousb.

● 1930 onwards
○ Before 1930

KM
MILES

C 115/8

HYPERICUM
TETRAPTERUM
Fr.

Square-stemmed
St. John's Wort

● All records

KM
MILES

C 115/9

HYPERICUM
HUMIFUSUM
L

Trailing
St. John's Wort

● All records

KM
MILES

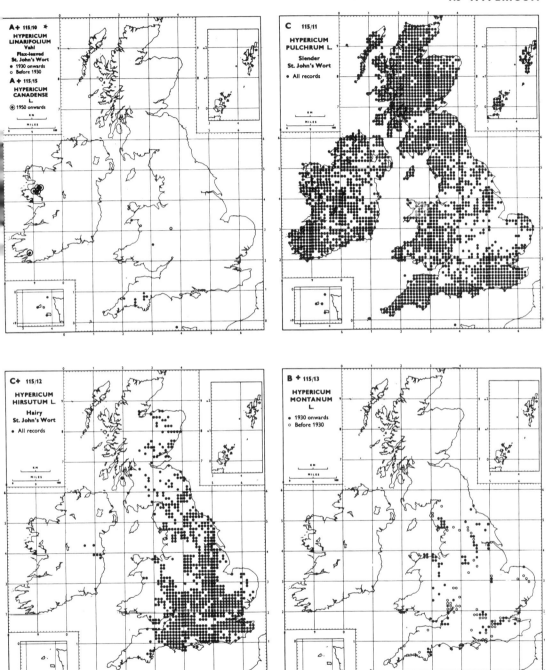

A+ 115/10 ✱

HYPERICUM
LINARIFOLIUM
Vahl

Flax-leaved
St. John's Wort
● 1930 onwards
○ Before 1930

A + 115/15

HYPERICUM
CANADENSE
L.

◉ 1950 onwards

C 115/11

HYPERICUM
PULCHRUM L.

Slender
St. John's Wort
● All records

C+ 115/12

HYPERICUM
HIRSUTUM L.

Hairy
St. John's Wort
● All records

B + 115/13

HYPERICUM
MONTANUM
L.

● 1930 onwards
○ Before 1930

GUTTIFERAE

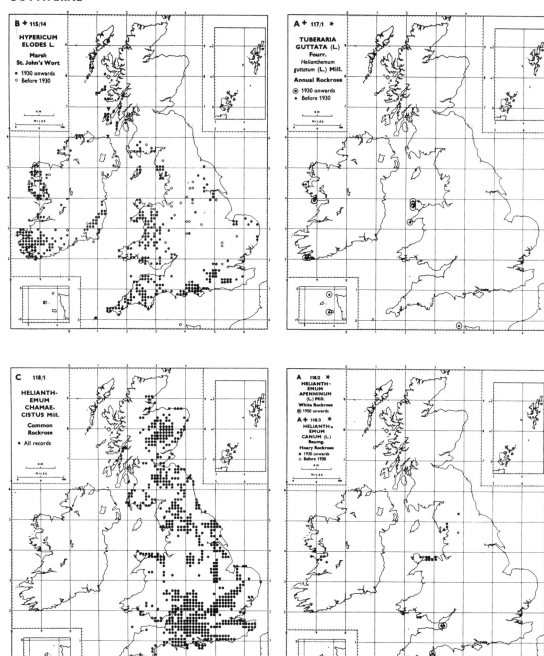

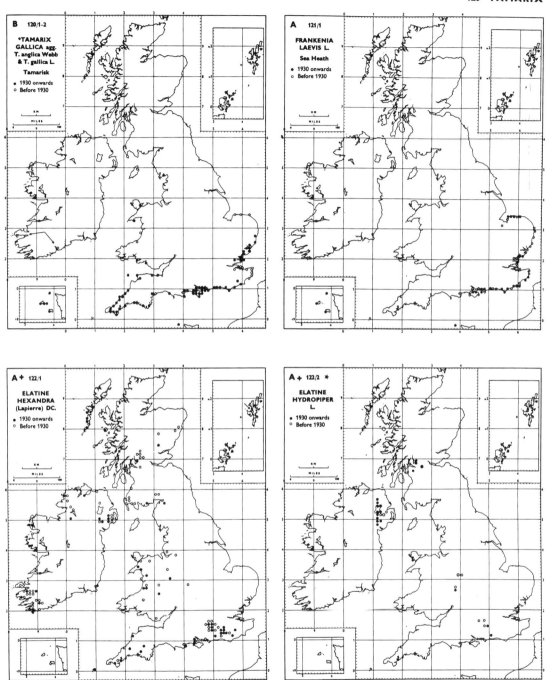

B 120/1-2

*TAMARIX
GALLICA agg.
T. anglica Webb
& T. gallica L.

Tamarisk

• 1930 onwards
○ Before 1930

A 121/1

FRANKENIA
LAEVIS L.

Sea Heath

• 1930 onwards
○ Before 1930

A + 122/1

ELATINE
HEXANDRA
(Lapierre) DC.

• 1930 onwards
○ Before 1930

A + 122/2 ✳

ELATINE
HYDROPIPER
L

• 1930 onwards
○ Before 1930

CARYOPHYLLACEAE

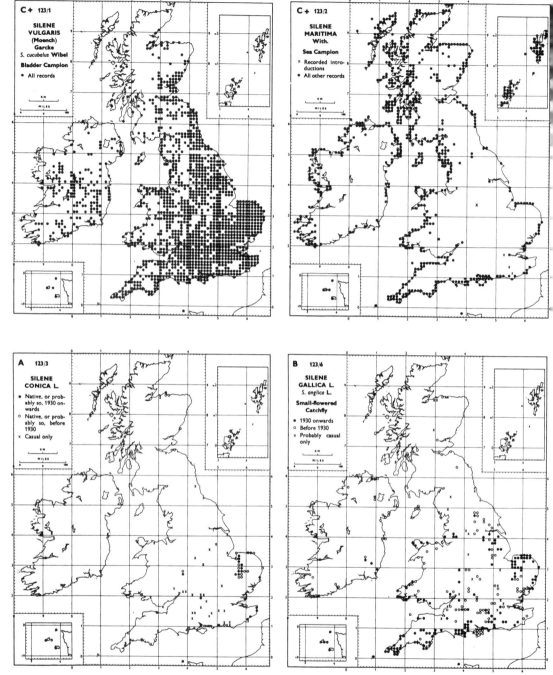

C+ 123/1

SILENE
VULGARIS
(Moench)
Garcke
S. cucubalus Wibel

Bladder Campion

● All records

C+ 123/2

SILENE
MARITIMA
With.

Sea Campion

× Recorded intro-
ductions
● All other records

A 123/3

SILENE
CONICA L.

● Native, or prob-
ably so, 1930 on-
wards
○ Native, or prob-
ably so, before
1930
× Casual only

B 123/6

SILENE
GALLICA L.
S. anglica L.

Small-flowered
Catchfly

● 1930 onwards
○ Before 1930
× Probably casual
only

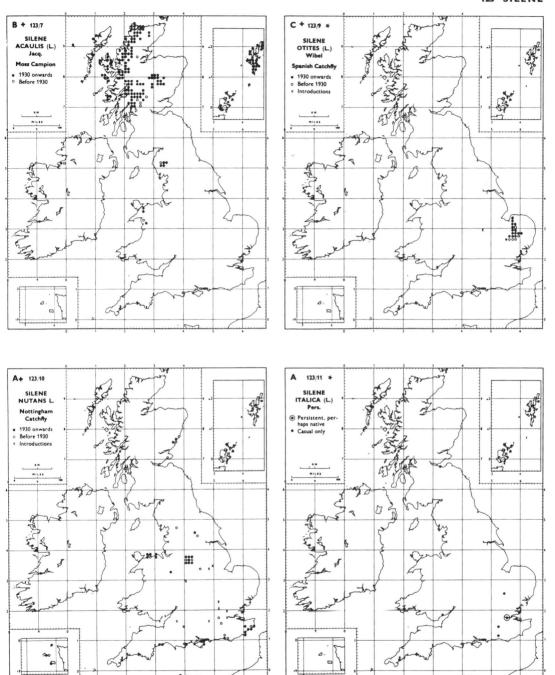

B + 123/7

SILENE ACAULIS (L.) Jacq.

Moss Campion

- 1930 onwards
- Before 1930

C + 123/9 ✳

SILENE OTITES (L.) Wibel

Spanish Catchfly

- 1930 onwards
- Before 1930
- × Introductions

A + 123/10

SILENE NUTANS L.

Nottingham Catchfly

- 1930 onwards
- Before 1930
- × Introductions

A 123/11 ✳

SILENE ITALICA (L.) Pers.

- ⊙ Persistent, perhaps native
- Casual only

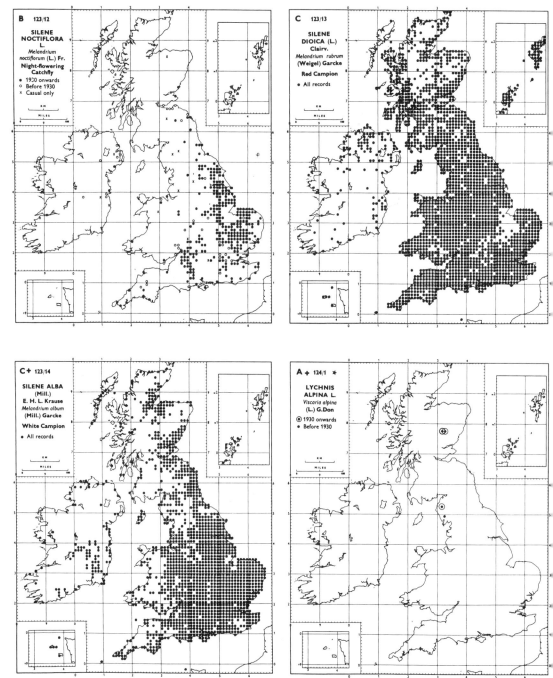

B 123/12

SILENE NOCTIFLORA L.
Melandrium noctiflorum (L.) Fr.
Night-flowering Catchfly

- 1930 onwards
- Before 1930
- × Casual only

KM
MILES

C 123/13

SILENE DIOICA (L.) Clairv.
Melandrium rubrum (Weigel) Garcke
Red Campion

- All records

KM
MILES

C + 123/14

SILENE ALBA (Mill.) E. H. L. Krause
Melandrium album (Mill.) Garcke
White Campion

- All records

KM
MILES

A + 124/1 ✳

LYCHNIS ALPINA L.
Viscaria alpina (L.) G.Don

- ⊙ 1930 onwards
- Before 1930

KM
MILES

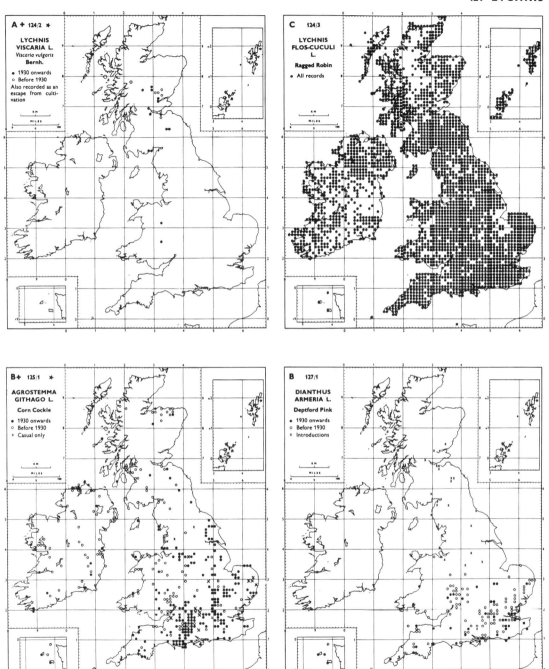

A + 124/2 ✱

**LYCHNIS
VISCARIA L.**
Viscaria vulgaris
Bernh.

● 1930 onwards
○ Before 1930
Also recorded as an
escape from culti-
vation

C ● 124/3

**LYCHNIS
FLOS-CUCULI
L.**

Ragged Robin

● All records

B + 125/1 ✱

**AGROSTEMMA
GITHAGO L.**

Corn Cockle

● 1930 onwards
○ Before 1930
× Casual only

B 127/1

**DIANTHUS
ARMERIA L.**

Deptford Pink

● 1930 onwards
○ Before 1930
× Introductions

65

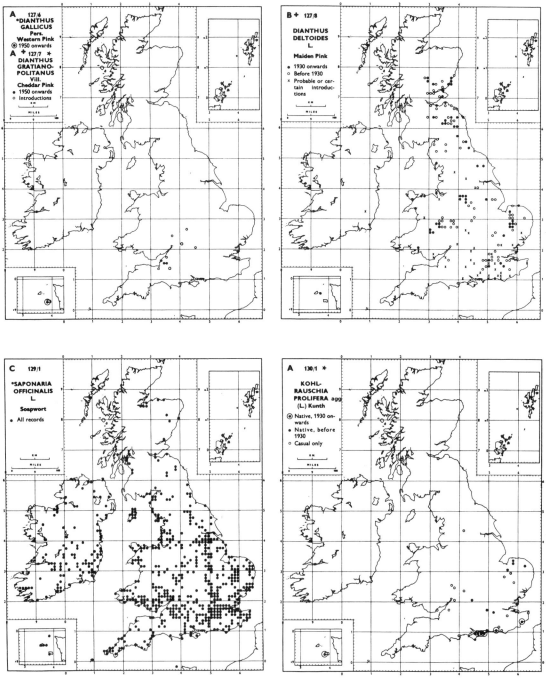

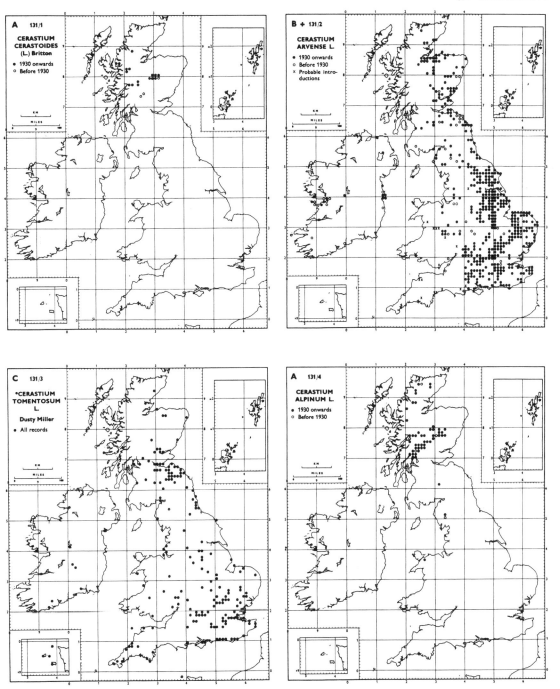

A 131/1

CERASTIUM
CERASTOIDES
(L.) Britton

• 1930 onwards
○ Before 1930

B ✛ 131/2

CERASTIUM
ARVENSE L.

• 1930 onwards
○ Before 1930
✕ Probable intro-
    ductions

C 131/3

*CERASTIUM
TOMENTOSUM
L

Dusty Miller

• All records

A 131/4

CERASTIUM
ALPINUM L.

• 1930 onwards
○ Before 1930

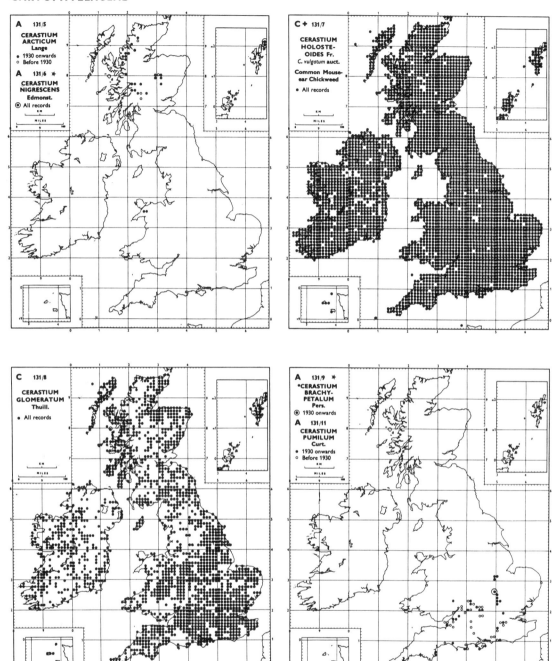

A   131/5

CERASTIUM
ARCTICUM
Lange
● 1930 onwards
○ Before 1930

A   131/6 ✶

CERASTIUM
NIGRESCENS
Edmonst.
◉ All records

C ✛ 131/7

CERASTIUM
HOLOSTE-
OIDES Fr.
C. vulgatum auct.

Common Mouse-
ear Chickweed
● All records

C   131/8

CERASTIUM
GLOMERATUM
Thuill.
● All records

A   131/9 ✶

*CERASTIUM
BRACHY-
PETALUM
Pers.
◉ 1930 onwards

A   131/11

CERASTIUM
PUMILUM
Curt.
● 1930 onwards
○ Before 1930

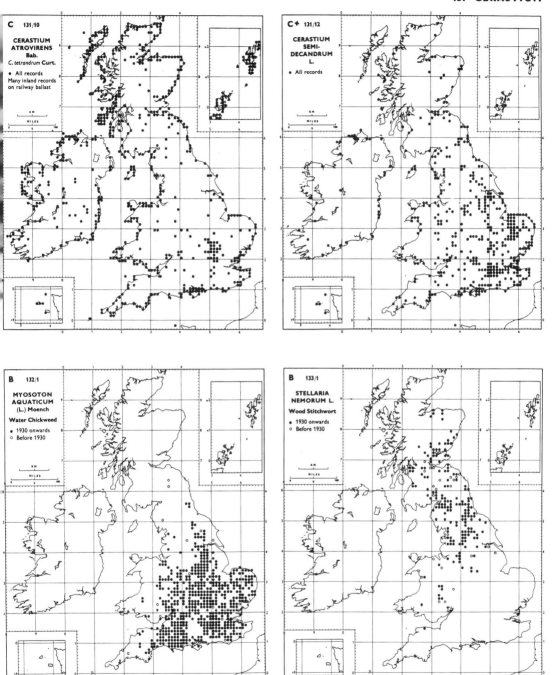

C 131/10

CERASTIUM ATROVIRENS
Bab.
C. tetrandrum Curt.

● All records
Many inland records on railway ballast

C✛ 131/12

CERASTIUM SEMI-DECANDRUM
L.

● All records

B 132/1

MYOSOTON AQUATICUM
(L.) Moench

Water Chickweed

● 1930 onwards
○ Before 1930

B 133/1

STELLARIA NEMORUM L.

Wood Stitchwort

● 1930 onwards
○ Before 1930

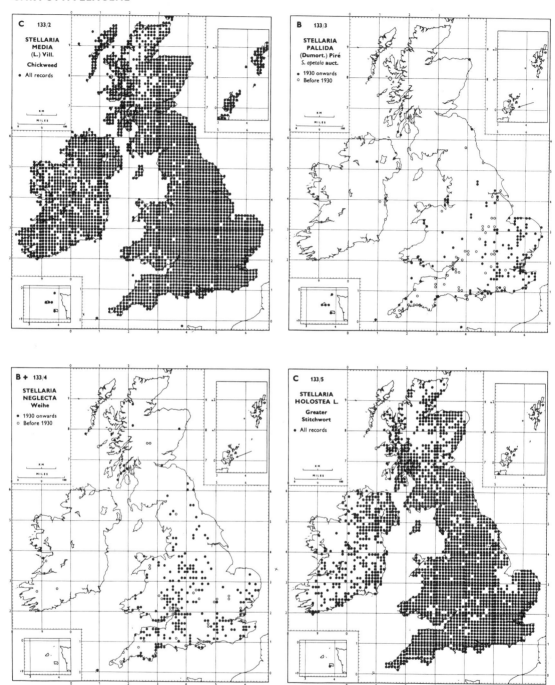

C 133/2

STELLARIA
MEDIA
(L.) Vill.

Chickweed

• All records

B 133/3

STELLARIA
PALLIDA
(Dumort.) Piré
S. apetala auct.

• 1930 onwards
○ Before 1930

B + 133/4

STELLARIA
NEGLECTA
Weihe

• 1930 onwards
○ Before 1930

C 133/5

STELLARIA
HOLOSTEA L.

Greater
Stitchwort

• All records

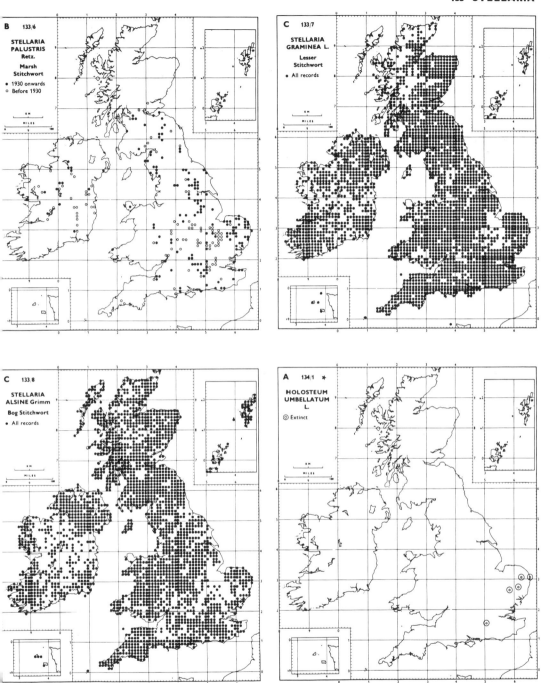

B 133/6

STELLARIA PALUSTRIS Retz.

Marsh Stitchwort

● 1930 onwards
○ Before 1930

C 133/7

STELLARIA GRAMINEA L.

Lesser Stitchwort

● All records

C 133/8

STELLARIA ALSINE Grimm

Bog Stitchwort

● All records

A 134/1 ✶

HOLOSTEUM UMBELLATUM L.

◎ Extinct

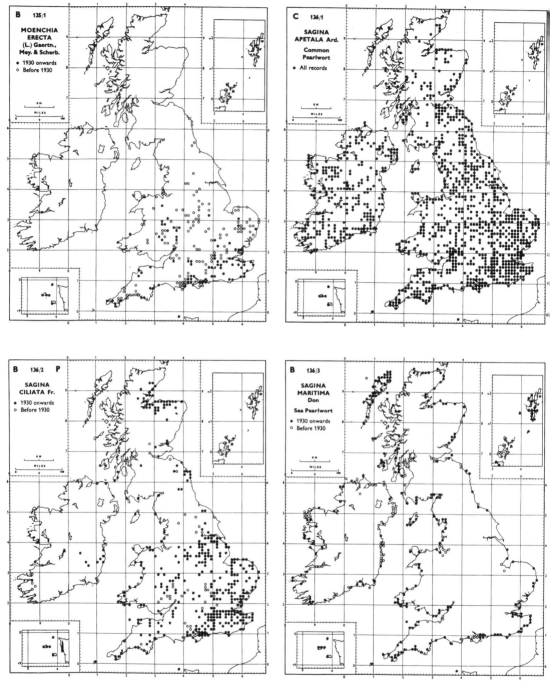

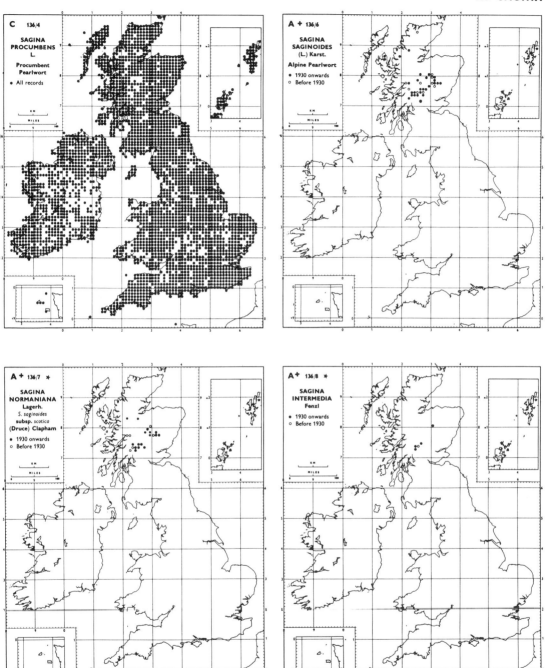

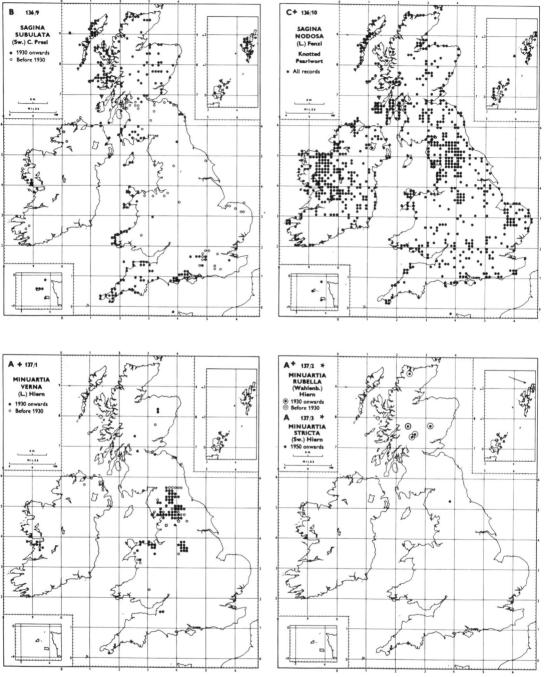

B 136/9

**SAGINA
SUBULATA**
(Sw.) C. Presl

• 1930 onwards
○ Before 1930

C+ 136/10

**SAGINA
NODOSA**
(L.) Fenzl

**Knotted
Pearlwort**

• All records

A + 137/1

**MINUARTIA
VERNA**
(L.) Hiern

• 1930 onwards
○ Before 1930

A+ 137/2 *

**MINUARTIA
RUBELLA**
(Wahlenb.)
Hiern

⊙ 1930 onwards
⊚ Before 1930

A 137/3 *

**MINUARTIA
STRICTA**
(Sw.) Hiern

• 1950 onwards

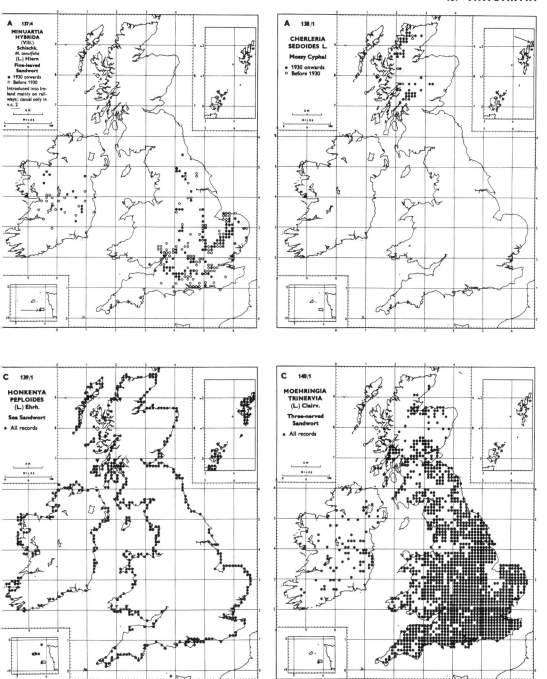

A 137/4
MINUARTIA HYBRIDA
(Vill.)
Schischk.
M. tenuifolia
(L.) Hiern
Fine-leaved Sandwort
● 1930 onwards
○ Before 1930
Introduced into Ireland mainly on railways; casual only in v.c. 2

A 138/1
CHERLERIA SEDOIDES L.
Mossy Cyphal
● 1930 onwards
○ Before 1930

C 139/1
HONKENYA PEPLOIDES
(L.) Ehrh.
Sea Sandwort
● All records

C 140/1
MOEHRINGIA TRINERVIA
(L.) Clairv.
Three-nerved Sandwort
● All records

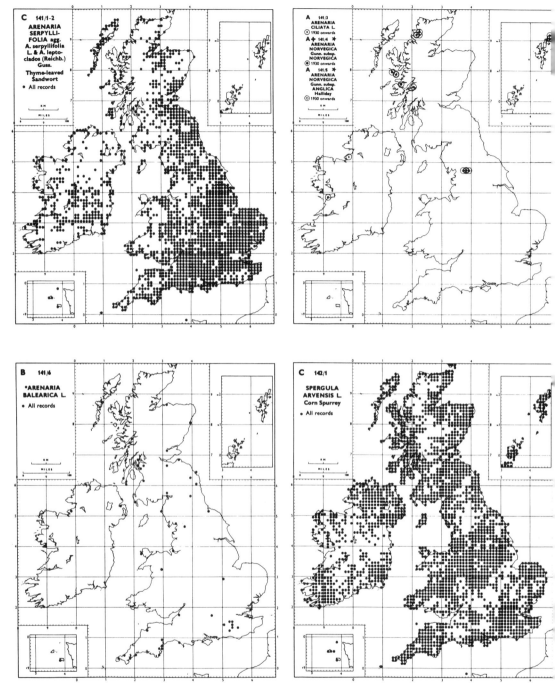

C 141/1-2
ARENARIA
SERPYLLI-
FOLIA agg.
A. serpyllifolia
L. & A. lepto-
clados (Reichb.)
Guss.
Thyme-leaved
Sandwort
• All records

A 141/3
ARENARIA
CILIATA L.
⊗ 1930 onwards
A ✦ 141/4 ✱
ARENARIA
NORVEGICA
Gunn. subsp.
NORVEGICA
◉ 1930 onwards
A 141/5 ✱
ARENARIA
NORVEGICA
Gunn. subsp.
ANGLICA
Halliday
◎ 1930 onwards

B 141/6
•ARENARIA
BALEARICA L.
• All records

C 142/1
SPERGULA
ARVENSIS L.
Corn Spurrey
• All records

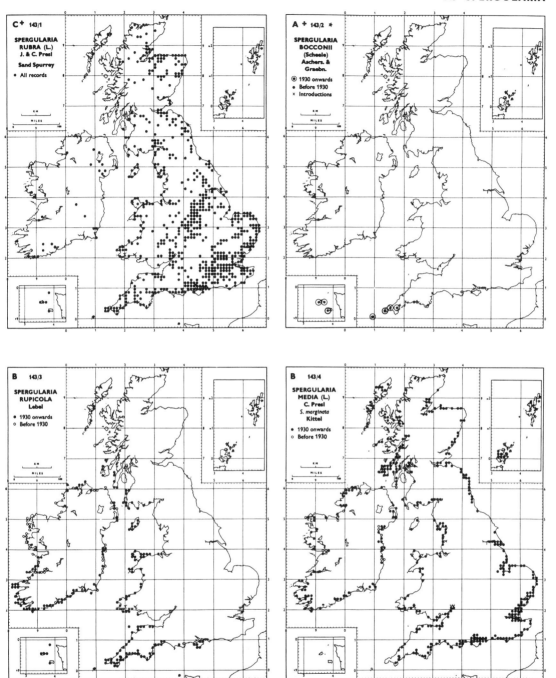

C+ 143/1

SPERGULARIA
RUBRA (L.)
J. & C. Presl

Sand Spurrey

• All records

A + 143/2 *

SPERGULARIA
BOCCONII
(Scheele)
Aschers. &
Graebn.

⊙ 1930 onwards
• Before 1930
× Introductions

B 143/3

SPERGULARIA
RUPICOLA
Lebel

• 1930 onwards
○ Before 1930

B 143/4

SPERGULARIA
MEDIA (L.)
C. Presl
S. marginata
Kittel

• 1930 onwards
○ Before 1930

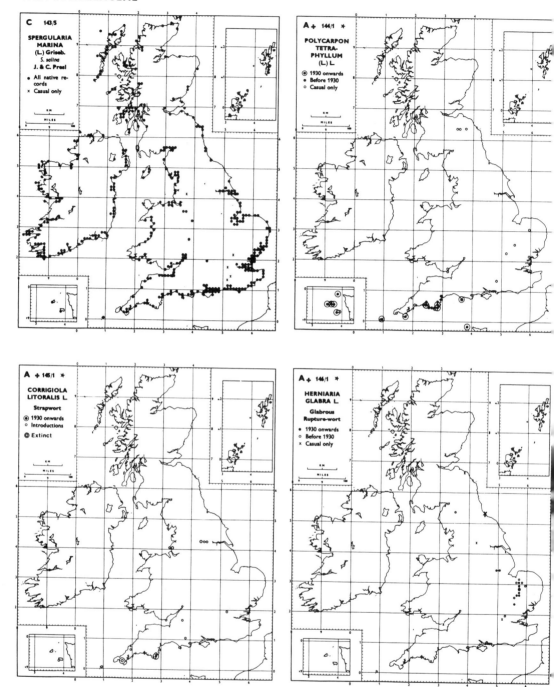

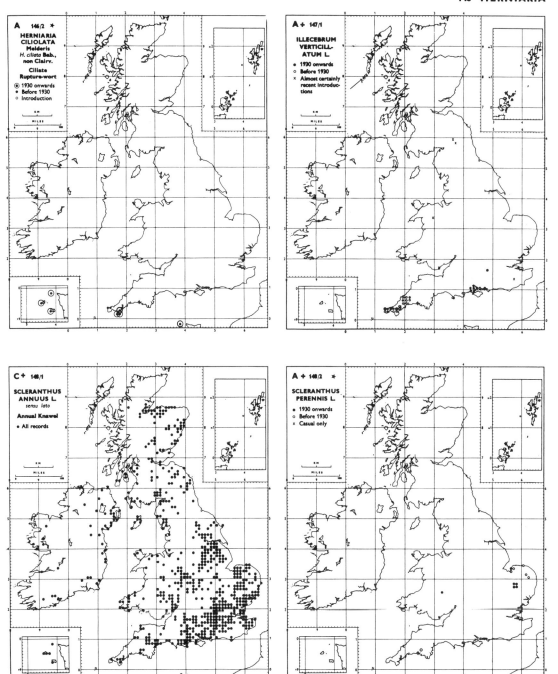

A 146/2 ✳
HERNIARIA
CILIOLATA
Melderis
H. ciliata Bab.,
non Clairv.
Ciliate
Rupture-wort
⊙ 1930 onwards
● Before 1930
○ Introduction

A + 147/1
ILLECEBRUM
VERTICILL-
ATUM L.
● 1930 onwards
○ Before 1930
✕ Almost certainly
recent introduc-
tions

C + 148/1
SCLERANTHUS
ANNUUS L.
sensu lato
Annual Knawel
● All records

A + 148/2 ✳
SCLERANTHUS
PERENNIS L.
● 1930 onwards
○ Before 1930
✕ Casual only

PORTULACACEAE

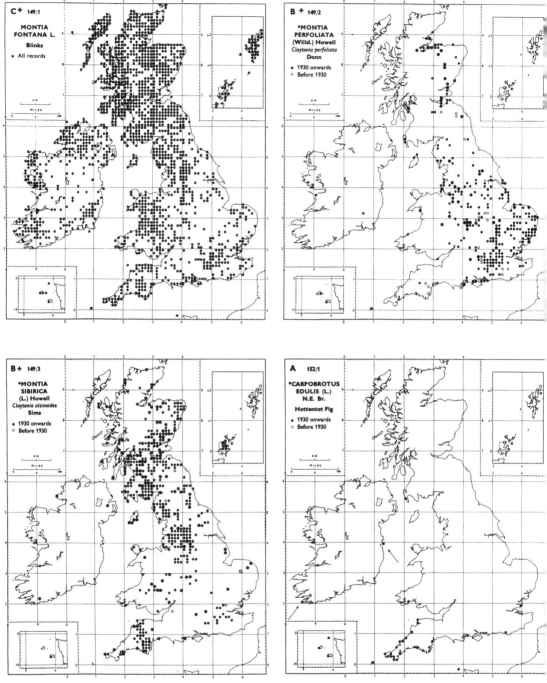

C+ 149/1
MONTIA FONTANA L.
Blinks
● All records

B+ 149/2
*MONTIA PERFOLIATA (Willd.) Howell
Claytonia perfoliata Donn
● 1930 onwards
○ Before 1930

B+ 149/3
*MONTIA SIBIRICA (L.) Howell
Claytonia alsinoides Sims
● 1930 onwards
○ Before 1930

A 152/1
*CARPOBROTUS EDULIS (L.) N.E. Br.
Hottentot Fig
● 1930 onwards
○ Before 1930

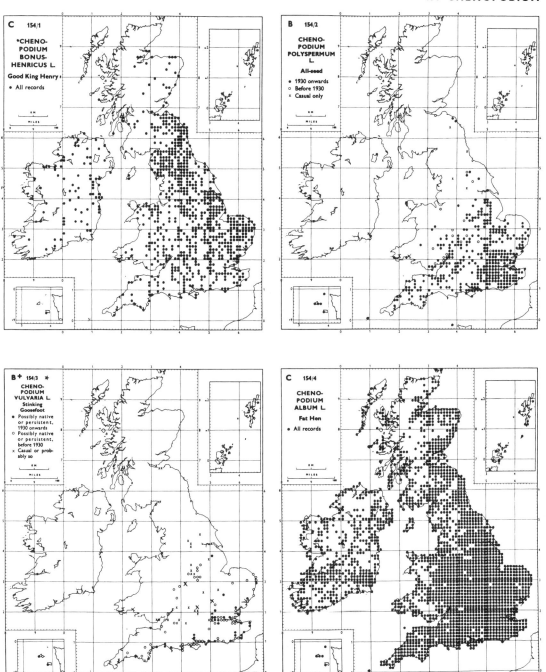

C 154/1

*CHENO-
PODIUM
BONUS-
HENRICUS L.

Good King Henry

• All records

B 154/2

CHENO-
PODIUM
POLYSPERMUM
L

All-seed

• 1930 onwards
○ Before 1930
× Casual only

B+ 154/3 ✳

CHENO-
PODIUM
VULVARIA L.

Stinking
Goosefoot

• Possibly native
  or persistent,
  1930 onwards
○ Possibly native
  or persistent,
  before 1930
× Casual or prob-
  ably so

C 154/4

CHENO-
PODIUM
ALBUM L.

Fat Hen

• All records

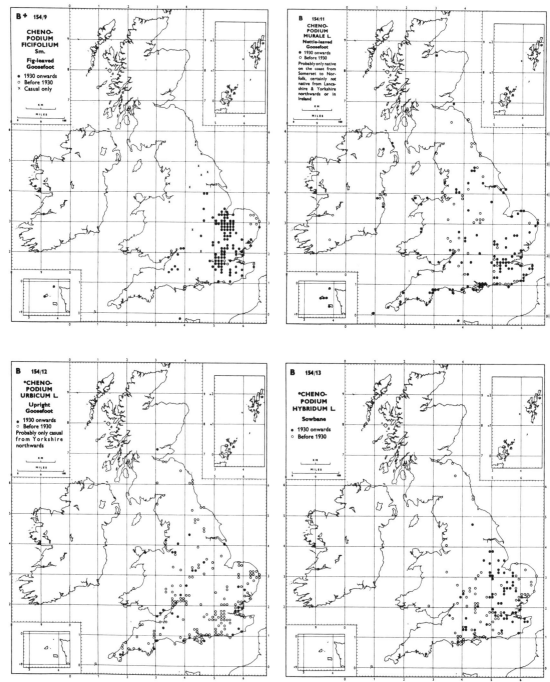

B + 154/9

CHENO-
PODIUM
FICIFOLIUM
Sm.

Fig-leaved
Goosefoot

• 1930 onwards
○ Before 1930
× Casual only

B 154/11

CHENO-
PODIUM
MURALE L.
Nettle-leaved
Goosefoot

• 1930 onwards
○ Before 1930
Probably only native
on the coast from
Somerset to Nor-
folk, certainly not
native from Lanca-
shire & Yorkshire
northwards or in
Ireland

B 154/12

*CHENO-
PODIUM
URBICUM L.

Upright
Goosefoot

• 1930 onwards
○ Before 1930
Probably only casual
from Yorkshire
northwards

B 154/13

*CHENO-
PODIUM
HYBRIDUM L.

Sowbane

• 1930 onwards
○ Before 1930

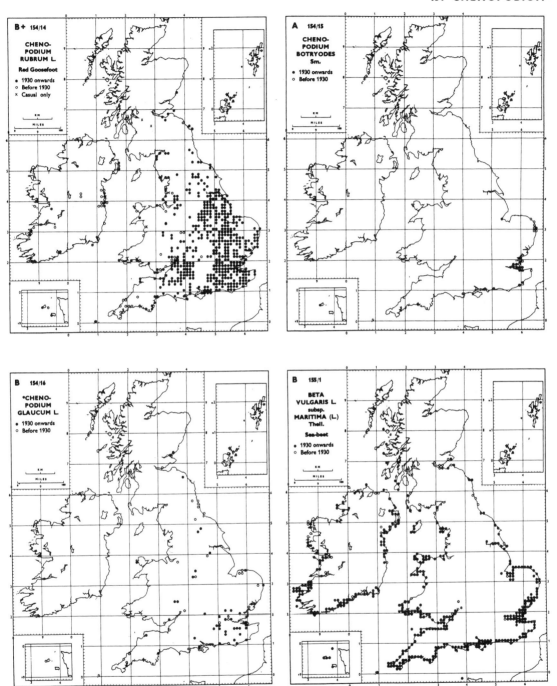

B+ 154/14

CHENO-
PODIUM
RUBRUM L.

Red Goosefoot

• 1930 onwards
○ Before 1930
× Casual only

A 154/15

CHENO-
PODIUM
BOTRYODES
Sm.

• 1930 onwards
○ Before 1930

B 154/16

*CHENO-
PODIUM
GLAUCUM L.

• 1930 onwards
○ Before 1930

B 155/1

BETA
VULGARIS L.
subsp.
MARITIMA (L.)
Thell.

Sea-beet

• 1930 onwards
○ Before 1930

83

CHENOPODIACEAE

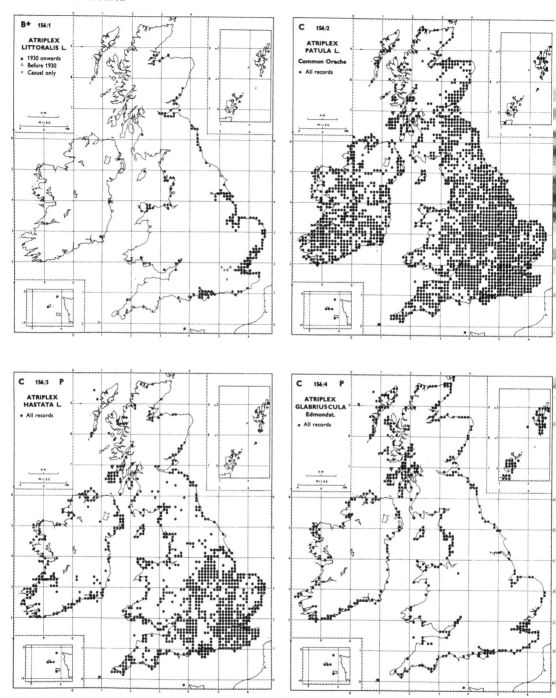

B+ 156/1

ATRIPLEX LITTORALIS L.

- 1930 onwards
- Before 1930
- Casual only

C 156/2

ATRIPLEX PATULA L.

Common Orache

- All records

C 156/3 **P**

ATRIPLEX HASTATA L.

- All records

C 156/4 **P**

ATRIPLEX GLABRIUSCULA Edmondst.

- All records

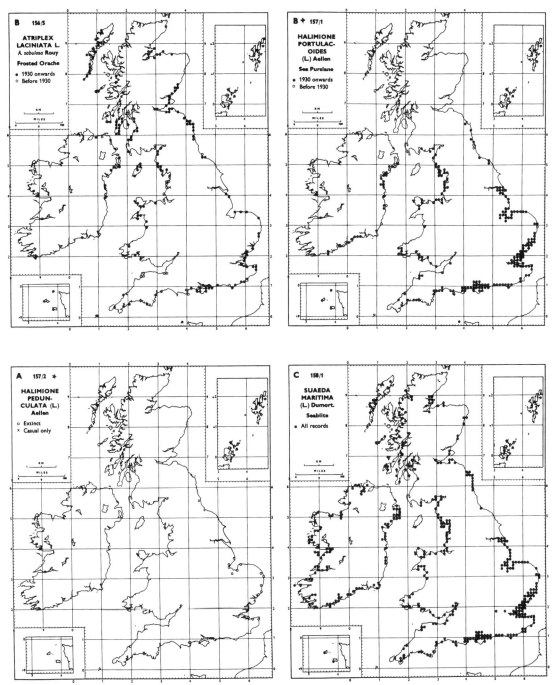

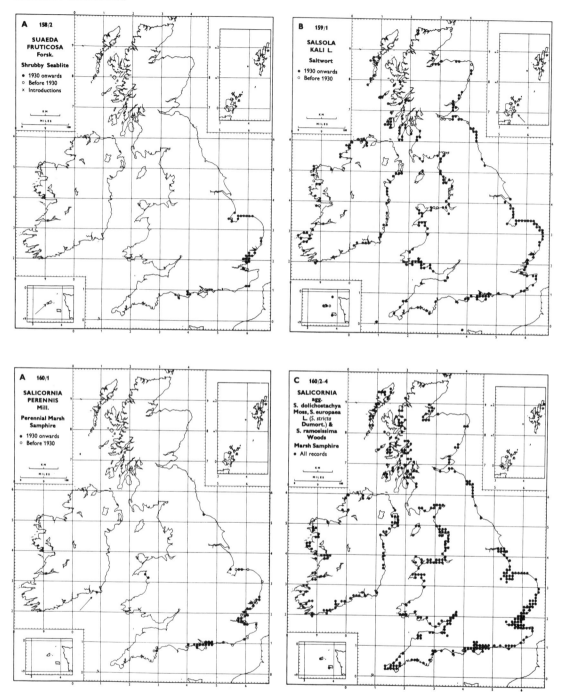

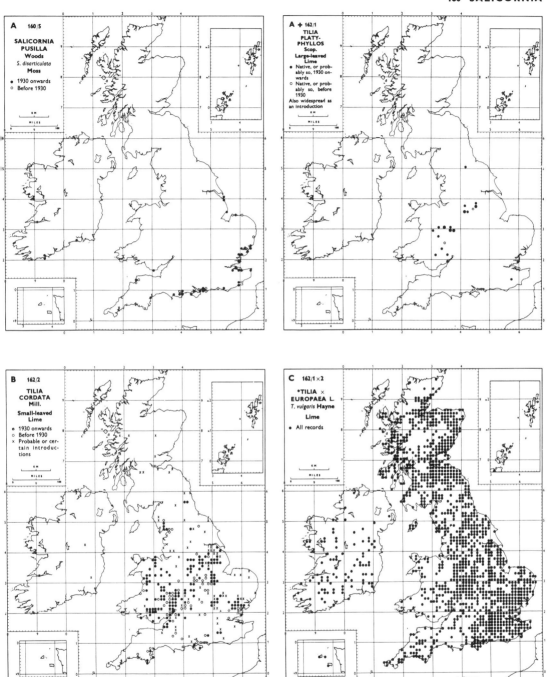

A 160/5

SALICORNIA
PUSILLA
Woods
S. disarticulata
Moss

● 1930 onwards
○ Before 1930

KM

MILES

A ✛ 162/1

TILIA
PLATY-
PHYLLOS
Scop.
**Large-leaved
Lime**
● Native, or prob-
ably so, 1930 on-
wards
○ Native, or prob-
ably so, before
1930
Also widespread as
an introduction

MILES

B 162/2

TILIA
CORDATA
Mill.

**Small-leaved
Lime**

● 1930 onwards
○ Before 1930
× Probable or cer-
tain introduc-
tions

KM

MILES

C 162/1 × 2

*TILIA ×
EUROPAEA L.
T. vulgaris* Hayne
Lime

● All records

KM

MILES

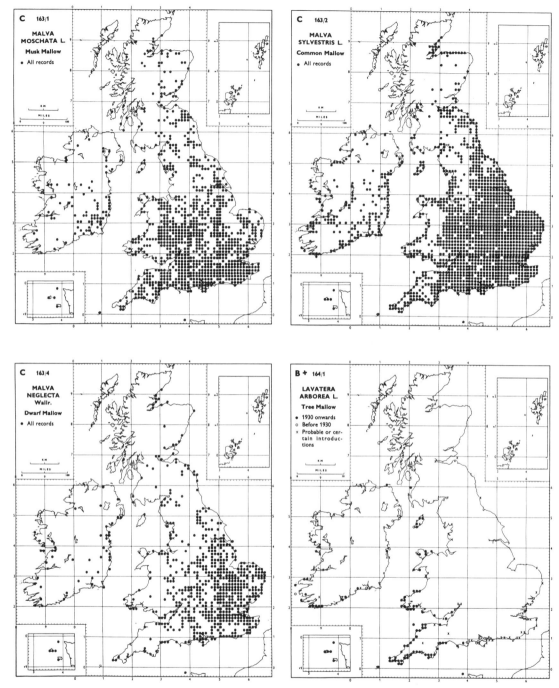

C 163/1
MALVA
MOSCHATA L.
Musk Mallow
• All records

C 163/2
MALVA
SYLVESTRIS L.
Common Mallow
• All records

C 163/4
MALVA
NEGLECTA
Wallr.
Dwarf Mallow
• All records

B + 164/1
LAVATERA
ARBOREA L.
Tree Mallow
• 1930 onwards
○ Before 1930
× Probable or cer-
  tain introduc-
  tions

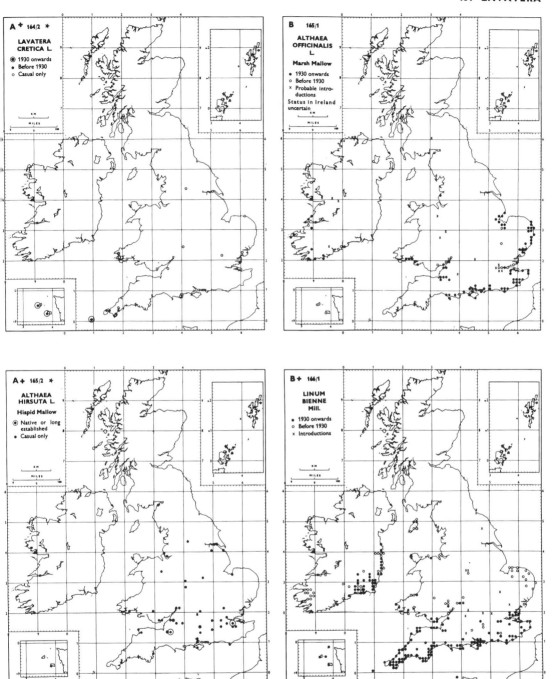

A + 164/2 *

LAVATERA
CRETICA L.

⊙ 1930 onwards
• Before 1930
○ Casual only

KM
MILES

B   165/1

ALTHAEA
OFFICINALIS
L.

Marsh Mallow

• 1930 onwards
○ Before 1930
× Probable intro-
   ductions
Status in Ireland
uncertain

KM
MILES

A + 165/2 *

ALTHAEA
HIRSUTA L.

Hispid Mallow

⊙ Native or long
   established
• Casual only

KM
MILES

B + 166/1

LINUM
BIENNE
Mill.

• 1930 onwards
○ Before 1930
× Introductions

KM
MILES

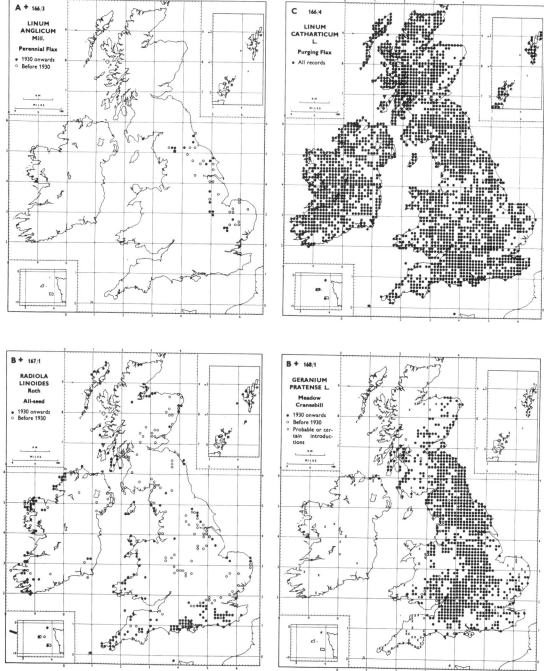

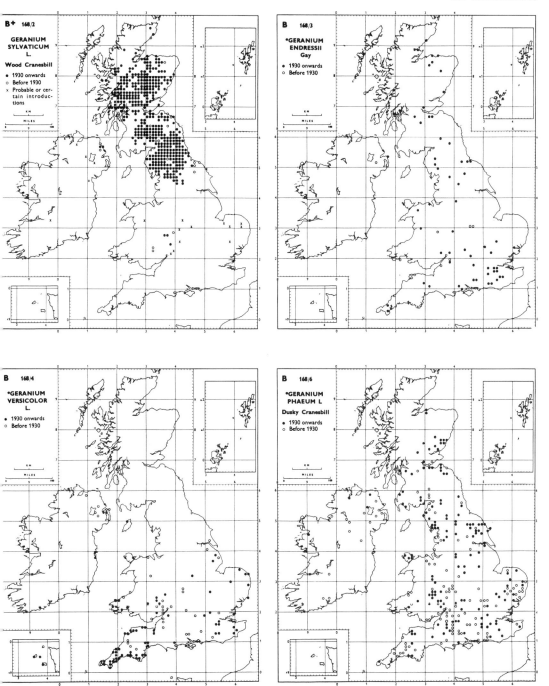

B+ 168/2

GERANIUM
SYLVATICUM
L.

Wood Cranesbill

● 1930 onwards
○ Before 1930
× Probable or cer-
  tain introduc-
  tions

KM
MILES

B 168/3

*GERANIUM
ENDRESSII
Gay

● 1930 onwards
○ Before 1930

KM
MILES

B 168/4

*GERANIUM
VERSICOLOR
L.

● 1930 onwards
○ Before 1930

KM
MILES

B 168/6

*GERANIUM
PHAEUM L

Dusky Cranesbill

● 1930 onwards
○ Before 1930

KM
MILES

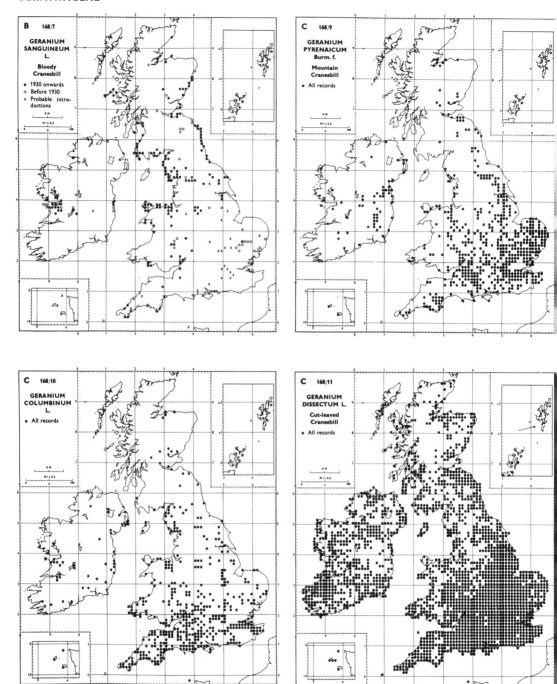

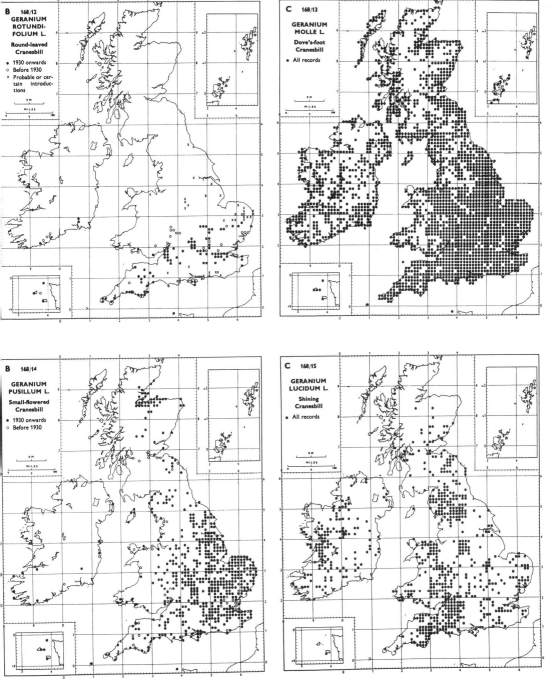

B 168/12
GERANIUM ROTUNDI-
FOLIUM L.

Round-leaved
Cranesbill

- 1930 onwards
- Before 1930
- Probable or cer-
  tain introduc-
  tions

C 168/13

GERANIUM
MOLLE L.

Dove's-foot
Cranesbill

- All records

B 168/14

GERANIUM
PUSILLUM L.

Small-flowered
Cranesbill

- 1930 onwards
- Before 1930

C 168/15

GERANIUM
LUCIDUM L.

Shining
Cranesbill

- All records

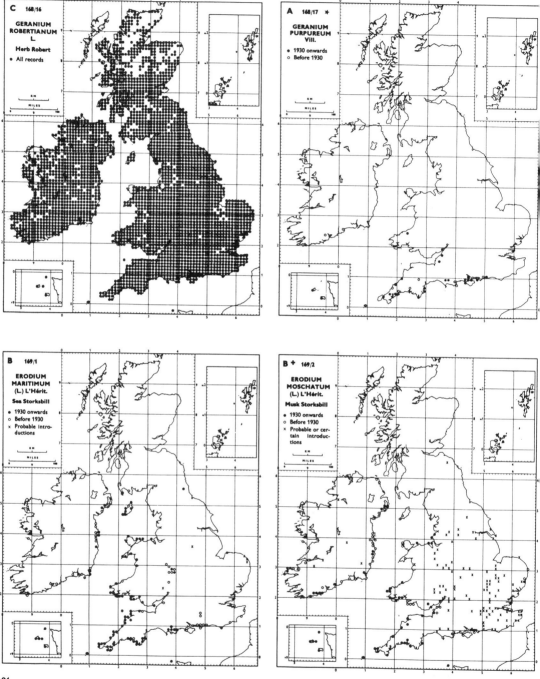

C 168/16

GERANIUM
ROBERTIANUM
L

Herb Robert

• All records

A 168/17 ✳

GERANIUM
PURPUREUM
VIII.

• 1930 onwards
○ Before 1930

B 169/1

ERODIUM
MARITIMUM
(L.) L'Hérit.

Sea Storksbill

• 1930 onwards
○ Before 1930
✕ Probable intro-
ductions

B ✛ 169/2

ERODIUM
MOSCHATUM
(L.) L'Hérit.

Musk Storksbill

• 1930 onwards
○ Before 1930
✕ Probable or cer-
tain introduc-
tions

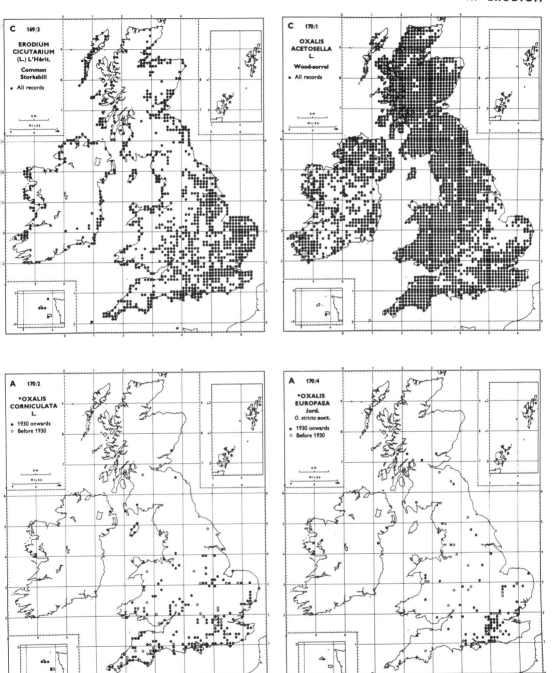

C   169/3

**ERODIUM
CICUTARIUM**
(L.) L'Hérit.

**Common
Storksbill**

• All records

C   170/1

**OXALIS
ACETOSELLA**
L.

Wood-sorrel

• All records

A   170/2

**•OXALIS
CORNICULATA**
L

• 1930 onwards
○ Before 1930

A   170/4

**•OXALIS
EUROPAEA**
Jord.
O. stricta auct.

• 1930 onwards
○ Before 1930

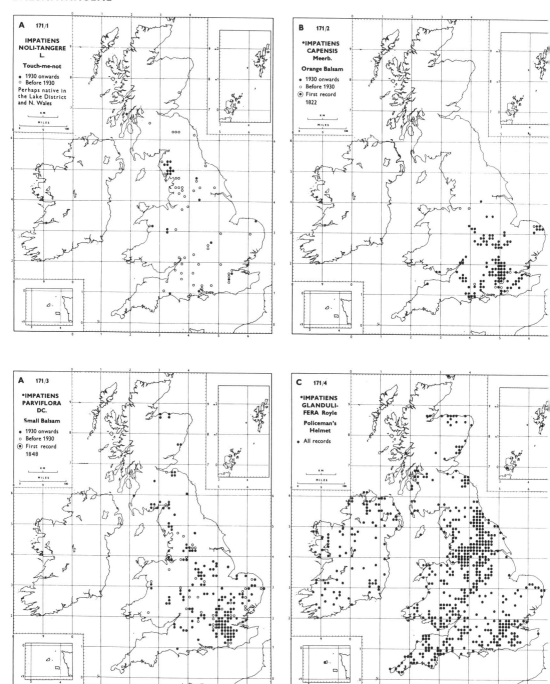

A 171/1

IMPATIENS
NOLI-TANGERE
L.

Touch-me-not

• 1930 onwards
○ Before 1930
Perhaps native in
the Lake District
and N. Wales

B 171/2

*IMPATIENS
CAPENSIS
Meerb.

Orange Balsam

• 1930 onwards
○ Before 1930
⊙ First record
1822

A 171/3

*IMPATIENS
PARVIFLORA
DC.

Small Balsam

• 1930 onwards
○ Before 1930
⊙ First record
1848

C 171/4

*IMPATIENS
GLANDULI-
FERA Royle

Policeman's
Helmet

• All records

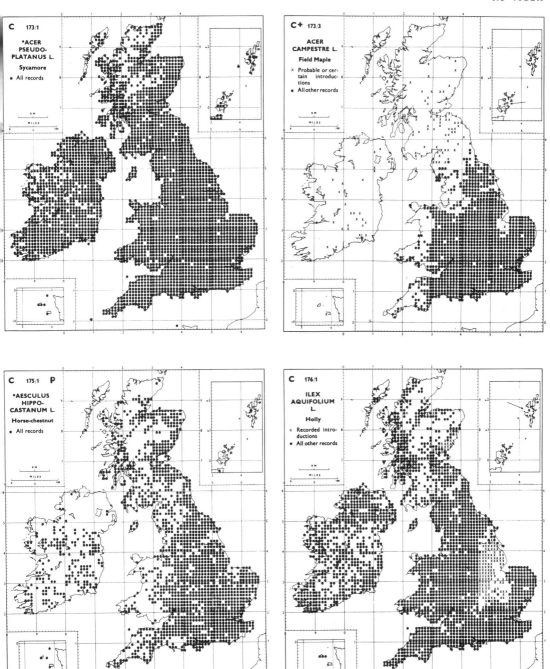

C 173/1

*ACER
PSEUDO-
PLATANUS L.

Sycamore

● All records

C+ 173/3

ACER
CAMPESTRE L.

Field Maple

× Probable or certain introductions
● All other records

C 175/1 P

*AESCULUS
HIPPO-
CASTANUM L.

Horse-chestnut

● All records

C 176/1

ILEX
AQUIFOLIUM
L.

Holly

× Recorded introductions
● All other records

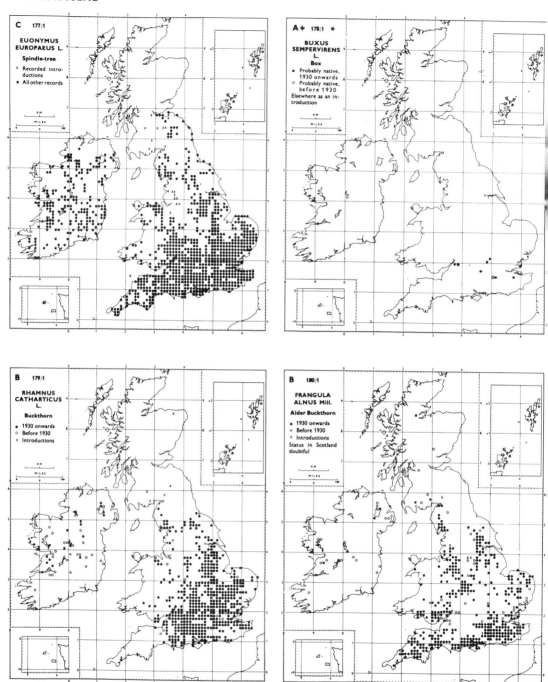

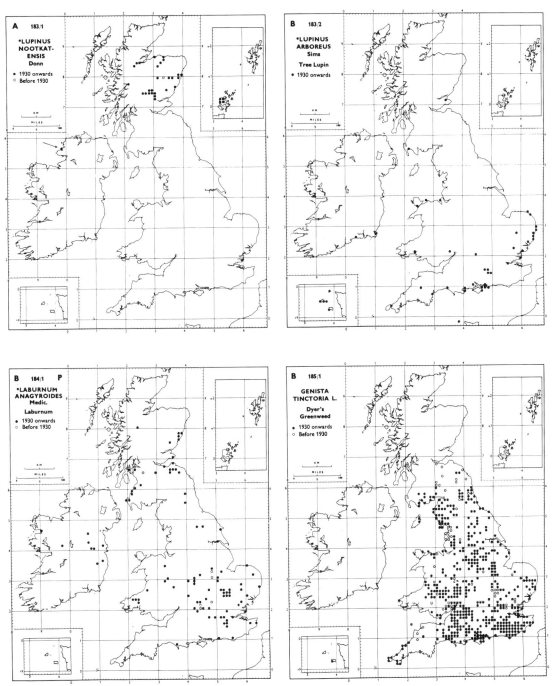

A 183/1
*LUPINUS
NOOTKAT-
ENSIS
Donn

• 1930 onwards
○ Before 1930

B 183/2
*LUPINUS
ARBOREUS
Sims

Tree Lupin

• 1930 onwards

B 184/1 P
*LABURNUM
ANAGYROIDES
Medic.

Laburnum

• 1930 onwards
○ Before 1930

B 185/1
GENISTA
TINCTORIA L.

Dyer's
Greenweed

• 1930 onwards
○ Before 1930

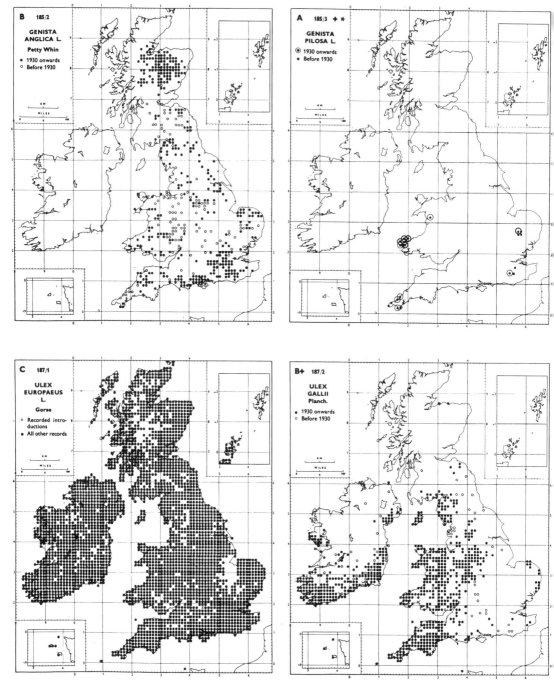

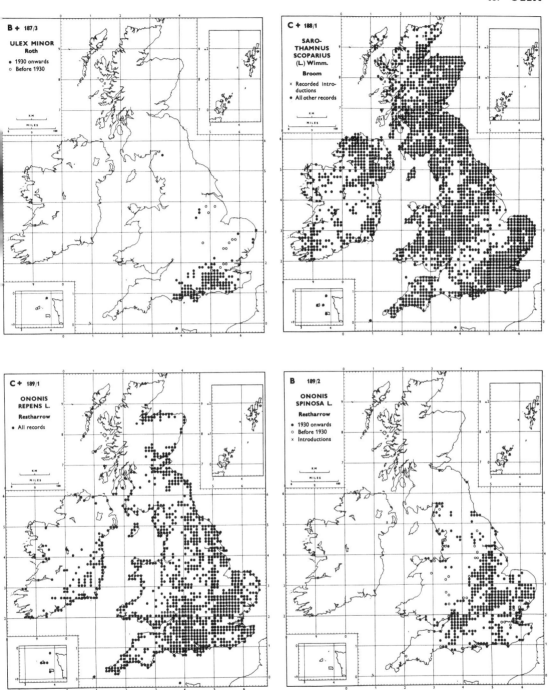

B + 187/3

ULEX MINOR
Roth

• 1930 onwards
○ Before 1930

C + 188/1

SARO-
THAMNUS
SCOPARIUS
(L.) Wimm.

Broom

× Recorded intro-
ductions
• All other records

C + 189/1

ONONIS
REPENS L.

Restharrow

• All records

B 189/2

ONONIS
SPINOSA L.

Restharrow

• 1930 onwards
○ Before 1930
× Introductions

LEGUMINOSAE

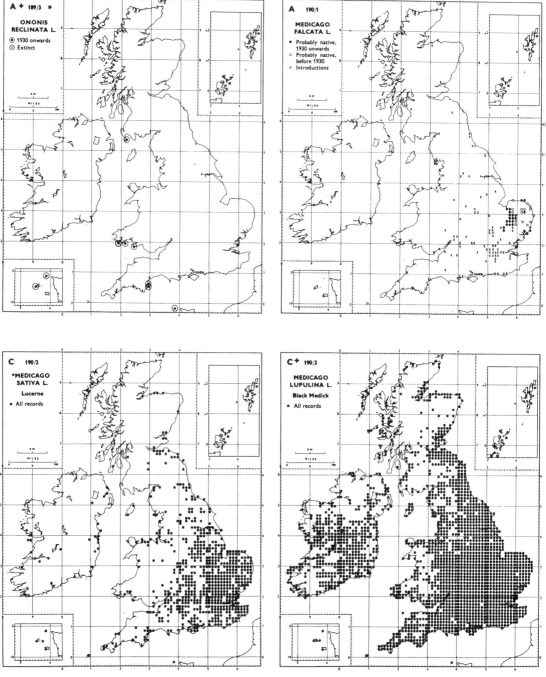

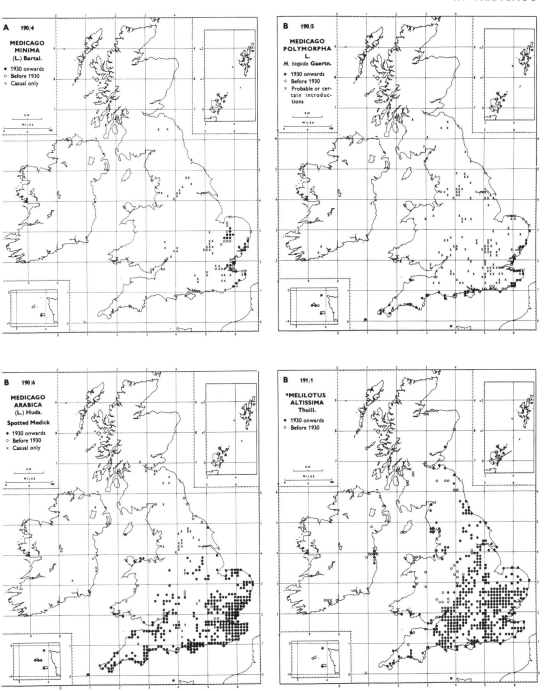

A 190/4

MEDICAGO MINIMA
(L.) Bartal.

- 1930 onwards
- ○ Before 1930
- × Casual only

B 190/5

MEDICAGO POLYMORPHA
L

M. hispida **Gaertn.**

- 1930 onwards
- ○ Before 1930
- × Probable or certain introductions

B 190/6

MEDICAGO ARABICA
(L.) Huds.

Spotted Medick

- 1930 onwards
- ○ Before 1930
- × Casual only

B 191/1

***MELILOTUS ALTISSIMA**
Thuill.

- 1930 onwards
- ○ Before 1930

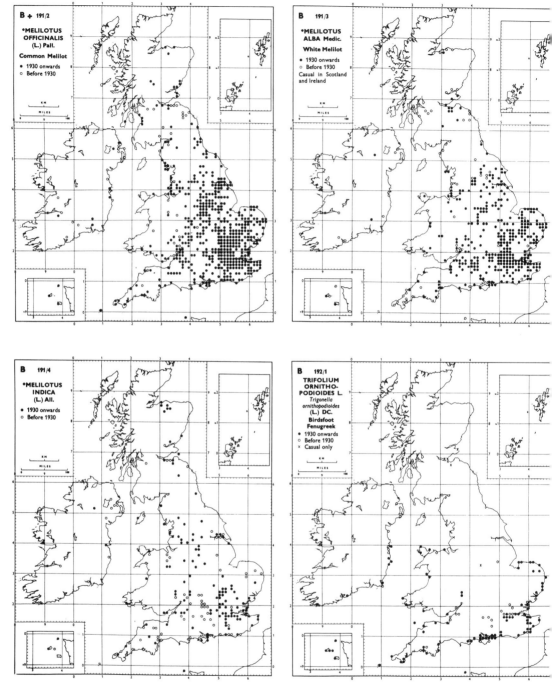

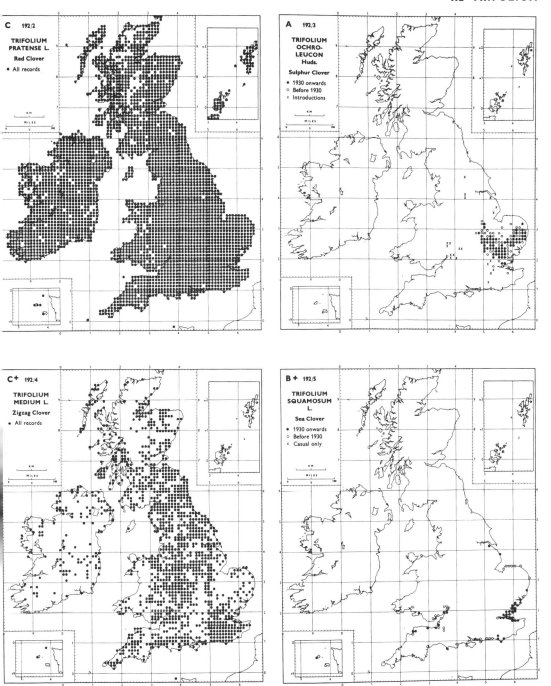

C 192/2

TRIFOLIUM PRATENSE L.

Red Clover

● All records

A 192/3

TRIFOLIUM OCHRO-LEUCON Huds.

Sulphur Clover

● 1930 onwards
○ Before 1930
× Introductions

C⁺ 192/4

TRIFOLIUM MEDIUM L.

Zigzag Clover

● All records

B⁺ 192/5

TRIFOLIUM SQUAMOSUM L.

Sea Clover

● 1930 onwards
○ Before 1930
× Casual only

LEGUMINOSAE

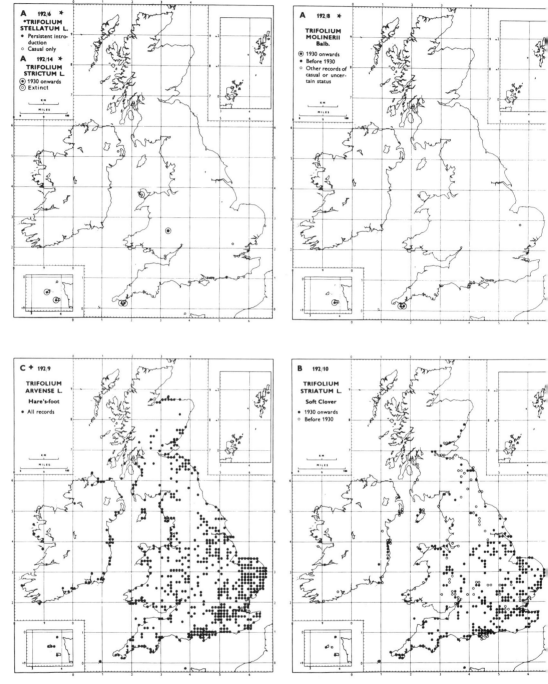

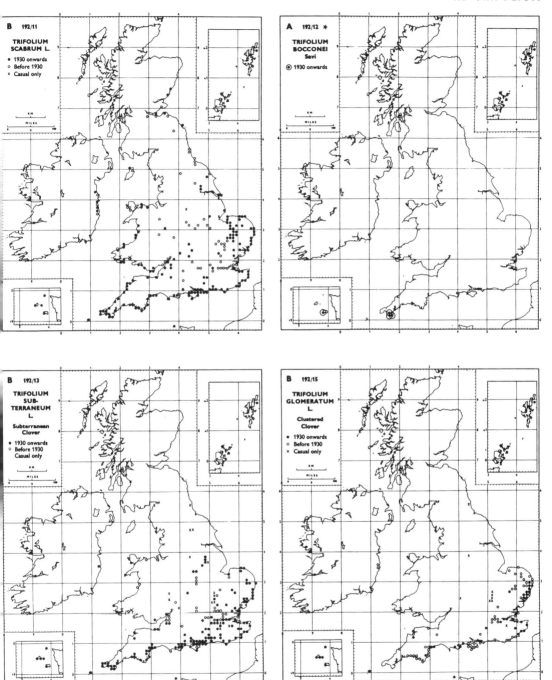

B 192/11

TRIFOLIUM
SCABRUM L.

• 1930 onwards
○ Before 1930
× Casual only

A 192/12 ✱

TRIFOLIUM
BOCCONEI
Savi

◉ 1930 onwards

B 192/13

TRIFOLIUM
SUB-
TERRANEUM
L.

Subterranean
Clover

• 1930 onwards
○ Before 1930
  Casual only

B 192/15

TRIFOLIUM
GLOMERATUM
L.

Clustered
Clover

• 1930 onwards
○ Before 1930
× Casual only

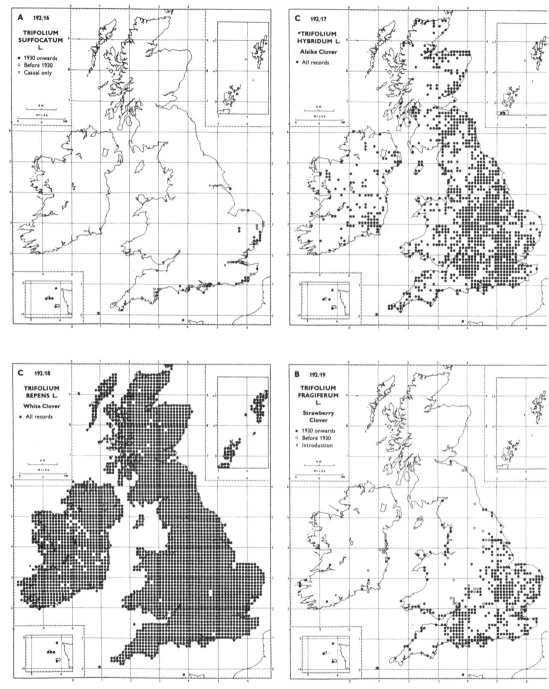

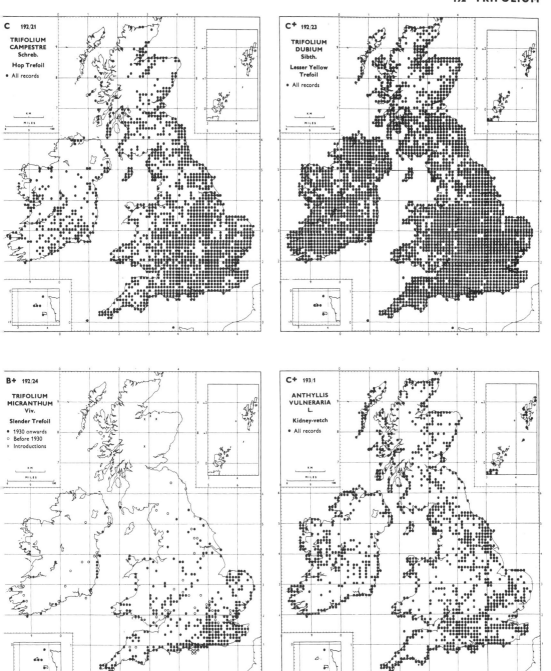

C 192/21

TRIFOLIUM
CAMPESTRE
Schreb.

Hop Trefoil

● All records

C+ 192/23

TRIFOLIUM
DUBIUM
Sibth.

Lesser Yellow
Trefoil

● All records

B+ 192/24

TRIFOLIUM
MICRANTHUM
Viv.

Slender Trefoil

● 1930 onwards
○ Before 1930
× Introductions

C+ 193/1

ANTHYLLIS
VULNERARIA
L.

Kidney-vetch

● All records

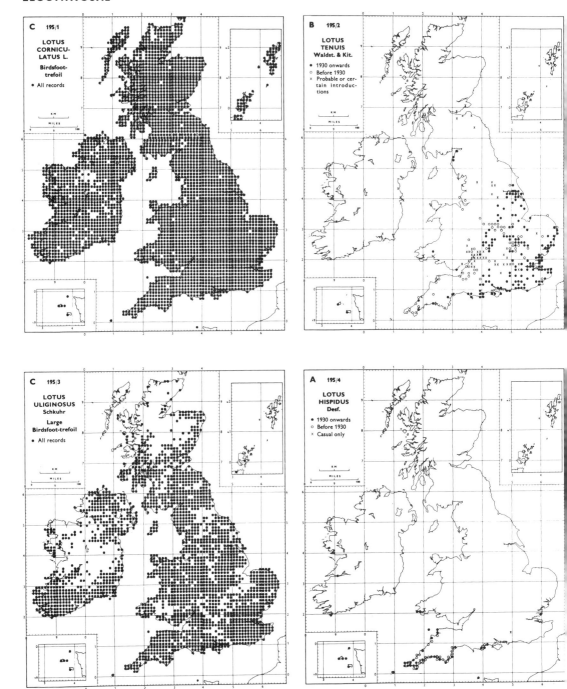

C 195/1
LOTUS CORNICU-LATUS L.
Birdsfoot-trefoil
• All records

B 195/2
LOTUS TENUIS Waldst. & Kit.
• 1930 onwards
○ Before 1930
× Probable or certain introductions

C 195/3
LOTUS ULIGINOSUS Schkuhr
Large Birdsfoot-trefoil
• All records

A 195/4
LOTUS HISPIDUS Desf.
• 1930 onwards
○ Before 1930
× Casual only

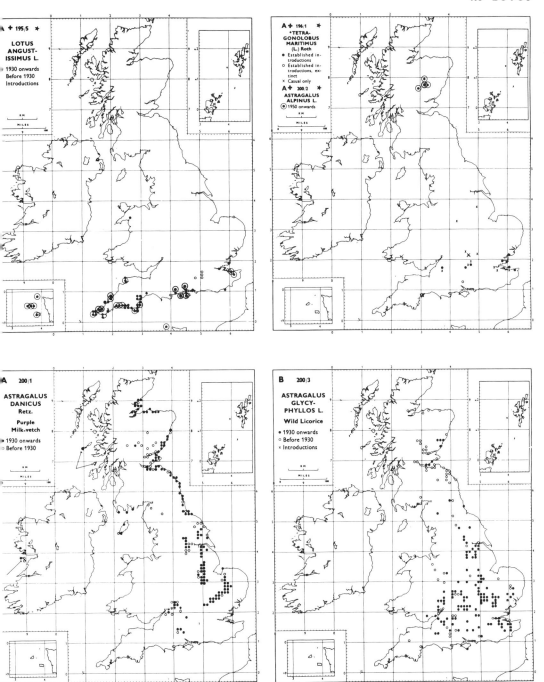

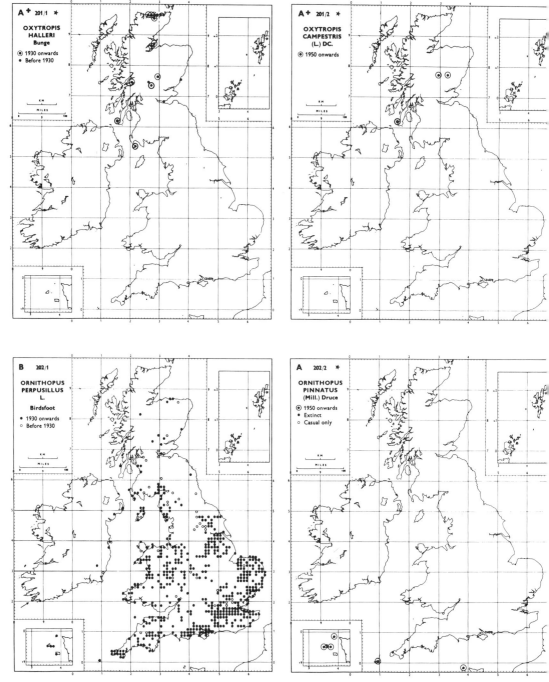

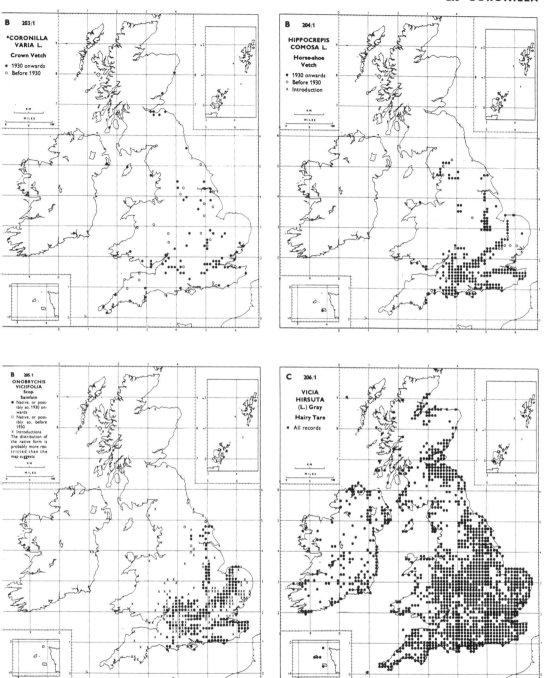

B 203/1

*CORONILLA
VARIA L.

Crown Vetch

● 1930 onwards
○ Before 1930

B 204/1

HIPPOCREPIS
COMOSA L.

Horse-shoe
Vetch

● 1930 onwards
○ Before 1930
× Introduction

B 205/1

ONOBRYCHIS
VICIIFOLIA
Scop.
Sainfoin

● Native, or poss-
ibly so, 1930 on-
wards
○ Native, or poss-
ibly so, before
1930
× Introductions
The distribution of
the native form is
probably more res-
tricted than the
map suggests

C 206/1

VICIA
HIRSUTA
(L.) Gray

Hairy Tare

● All records

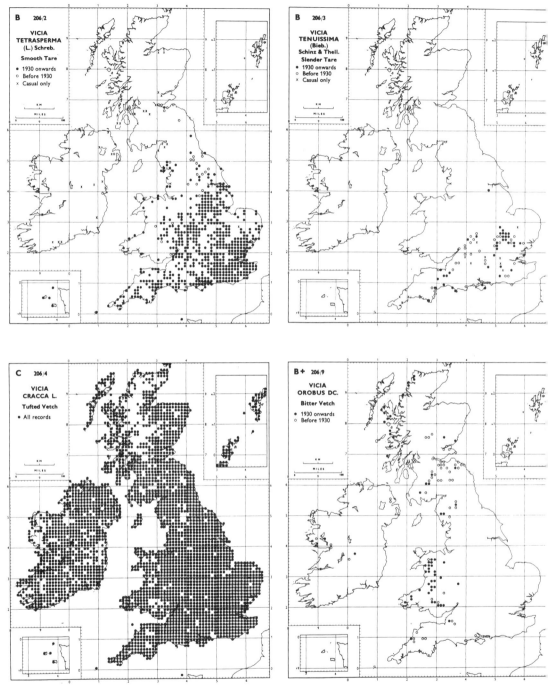

B 206/2

VICIA
TETRASPERMA
(L.) Schreb.

Smooth Tare

- 1930 onwards
∘ Before 1930
× Casual only

B 206/3

VICIA
TENUISSIMA
(Bieb.)
Schinz & Thell.

Slender Tare

- 1930 onwards
∘ Before 1930
× Casual only

C 206/4

VICIA
CRACCA L.

Tufted Vetch

- All records

B+ 206/9

VICIA
OROBUS DC.

Bitter Vetch

- 1930 onwards
∘ Before 1930

114

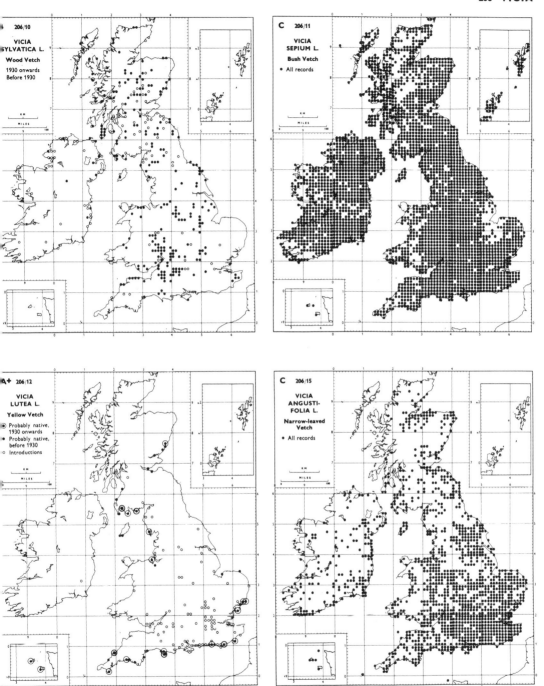

206/10
VICIA
SYLVATICA L.
Wood Vetch
○ 1930 onwards
○ Before 1930

C 206/11
VICIA
SEPIUM L.
Bush Vetch
● All records

206/12
VICIA
LUTEA L.
Yellow Vetch
◉ Probably native, 1930 onwards
● Probably native, before 1930
○ Introductions

C 206/15
VICIA
ANGUSTI-
FOLIA L.
Narrow-leaved
Vetch
● All records

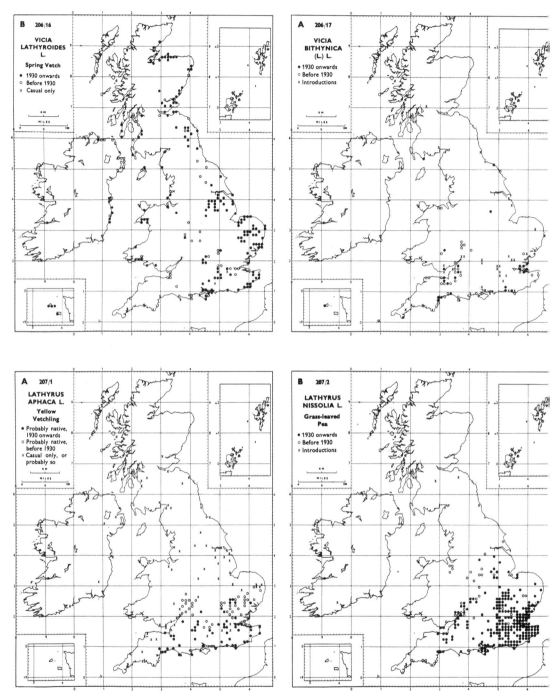

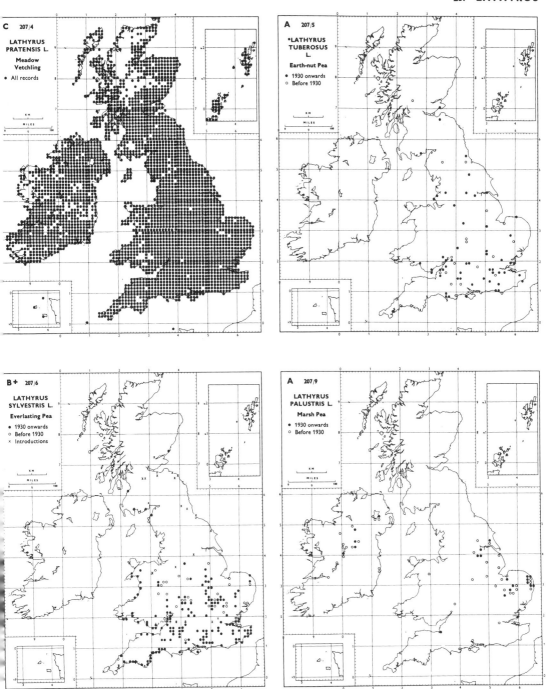

C 207/4

LATHYRUS
PRATENSIS L.

Meadow
Vetchling

• All records

A 207/5

•LATHYRUS
TUBEROSUS
L.

Earth-nut Pea

• 1930 onwards
○ Before 1930

B+ 207/6

LATHYRUS
SYLVESTRIS L.

Everlasting Pea

• 1930 onwards
○ Before 1930
× Introductions

A 207/9

LATHYRUS
PALUSTRIS L.

Marsh Pea

• 1930 onwards
○ Before 1930

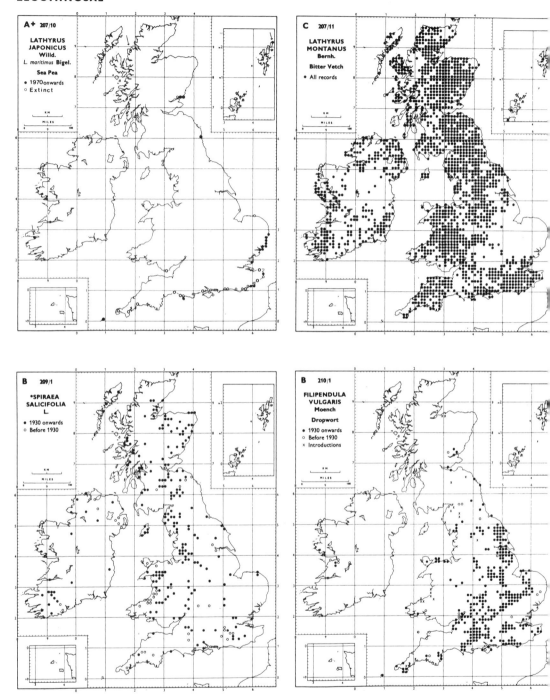

A+ 207/10

LATHYRUS
JAPONICUS
Willd.
L. maritimus Bigel.

Sea Pea

● 1970 onwards
○ Extinct

C 207/11

LATHYRUS
MONTANUS
Bernh.

Bitter Vetch

● All records

B 209/1

*SPIRAEA
SALICIFOLIA
L.

● 1930 onwards
○ Before 1930

B 210/1

FILIPENDULA
VULGARIS
Moench

Dropwort

● 1930 onwards
○ Before 1930
✕ Introductions

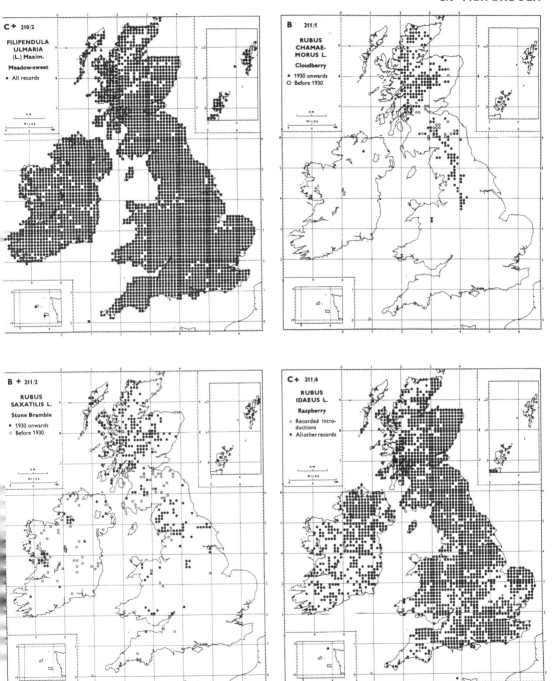

C + 210/2

FILIPENDULA
ULMARIA
(L.) Maxim.

Meadow-sweet

● All records

B 211/1

RUBUS
CHAMAE-
MORUS L.

Cloudberry

● 1930 onwards
○ Before 1930

B + 211/2

RUBUS
SAXATILIS L.

Stone Bramble

● 1930 onwards
○ Before 1930

C + 211/6

RUBUS
IDAEUS L.

Raspberry

× Recorded Intro-
   ductions
● All other records

ROSACEAE

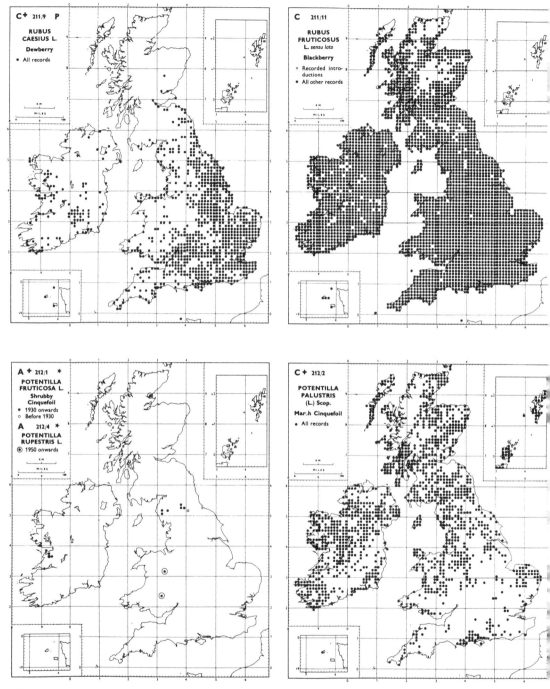

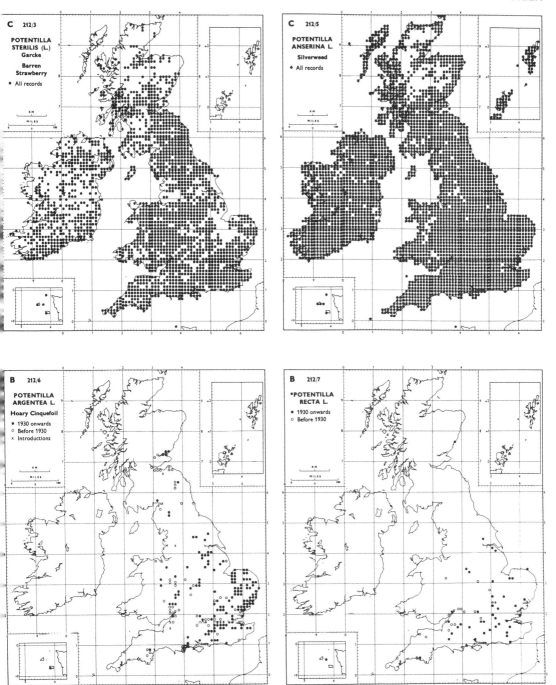

C 212/3

POTENTILLA
STERILIS (L.)
Garcke

Barren
Strawberry

• All records

C 212/5

POTENTILLA
ANSERINA L.

Silverweed

• All records

B 212/6

POTENTILLA
ARGENTEA L.

Hoary Cinquefoil

• 1930 onwards
○ Before 1930
× Introductions

B 212/7

•POTENTILLA
RECTA L.

• 1930 onwards
○ Before 1930

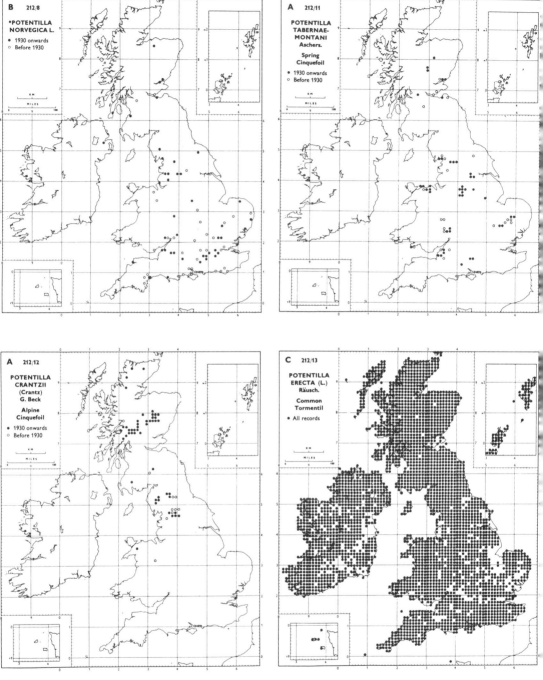

B 212/8

*POTENTILLA
NORVEGICA L.

• 1930 onwards
○ Before 1930

A 212/11

POTENTILLA
TABERNAE-
MONTANI
Aschers.

Spring
Cinquefoil

• 1930 onwards
○ Before 1930

A 212/12

POTENTILLA
CRANTZII
(Crantz)
G. Beck

Alpine
Cinquefoil

• 1930 onwards
○ Before 1930

C 212/13

POTENTILLA
ERECTA (L.)
Räusch.

Common
Tormentil

• All records

122

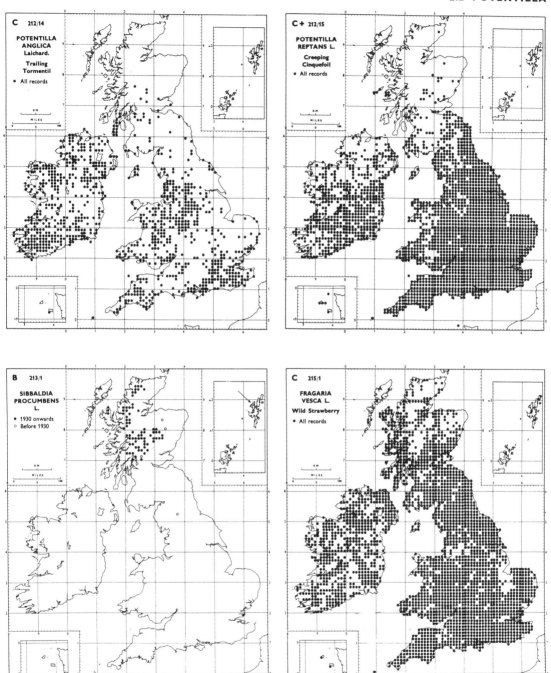

C 212/14

POTENTILLA
ANGLICA
Laichard.

Trailing
Tormentil

• All records

C+ 212/15

POTENTILLA
REPTANS L.

Creeping
Cinquefoil

• All records

B 213/1

SIBBALDIA
PROCUMBENS
L.

• 1930 onwards
○ Before 1930

C 215/1

FRAGARIA
VESCA L.

Wild Strawberry

• All records

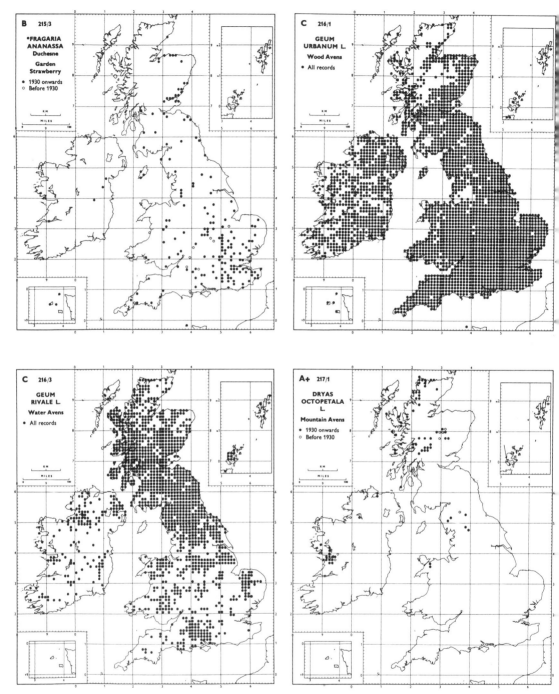

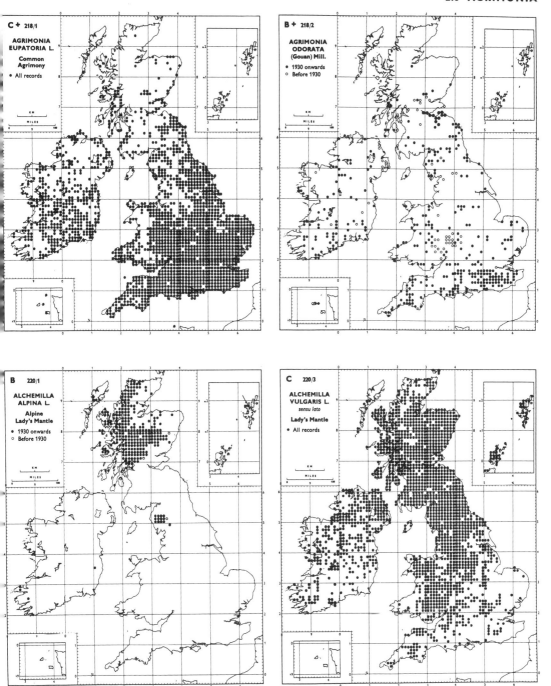

C + 218/1
AGRIMONIA
EUPATORIA L.
Common
Agrimony
• All records

B + 218/2
AGRIMONIA
ODORATA
(Gouan) Mill.
• 1930 onwards
○ Before 1930

B 220/1
ALCHEMILLA
ALPINA L.
Alpine
Lady's Mantle
• 1930 onwards
○ Before 1930

C 220/3
ALCHEMILLA
VULGARIS L.
sensu lato
Lady's Mantle
• All records

125

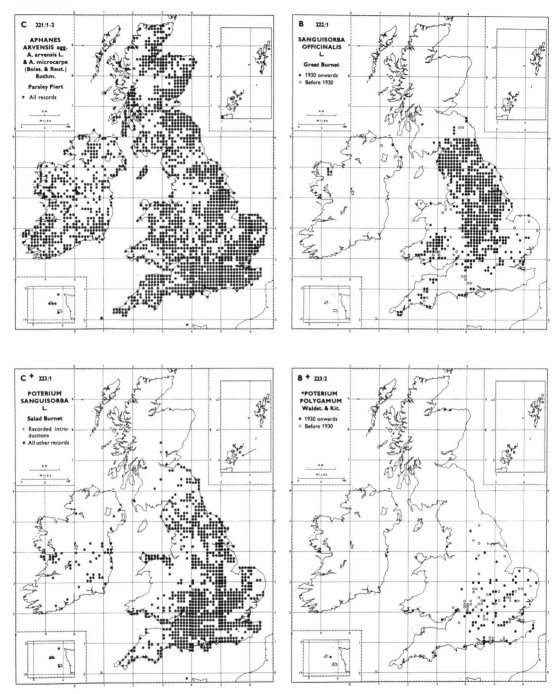

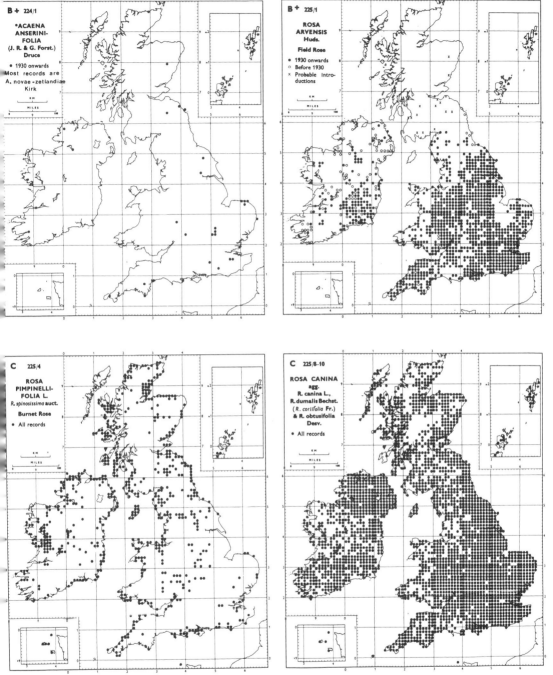

B + 224/1

*ACAENA
ANSERINI-
FOLIA
(J. R. & G. Forst.)
Druce

● 1930 onwards
Most records are
A. novae-zetlandiae
Kirk

KM
MILES

B + 225/1

ROSA
ARVENSIS
Huds.

Field Rose

● 1930 onwards
○ Before 1930
× Probable intro-
ductions

KM
MILES

C 225/4

ROSA
PIMPINELLI-
FOLIA L.
R. spinosissima auct.

Burnet Rose

● All records

KM
MILES

C 225/8-10

ROSA CANINA
agg.
R. canina L.,
R. dumalis Bechst.
(R. coriifolia Fr.)
& R. obtusifolia
Desv.

● All records

KM
MILES

ROSACEAE

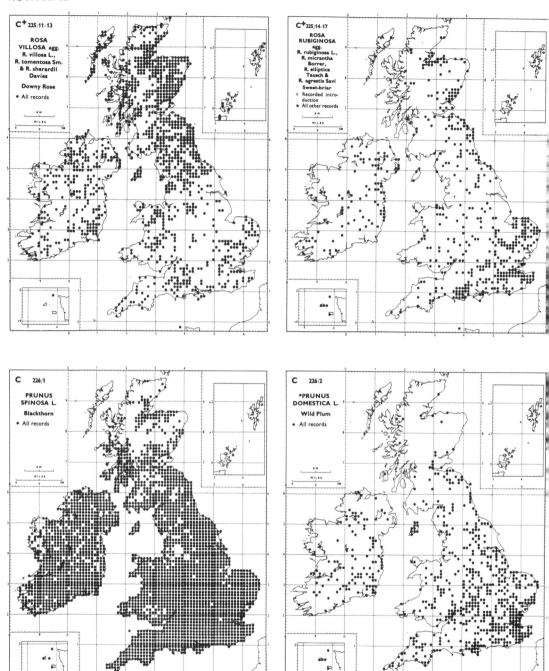

C⁺ 225/11-13

**ROSA
VILLOSA** agg.
R. villosa L.,
R. tomentosa Sm.
& R. sherardii
Davies
Downy Rose
● All records

C⁺ 225/14-17

**ROSA
RUBIGINOSA**
agg.
R. rubiginosa L.,
R. micrantha
Borrer,
R. elliptica
Tausch &
R. agrestis Savi
Sweet-briar
× Recorded intro-
duction
● All other records

C 226/1

**PRUNUS
SPINOSA** L.
Blackthorn
● All records

C 226/2

**●PRUNUS
DOMESTICA** L.
Wild Plum
● All records

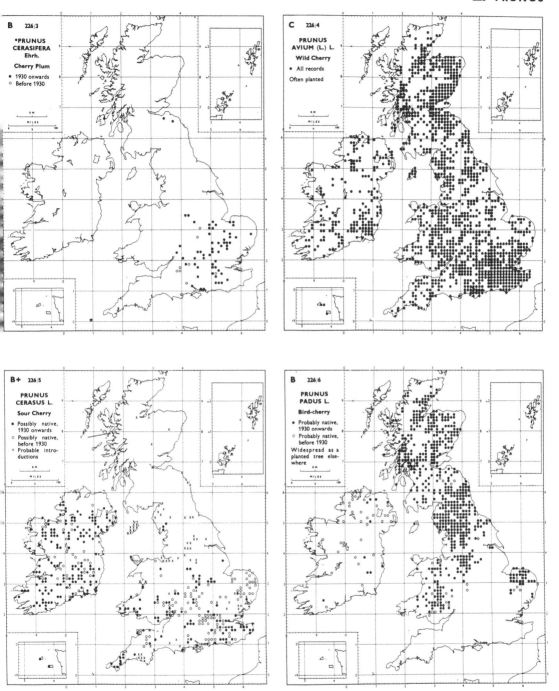

B 226/3

*PRUNUS
CERASIFERA
Ehrh.

Cherry Plum

• 1930 onwards
○ Before 1930

C 226/4

PRUNUS
AVIUM (L.) L.

Wild Cherry

• All records

Often planted

B+ 226/5

PRUNUS
CERASUS L.

Sour Cherry

• Possibly native,
1930 onwards
○ Possibly native,
before 1930
× Probable intro-
ductions

B 226/6

PRUNUS
PADUS L.

Bird-cherry

• Probably native,
1930 onwards
○ Probably native,
before 1930
Widespread as a
planted tree else-
where

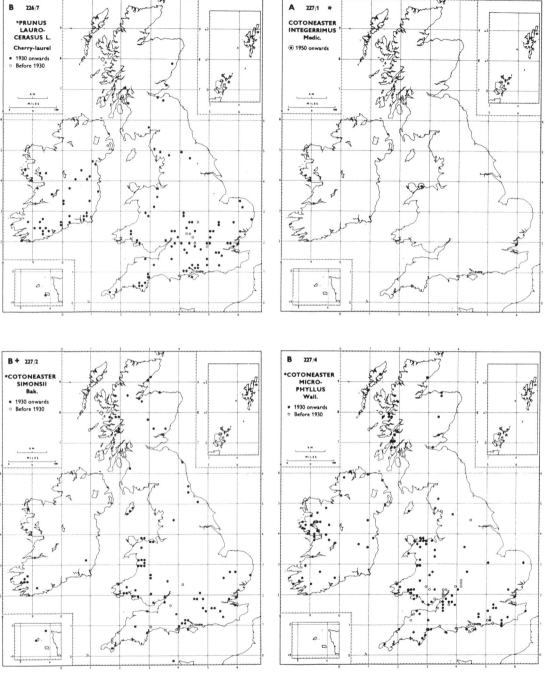

B 226/7
*PRUNUS
LAURO-
CERASUS L.
Cherry-laurel
● 1930 onwards
○ Before 1930

A 227/1 ✳
COTONEASTER
INTEGERRIMUS
Medic.
◉ 1950 onwards

B ✛ 227/2
*COTONEASTER
SIMONSII
Bak.
● 1930 onwards
○ Before 1930

B 227/4
*COTONEASTER
MICRO-
PHYLLUS
Wall.
● 1930 onwards
○ Before 1930

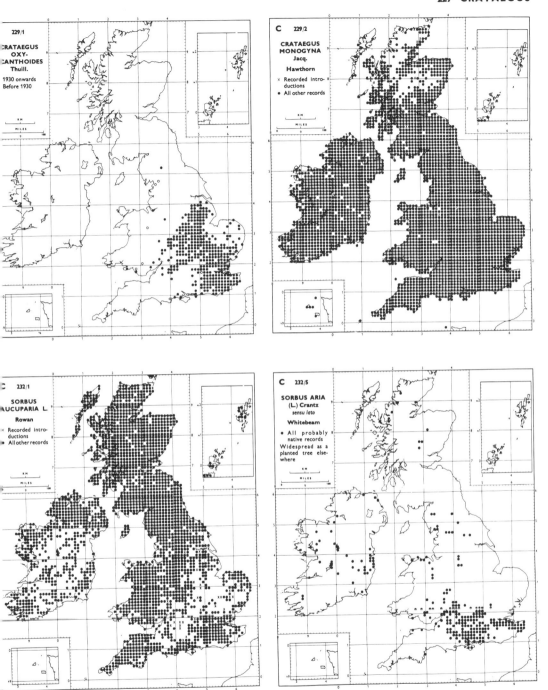

229/1
CRATAEGUS
OXY-
CANTHOIDES
Thuill.

1930 onwards
Before 1930

C 229/2
CRATAEGUS
MONOGYNA
Jacq.

Hawthorn

× Recorded intro-
ductions
● All other records

C 232/1

SORBUS
AUCUPARIA L.

Rowan

× Recorded intro-
ductions
● All other records

C 232/5

SORBUS ARIA
(L.) Crantz
sensu lato

Whitebeam

● All probably
native records
Widespread as a
planted tree else-
where

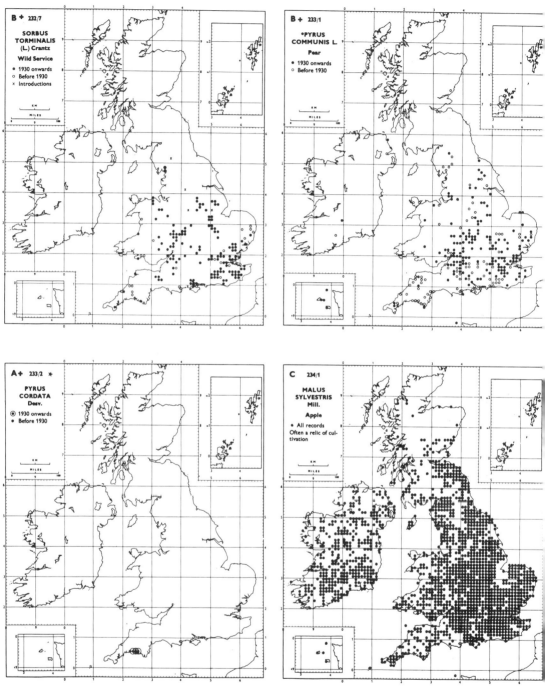

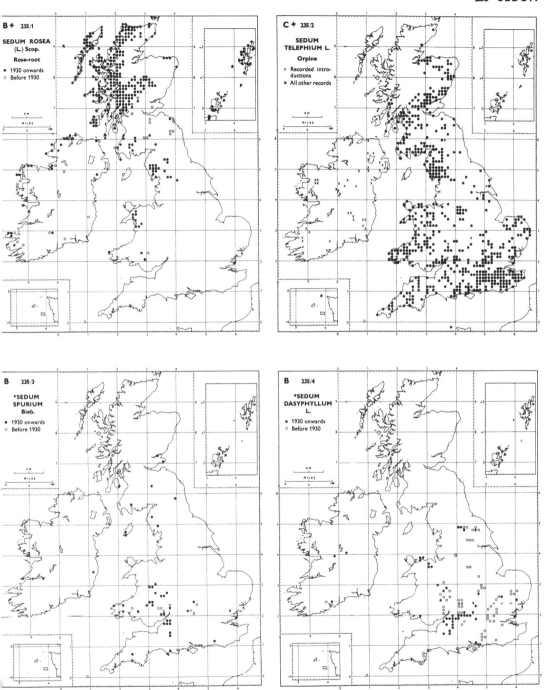

B + 235/1

SEDUM ROSEA
(L.) Scop.
Rose-root

• 1930 onwards
○ Before 1930

C + 235/2

SEDUM
TELEPHIUM L.
Orpine

× Recorded intro-
ductions
• All other records

B 235/3

*SEDUM
SPURIUM
Bieb.

• 1930 onwards
○ Before 1930

B 235/4

*SEDUM
DASYPHYLLUM
L.

• 1930 onwards
○ Before 1930

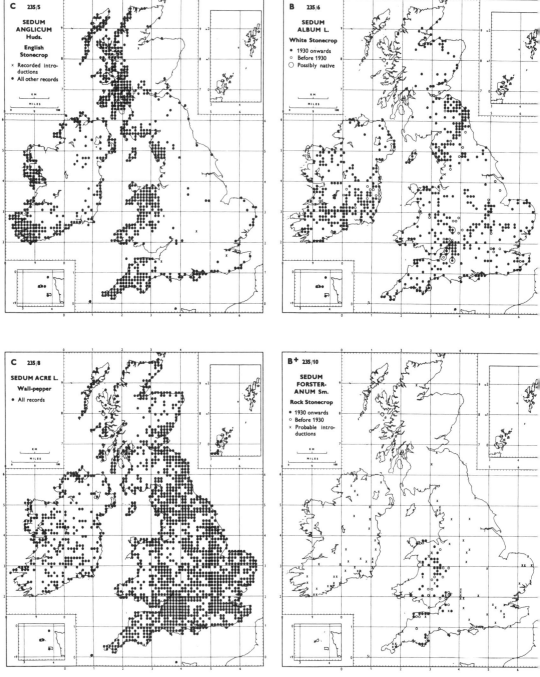

C 235/5

SEDUM
ANGLICUM
Huds.

English
Stonecrop

× Recorded Intro-
ductions
● All other records

B 235/6

SEDUM
ALBUM L.

White Stonecrop

● 1930 onwards
○ Before 1930
◯ Possibly native

C 235/8

SEDUM ACRE L.

Wall-pepper

● All records

B+ 235/10

SEDUM
FORSTER-
ANUM Sm.

Rock Stonecrop

● 1930 onwards
○ Before 1930
× Probable intro-
ductions

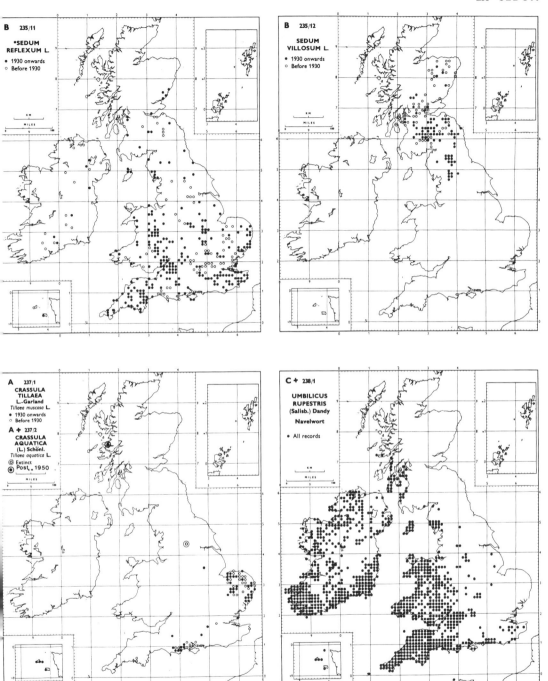

B 235/11
•SEDUM
REFLEXUM L.
• 1930 onwards
○ Before 1930

B 235/12
SEDUM
VILLOSUM L.
• 1930 onwards
○ Before 1930

A 237/1
CRASSULA
TILLAEA
L.-Garland
Tillaea muscosa L.
• 1930 onwards
○ Before 1930

A + 237/2
CRASSULA
AQUATICA
(L.) Schönl.
Tillaea aquatica L.
◉ Extinct
● Post. 1950

C + 238/1
UMBILICUS
RUPESTRIS
(Salisb.) Dandy
Navelwort
• All records

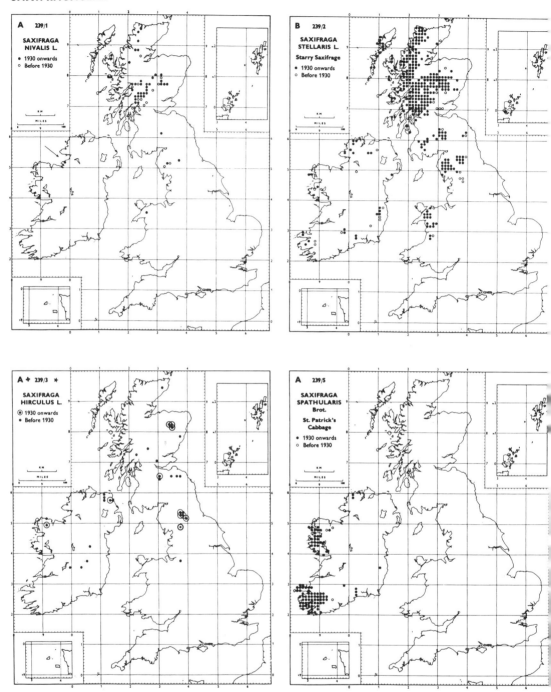

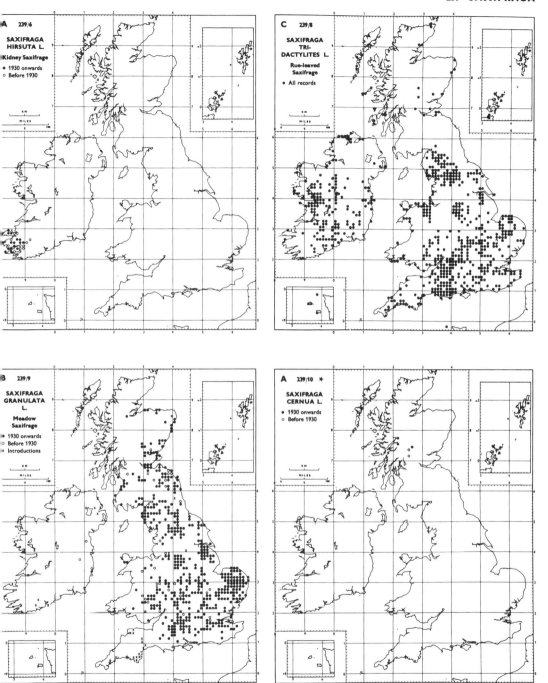

A 239/6

SAXIFRAGA
HIRSUTA L.

Kidney Saxifrage

● 1930 onwards
○ Before 1930

KM
MILES

C 239/8

SAXIFRAGA
TRI-
DACTYLITES L.

Rue-leaved
Saxifrage

● All records

KM
MILES

B 239/9

SAXIFRAGA
GRANULATA
L

Meadow
Saxifrage

▤ 1930 onwards
○ Before 1930
✕ Introductions

KM
MILES

A 239/10 ✱

SAXIFRAGA
CERNUA L.

● 1930 onwards
○ Before 1930

KM
MILES

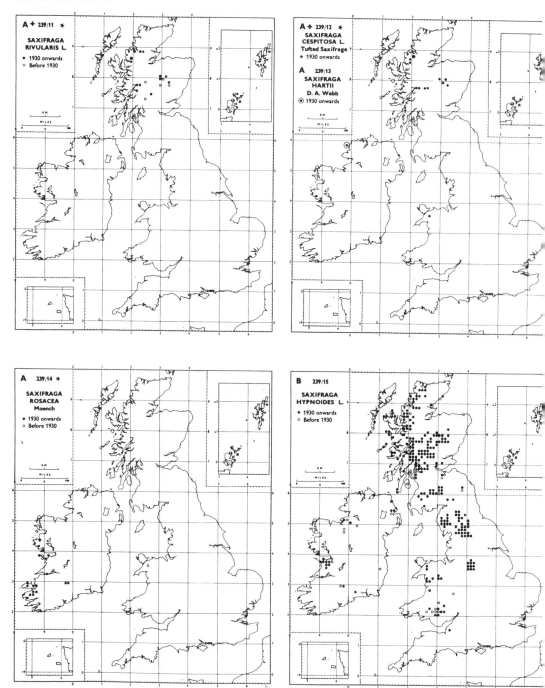

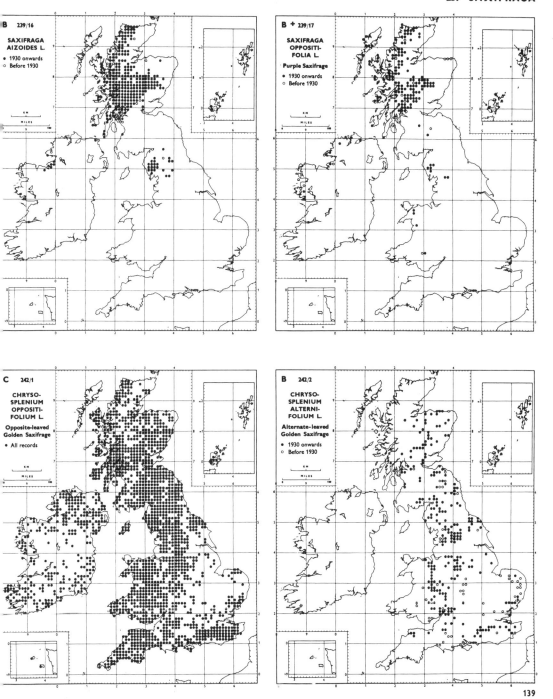

B 239/16

SAXIFRAGA
AIZOIDES L.

• 1930 onwards
○ Before 1930

B + 239/17

SAXIFRAGA
OPPOSITI-
FOLIA L.

Purple Saxifrage

• 1930 onwards
○ Before 1930

C 242/1

CHRYSO-
SPLENIUM
OPPOSITI-
FOLIUM L.

Opposite-leaved
Golden Saxifrage

• All records

B 242/2

CHRYSO-
SPLENIUM
ALTERNI-
FOLIUM L.

Alternate-leaved
Golden Saxifrage

• 1930 onwards
○ Before 1930

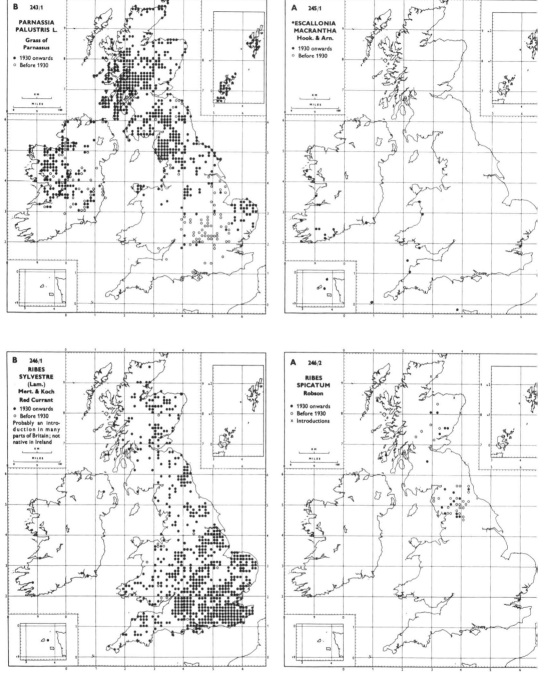

B 243/1

PARNASSIA
PALUSTRIS L.

Grass of
Parnassus

• 1930 onwards
○ Before 1930

A 245/1

*ESCALLONIA
MACRANTHA
Hook. & Arn.

• 1930 onwards
○ Before 1930

B 246/1

RIBES
SYLVESTRE
(Lam.)
Mert. & Koch
Red Currant

• 1930 onwards
○ Before 1930
Probably an intro-
duction in many
parts of Britain; not
native in Ireland

A 246/2

RIBES
SPICATUM
Robson

• 1930 onwards
○ Before 1930
× Introductions

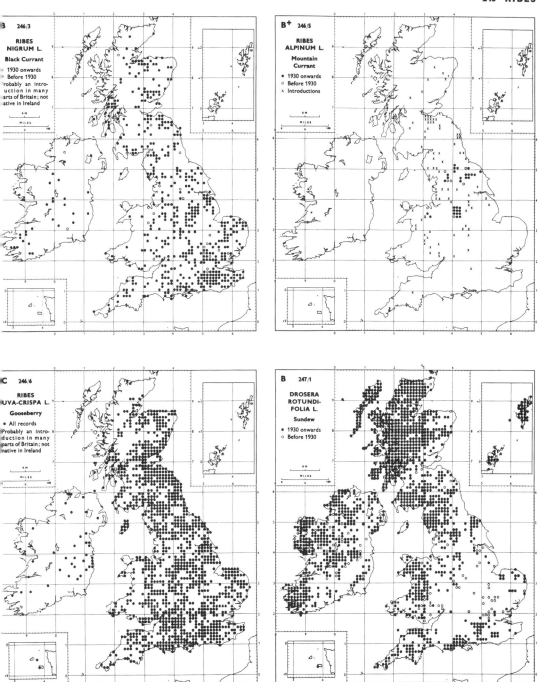

B 246/3

RIBES
NIGRUM L.

Black Currant

● 1930 onwards
● Before 1930
Probably an intro-
duction in many
parts of Britain; not
native in Ireland

B+ 246/5

RIBES
ALPINUM L.

Mountain
Currant

● 1930 onwards
○ Before 1930
x Introductions

C 246/6

RIBES
UVA-CRISPA L.

Gooseberry

● All records
Probably an intro-
duction in many
parts of Britain; not
native in Ireland

B 247/1

DROSERA
ROTUNDI-
FOLIA L.

Sundew

● 1930 onwards
○ Before 1930

DROSERACEAE

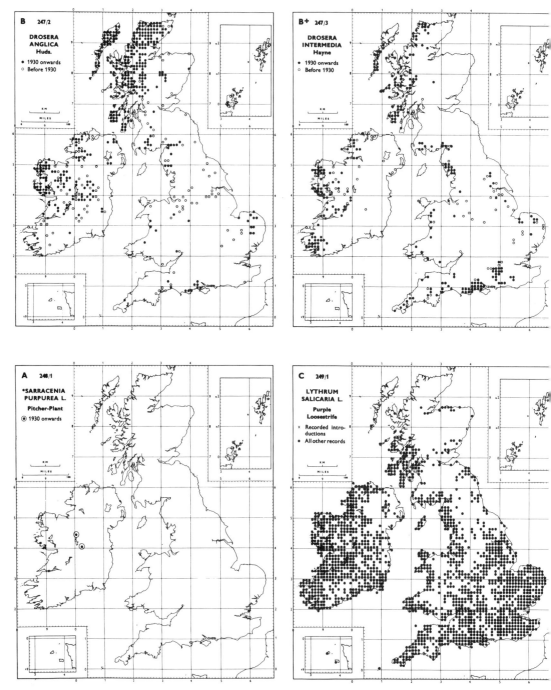

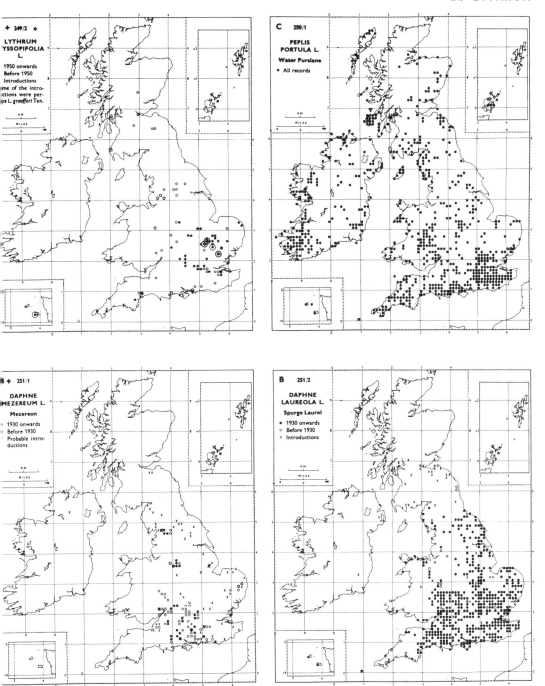

+ 249/2 ★

**LYTHRUM
YSSOPIFOLIA**
L

1950 onwards
Before 1950
Introductions
me of the intro-
ctions were per-
ps *L. groefferi* Ten.

C   250/1

**PEPLIS
PORTULA** L.

Water Purslane
● All records

● + 251/1

**DAPHNE
MEZEREUM** L.

Mezereon

● 1930 onwards
○ Before 1930
× Probable intro-
ductions

B   251/2

**DAPHNE
LAUREOLA** L.

Spurge Laurel

● 1930 onwards
○ Before 1930
× Introductions

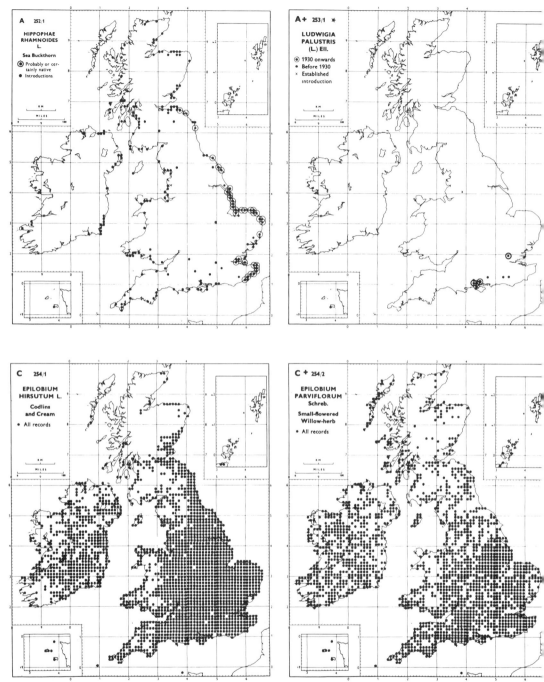

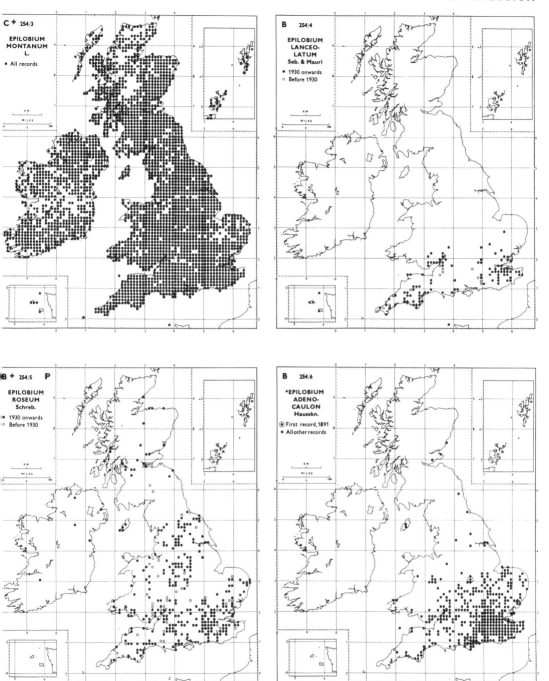

C + 254/3

EPILOBIUM
MONTANUM
L.

• All records

B 254/4

EPILOBIUM
LANCEO-
LATUM
Seb. & Mauri

• 1930 onwards
○ Before 1930

B + 254/5 P

EPILOBIUM
ROSEUM
Schreb.

• 1930 onwards
○ Before 1930

B 254/6

*EPILOBIUM
ADENO-
CAULON
Hausskn.

⊙ First record, 1891
• All other records

ONAGRACEAE

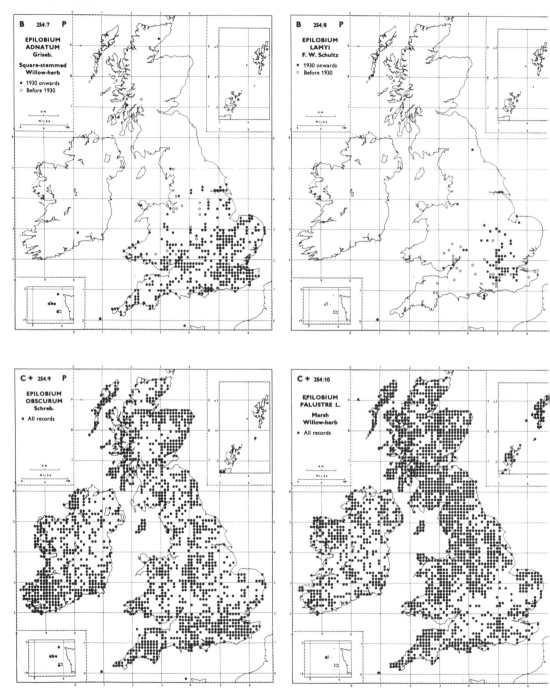

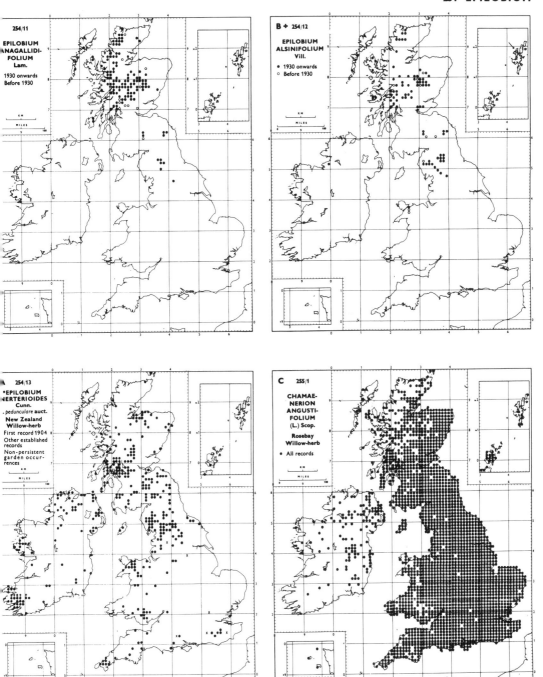

254/11
EPILOBIUM
ANAGALLIDI-
FOLIUM
Lam.

1930 onwards
Before 1930

KM
MILES

B + 254/12
EPILOBIUM
ALSINIFOLIUM
Vill.

• 1930 onwards
○ Before 1930

KM
MILES

254/13
*EPILOBIUM
NERTERIOIDES
Cunn.
. pedunculare auct.
New Zealand
Willow-herb
First record 1904
Other established
records
Non-persistent
garden occur-
rences

KM
MILES

C   255/1
CHAMAE-
NERION
ANGUSTI-
FOLIUM
(L.) Scop.
Rosebay
Willow-herb
• All records

KM
MILES

ONAGRACEAE

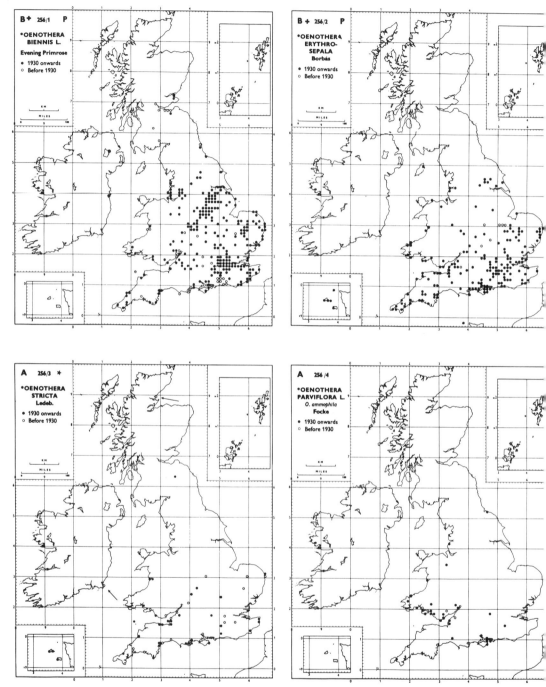

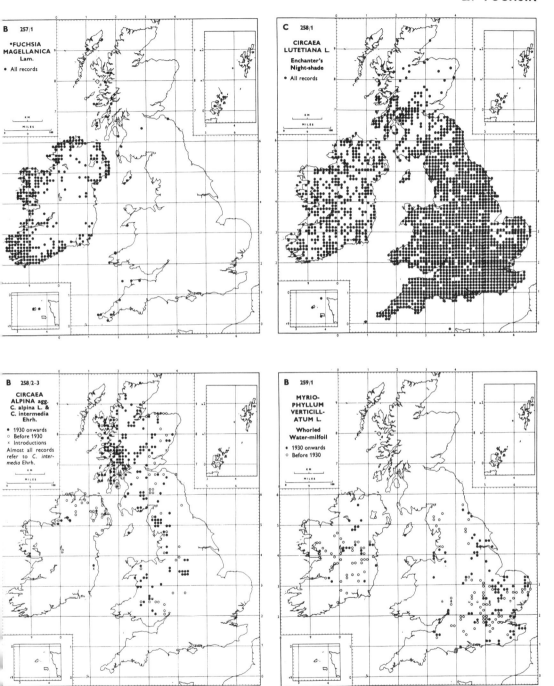

B 257/1
*FUCHSIA
MAGELLANICA
Lam.
● All records

C 258/1
CIRCAEA
LUTETIANA L.
Enchanter's
Night-shade
● All records

B 258/2-3
CIRCAEA
ALPINA agg.
C. alpina L. &
C. intermedia
Ehrh.
● 1930 onwards
○ Before 1930
× Introductions
Almost all records
refer to C. inter-
media Ehrh.

B 259/1
MYRIO-
PHYLLUM
VERTICILL-
ATUM L.
Whorled
Water-milfoil
● 1930 onwards
○ Before 1930

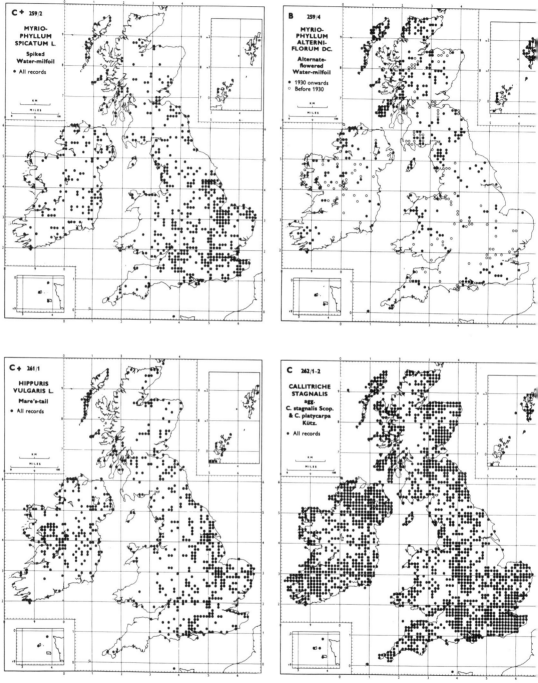

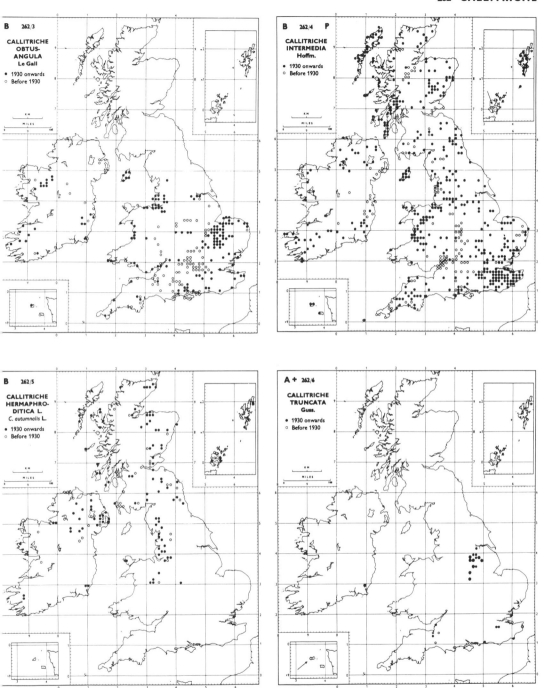

B 262/3

CALLITRICHE
OBTUS-
ANGULA
Le Gall

• 1930 onwards
○ Before 1930

B 262/4 P

CALLITRICHE
INTERMEDIA
Hoffm.

• 1930 onwards
○ Before 1930

B 262/5

CALLITRICHE
HERMAPHRO-
DITICA L.
C. autumnalis L.

• 1930 onwards
○ Before 1930

A + 262/6

CALLITRICHE
TRUNCATA
Guss.

• 1930 onwards
○ Before 1930

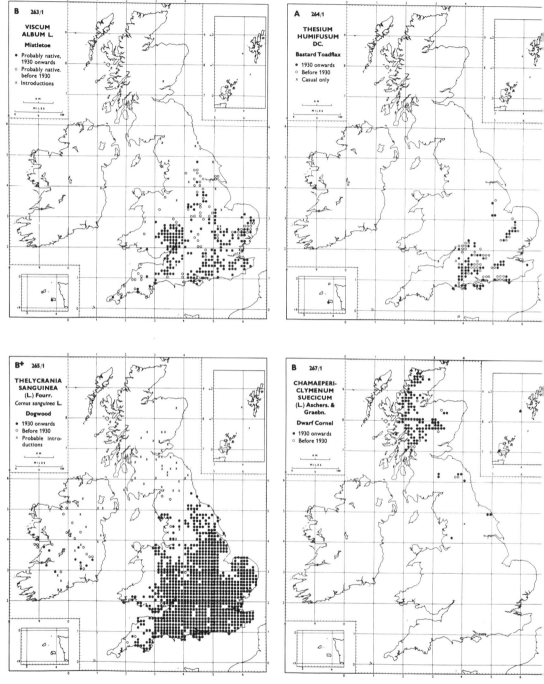

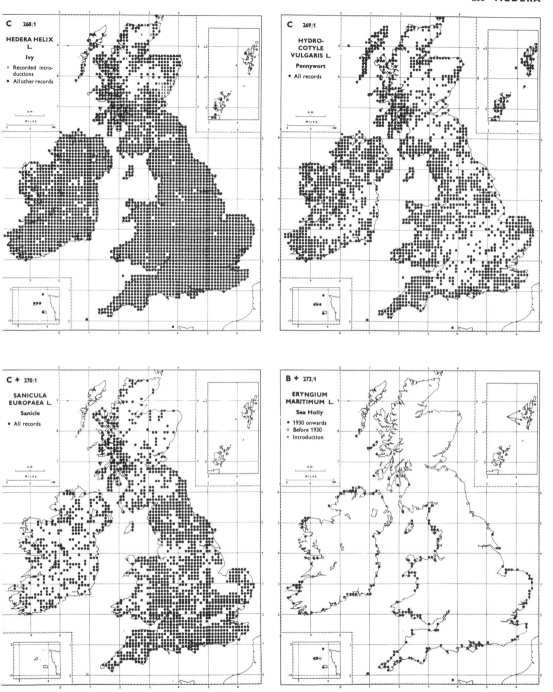

C 268/1

HEDERA HELIX
L.

Ivy

× Recorded intro-
ductions
• All other records

C 269/1

HYDRO-
COTYLE
VULGARIS L.

Pennywort

• All records

C + 270/1

SANICULA
EUROPAEA L.

Sanicle

• All records

B + 272/1

ERYNGIUM
MARITIMUM L.

Sea Holly

• 1930 onwards
○ Before 1930
× Introduction

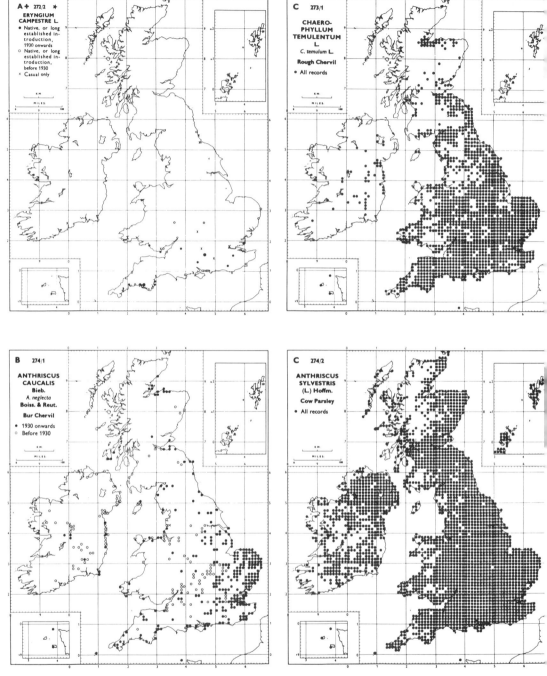

A + 272/2 ＊
ERYNGIUM
CAMPESTRE L.
• Native, or long
  established in-
  troduction,
  1930 onwards
○ Native, or long
  established in-
  troduction,
  before 1930
× Casual only

C 273/1
CHAERO-
PHYLLUM
TEMULENTUM
L.
C. temulum L.
Rough Chervil
• All records

B 274/1
ANTHRISCUS
CAUCALIS
Bieb.
A. neglecta
Boiss. & Reut.
Bur Chervil
• 1930 onwards
○ Before 1930

C 274/2
ANTHRISCUS
SYLVESTRIS
(L.) Hoffm.
Cow Parsley
• All records

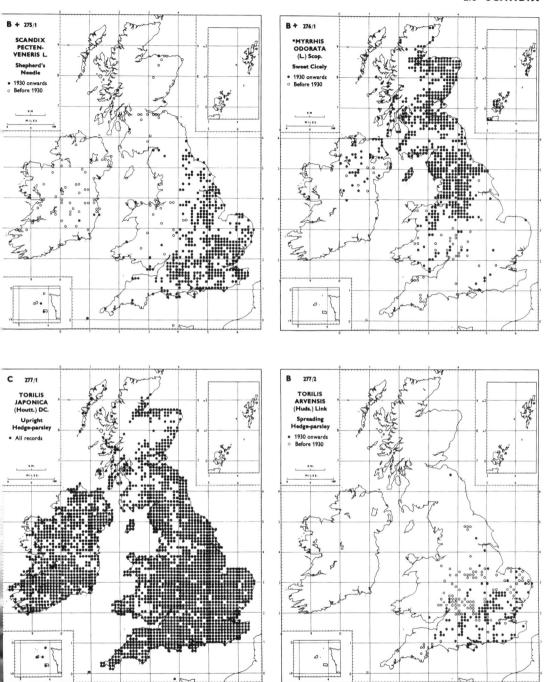

B + 275/1

SCANDIX
PECTEN-
VENERIS L.

Shepherd's
Needle

● 1930 onwards
○ Before 1930

B + 276/1

°MYRRHIS
ODORATA
(L.) Scop.

Sweet Cicely

● 1930 onwards
○ Before 1930

C 277/1

TORILIS
JAPONICA
(Houtt.) DC.

Upright
Hedge-parsley

● All records

B 277/2

TORILIS
ARVENSIS
(Huds.) Link

Spreading
Hedge-parsley

● 1930 onwards
○ Before 1930

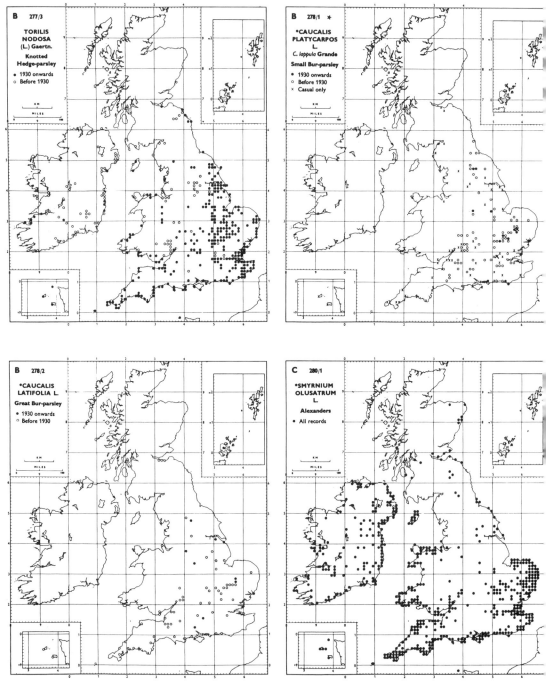

B 277/3

TORILIS
NODOSA
(L.) Gaertn.
Knotted
Hedge-parsley

- 1930 onwards
- Before 1930

B 278/1 ✶

•CAUCALIS
PLATYCARPOS
L
C. lappula Grande
Small Bur-parsley

- 1930 onwards
- Before 1930
× Casual only

B 278/2

•CAUCALIS
LATIFOLIA L.
Great Bur-parsley

- 1930 onwards
- Before 1930

C 280/1

•SMYRNIUM
OLUSATRUM
L
Alexanders

- All records

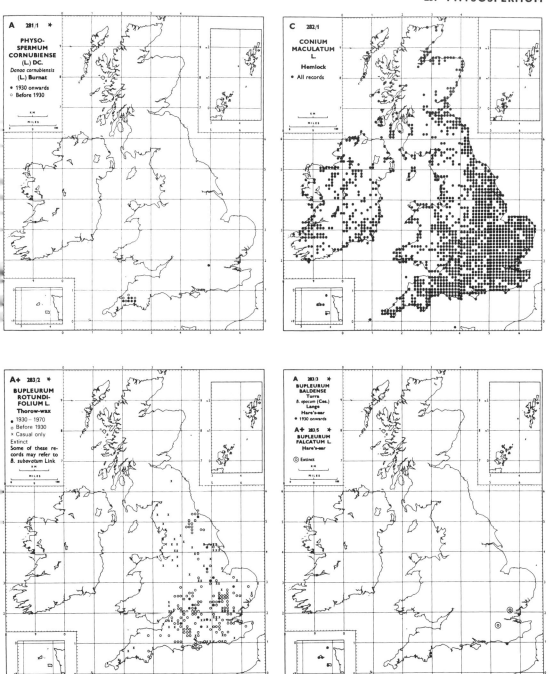

UMBELLIFERAE

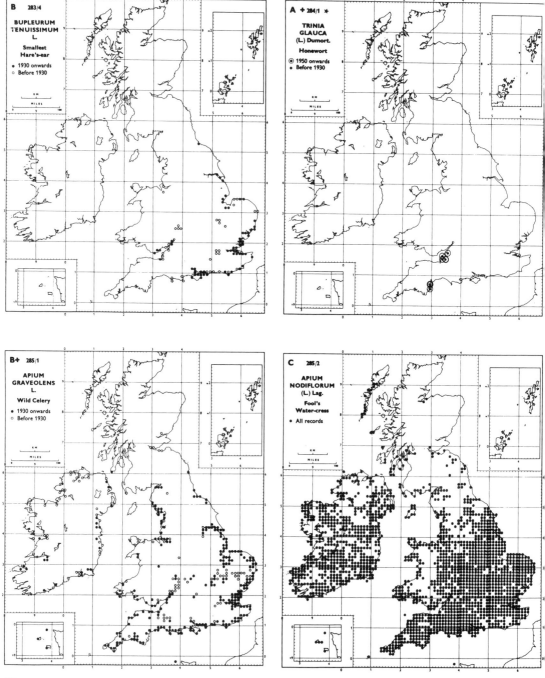

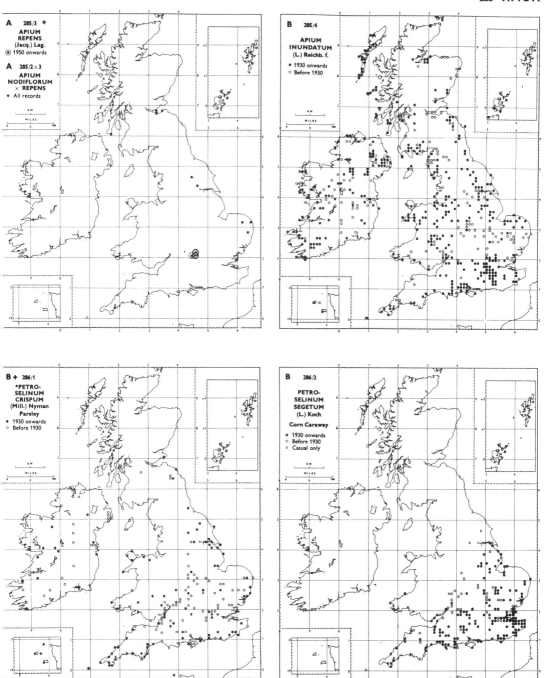

A 285/3 ✳
APIUM
REPENS
(Jacq.) Lag.
⊙ 1950 onwards

A 285/2 × 3
APIUM
NODIFLORUM
× REPENS
• All records

B 285/4
APIUM
INUNDATUM
(L.) Reichb. f.
• 1930 onwards
○ Before 1930

B + 286/1
*PETRO-
SELINUM
CRISPUM
(Mill.) Nyman
Parsley
• 1930 onwards
○ Before 1930

B 286/2
PETRO-
SELINUM
SEGETUM
(L.) Koch
Corn Caraway
• 1930 onwards
○ Before 1930
× Casual only

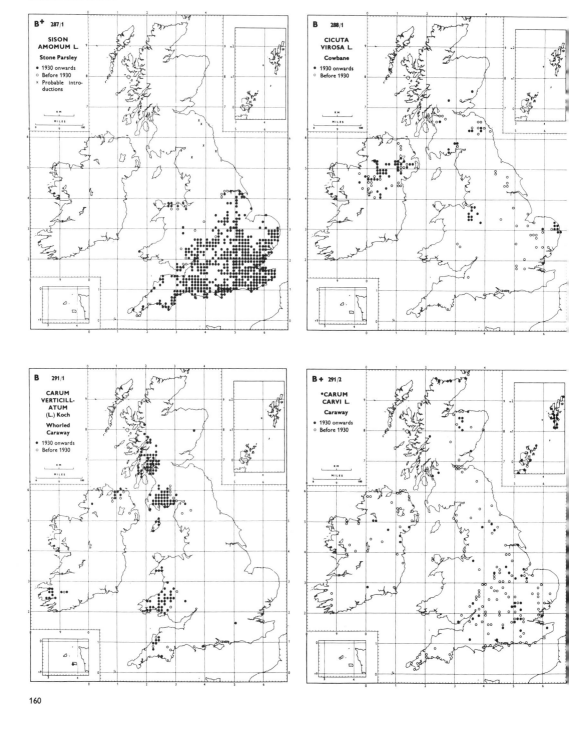

B+ 287/1

SISON
AMOMUM L.

Stone Parsley

• 1930 onwards
○ Before 1930
× Probable intro-
ductions

B 288/1

CICUTA
VIROSA L.

Cowbane

• 1930 onwards
○ Before 1930

B 291/1

CARUM
VERTICILL-
ATUM
(L.) Koch

Whorled
Caraway

• 1930 onwards
○ Before 1930

B+ 291/2

*CARUM
CARVI L.

Caraway

• 1930 onwards
○ Before 1930

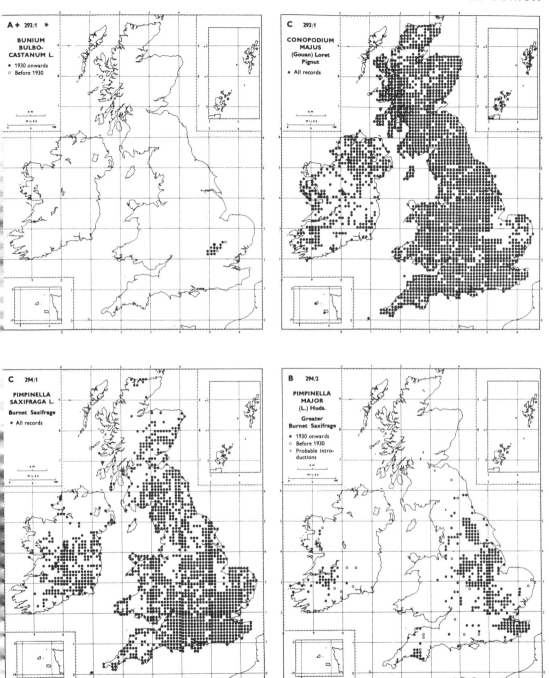

A + 292/1 *

BUNIUM
BULBO-
CASTANUM L.

• 1930 onwards
○ Before 1930

C   293/1

CONOPODIUM
MAJUS
(Gouan) Loret
Pignut

• All records

C   294/1

PIMPINELLA
SAXIFRAGA L.

Burnet Saxifrage

• All records

B   294/2

PIMPINELLA
MAJOR
(L.) Huds.

Greater
Burnet Saxifrage

• 1930 onwards
○ Before 1930
× Probable intro-
    ductions

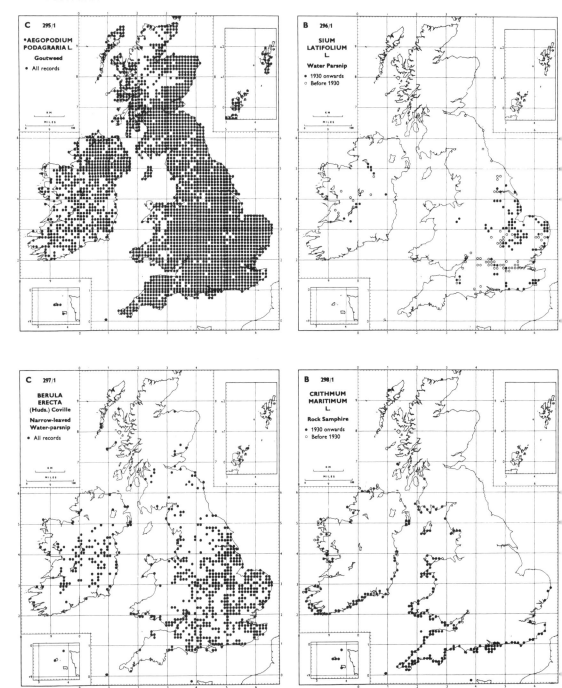

C   295/1

•AEGOPODIUM
PODAGRARIA L.

Goutweed

• All records

B   296/1

SIUM
LATIFOLIUM
L.

Water Parsnip

• 1930 onwards
○ Before 1930

C   297/1

BERULA
ERECTA
(Huds.) Coville

Narrow-leaved
Water-parsnip

• All records

B   298/1

CRITHMUM
MARITIMUM
L.

Rock Samphire

• 1930 onwards
○ Before 1930

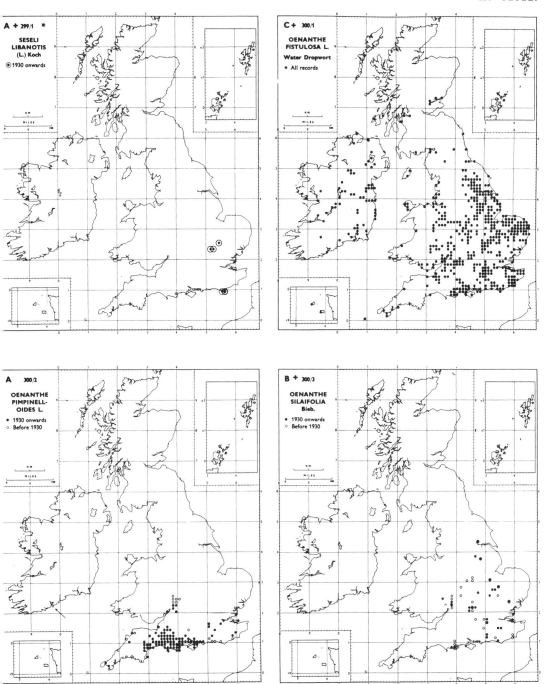

A + 299/1 *

SESELI
LIBANOTIS
(L.) Koch

⊙ 1930 onwards

C + 300/1

OENANTHE
FISTULOSA L.
Water Dropwort
• All records

A 300/2

OENANTHE
PIMPINELL-
OIDES L.
• 1930 onwards
○ Before 1930

B + 300/3

OENANTHE
SILAIFOLIA
Bieb.
• 1930 onwards
○ Before 1930

163

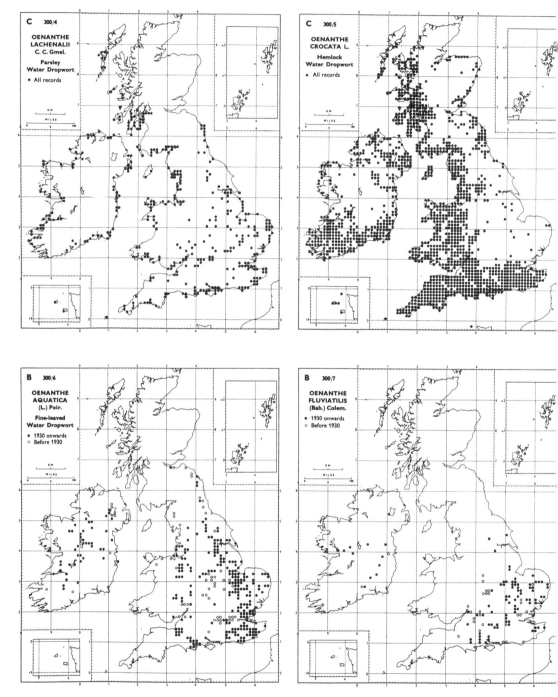

C 300/4

OENANTHE
LACHENALII
C. C. Gmel.

Parsley
Water Dropwort

• All records

C 300/5

OENANTHE
CROCATA L.

Hemlock
Water Dropwort

• All records

B 300/6

OENANTHE
AQUATICA
(L.) Poir.

Fine-leaved
Water Dropwort

• 1930 onwards
○ Before 1930

B 300/7

OENANTHE
FLUVIATILIS
(Bab.) Colem.

• 1930 onwards
○ Before 1930

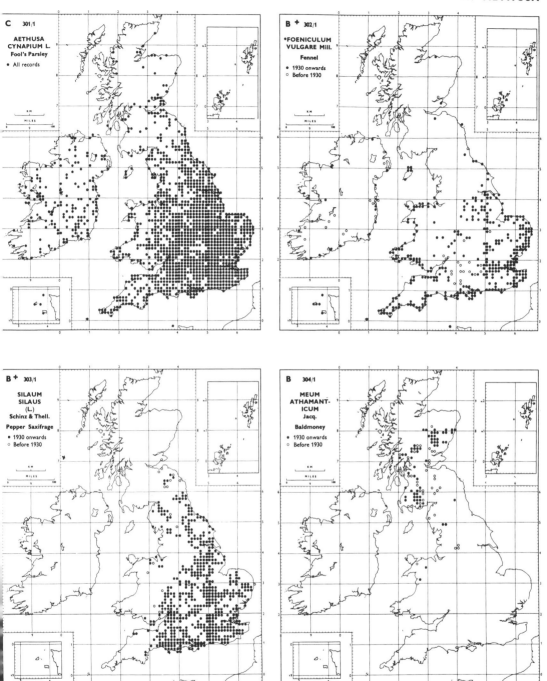

C 301/1

AETHUSA
CYNAPIUM L.
Fool's Parsley

● All records

B + 302/1

*FOENICULUM
VULGARE Mill.

Fennel

● 1930 onwards
○ Before 1930

B + 303/1

SILAUM
SILAUS
(L.)
Schinz & Thell.

Pepper Saxifrage

● 1930 onwards
○ Before 1930

B 304/1

MEUM
ATHAMANT-
ICUM
Jacq.

Baldmoney

● 1930 onwards
○ Before 1930

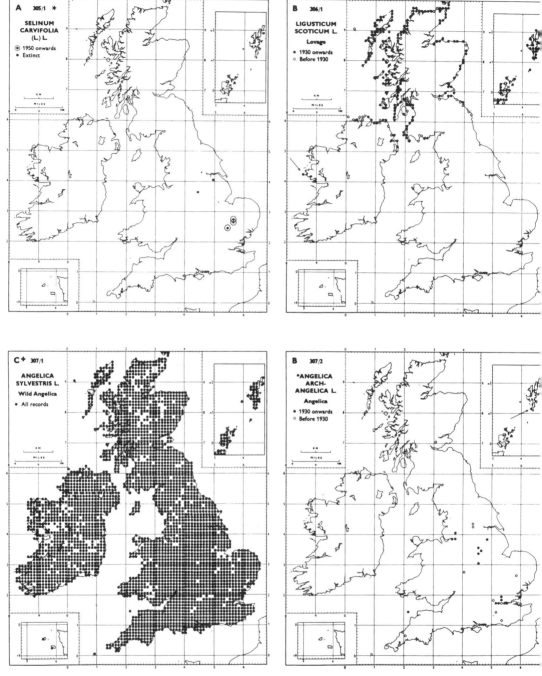

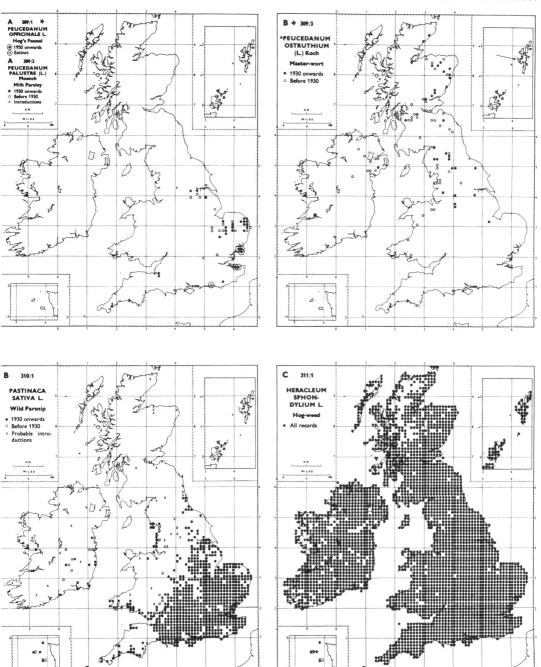

A 309/1 ✳
PEUCEDANUM
OFFICINALE L.
Hog's Fennel
⊙ 1950 onwards
⊘ Extinct

A 309/2
PEUCEDANUM
PALUSTRE (L.)
Moench
Milk Parsley
● 1930 onwards
○ Before 1930
✕ Introductions

B ✚ 309/3
●PEUCEDANUM
OSTRUTHIUM
(L.) Koch
Master-wort
● 1930 onwards
○ Before 1930

B 310/1
PASTINACA
SATIVA L.
Wild Parsnip
● 1930 onwards
○ Before 1930
✕ Probable intro-
ductions

C 311/1
HERACLEUM
SPHON-
DYLIUM L.
Hog-weed
● All records

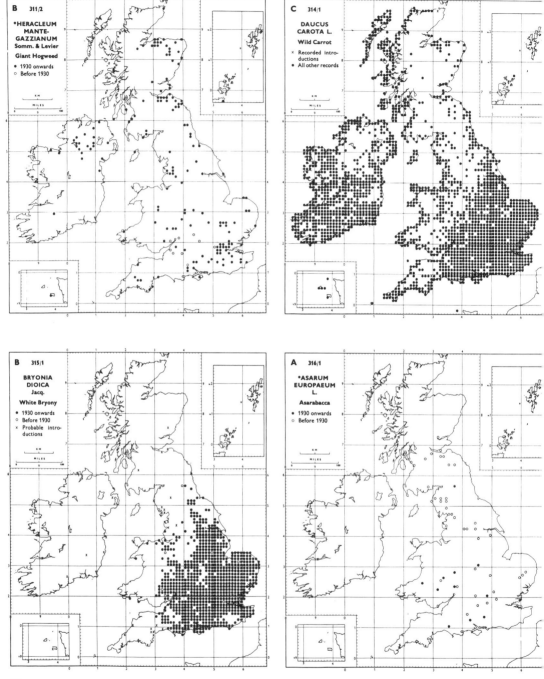

B 311/2

*HERACLEUM
MANTE-
GAZZIANUM
Somm. & Levier
Giant Hogweed

● 1930 onwards
○ Before 1930

C 314/1

DAUCUS
CAROTA L.

Wild Carrot

× Recorded intro-
ductions
● All other records

B 315/1

BRYONIA
DIOICA
Jacq.

White Bryony

● 1930 onwards
○ Before 1930
× Probable intro-
ductions

A 316/1

*ASARUM
EUROPAEUM
L.

Asarabacca

● 1930 onwards
○ Before 1930

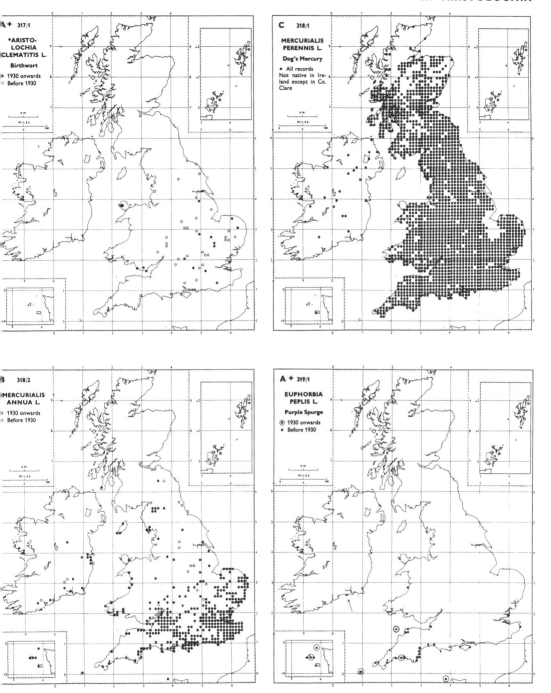

A + 317/1

*ARISTO-
LOCHIA
CLEMATITIS L.

Birthwort

● 1930 onwards
○ Before 1930

C 318/1

MERCURIALIS
PERENNIS L.

Dog's Mercury

● All records
Not native in Ire-
land except in Co.
Clare

B 318/2

MERCURIALIS
ANNUA L.

● 1930 onwards
○ Before 1930

A + 319/1

EUPHORBIA
PEPLIS L.

Purple Spurge

◉ 1930 onwards
● Before 1930

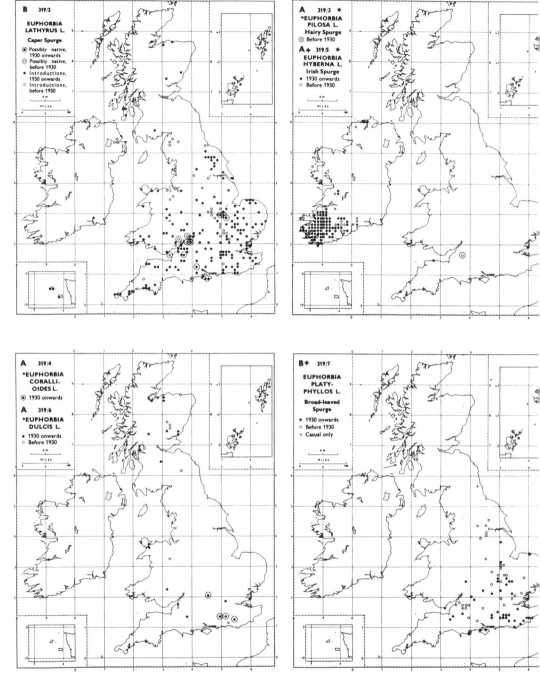

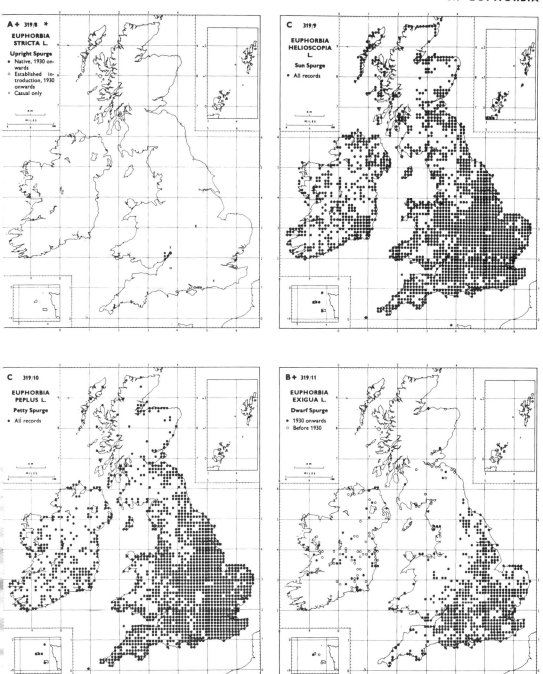

A+ 319/8 *

EUPHORBIA STRICTA L.

Upright Spurge

- Native, 1930 onwards
- ○ Established Introduction, 1930 onwards
- × Casual only

C 319/9

EUPHORBIA HELIOSCOPIA L.

Sun Spurge

- All records

C 319/10

EUPHORBIA PEPLUS L.

Petty Spurge

- All records

B+ 319/11

EUPHORBIA EXIGUA L.

Dwarf Spurge

- 1930 onwards
- ○ Before 1930

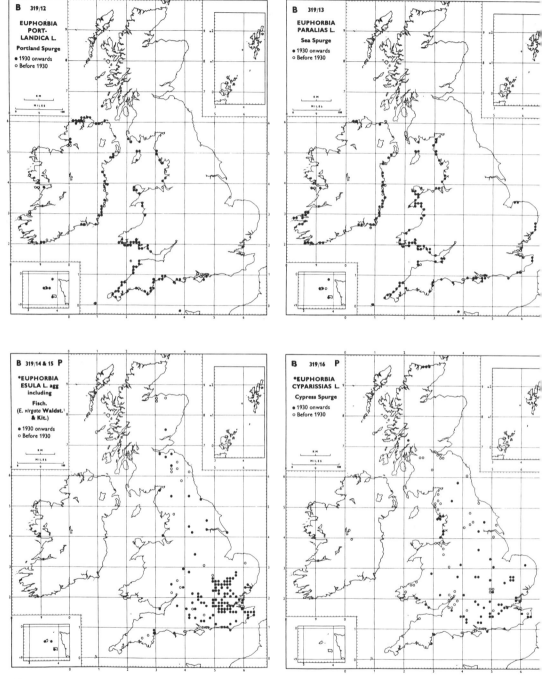

B 319/12

EUPHORBIA PORT-LANDICA L.

Portland Spurge

● 1930 onwards
○ Before 1930

B 319/13

EUPHORBIA PARALIAS L.

Sea Spurge

● 1930 onwards
○ Before 1930

B 319/14 & 15 P

*EUPHORBIA ESULA L. agg
including

Fisch.
(E. virgata Waldst.
& Kit.)

● 1930 onwards
○ Before 1930

B 319/16 P

*EUPHORBIA CYPARISSIAS L.

Cypress Spurge

● 1930 onwards
○ Before 1930

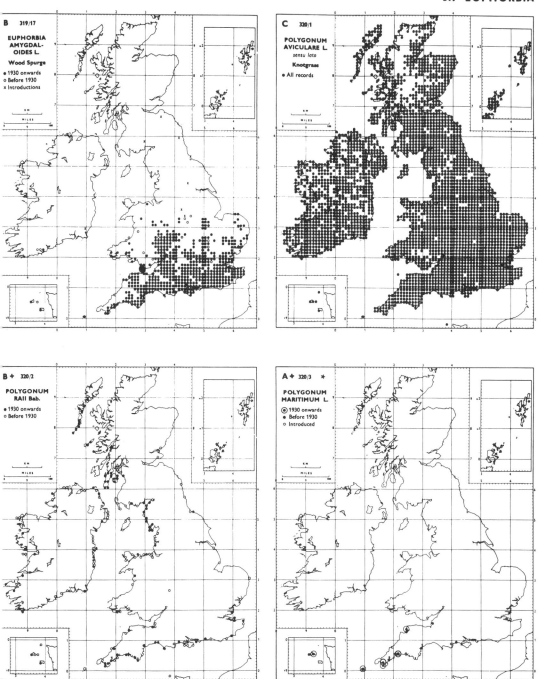

B 319/17

EUPHORBIA AMYGDAL-OIDES L.

Wood Spurge

• 1930 onwards
○ Before 1930
× Introductions

C 320/1

POLYGONUM AVICULARE L.
sensu lato

Knotgrass

• All records

B + 320/2

POLYGONUM RAII Bab.

• 1930 onwards
○ Before 1930

A + 320/3 ✶

POLYGONUM MARITIMUM L.

◉ 1930 onwards
• Before 1930
○ Introduced

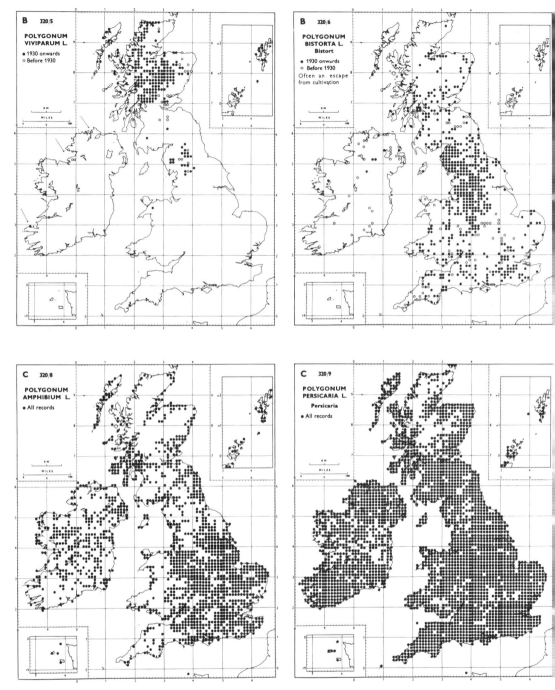

B   320/5

POLYGONUM
VIVIPARUM L.

● 1930 onwards
○ Before 1930

B   320/6

POLYGONUM
BISTORTA L.
Bistort

● 1930 onwards
○ Before 1930
Often an escape
from cultivation

C   320/8

POLYGONUM
AMPHIBIUM L.

● All records

C   320/9

POLYGONUM
PERSICARIA L.

Persicaria

● All records

174

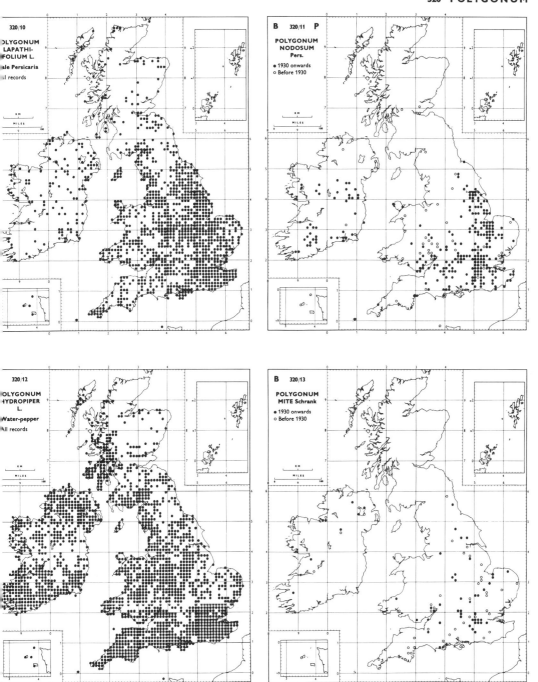

320/10

POLYGONUM
LAPATHI-
FOLIUM L.
Pale Persicaria
All records

B   320/11   P

POLYGONUM
NODOSUM
Pers.

• 1930 onwards
○ Before 1930

320/12

POLYGONUM
HYDROPIPER
L.
Water-pepper
All records

B   320/13

POLYGONUM
MITE Schrank

• 1930 onwards
○ Before 1930

175

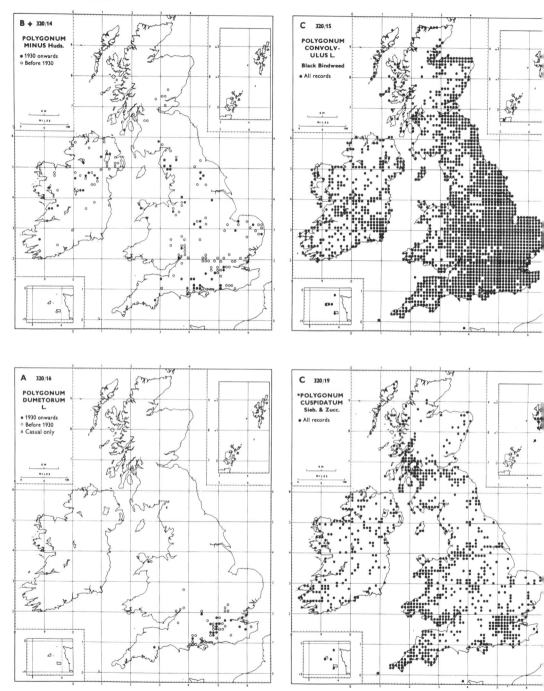

B + 320/14
POLYGONUM
MINUS Huds.
• 1930 onwards
○ Before 1930

C 320/15
POLYGONUM
CONVOLV-
ULUS L.
Black Bindweed
• All records

A 320/16
POLYGONUM
DUMETORUM
L.
• 1930 onwards
○ Before 1930
× Casual only

C 320/19
*POLYGONUM
CUSPIDATUM
Sieb. & Zucc.
• All records

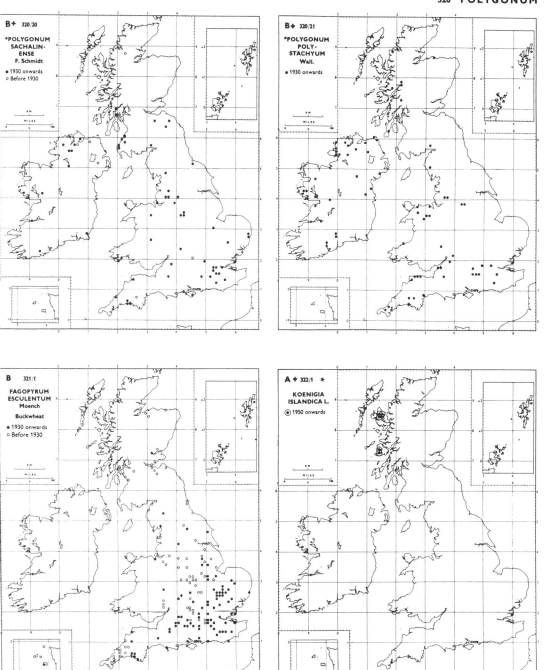

B+ 320/20
*POLYGONUM
SACHALIN-
ENSE
F. Schmidt
● 1930 onwards
○ Before 1930

B+ 320/21
*POLYGONUM
POLY-
STACHYUM
Wall.
● 1930 onwards

B 321/1
FAGOPYRUM
ESCULENTUM
Moench
Buckwheat
● 1930 onwards
○ Before 1930

A+ 322/1 ✳
KOENIGIA
ISLANDICA L.
◉ 1950 onwards

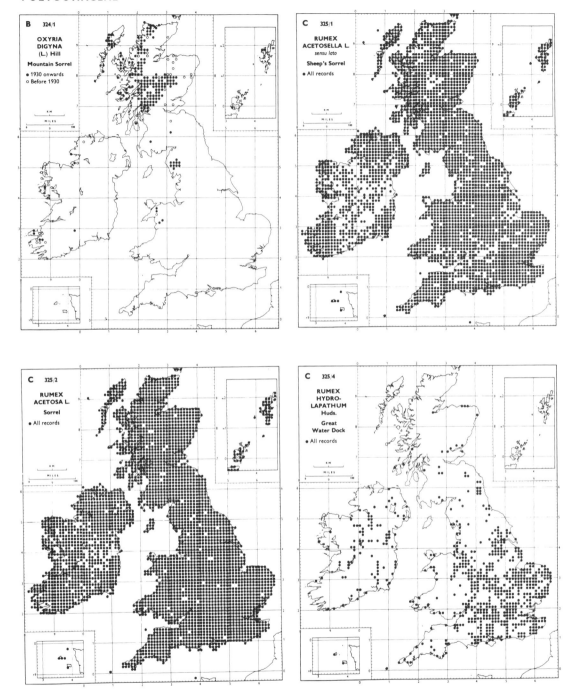

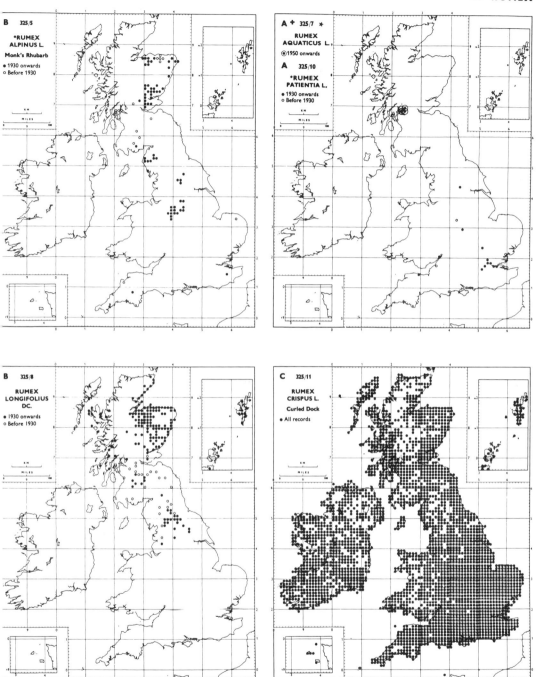

B 325/5

*RUMEX ALPINUS L.
Monk's Rhubarb
● 1930 onwards
○ Before 1930

A + 325/7 ✶
RUMEX AQUATICUS L.
⊚1950 onwards
A 325/10
*RUMEX PATIENTIA L.
● 1930 onwards
○ Before 1930

B 325/8
RUMEX LONGIFOLIUS DC.
● 1930 onwards
○ Before 1930

C 325/11
RUMEX CRISPUS L.
Curled Dock
● All records

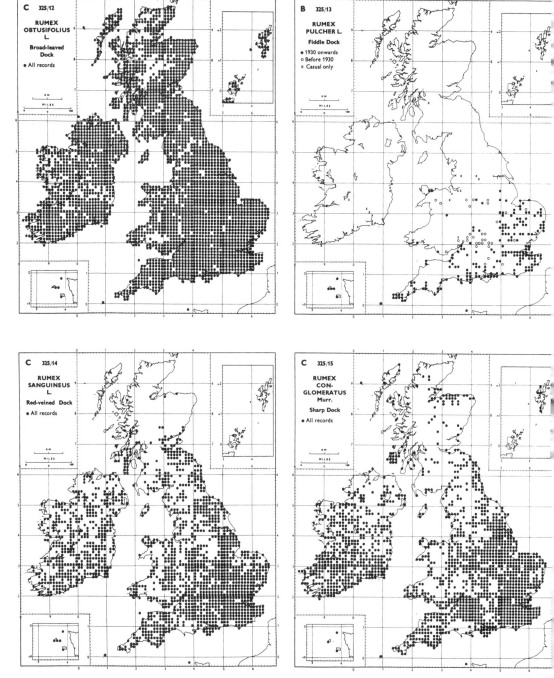

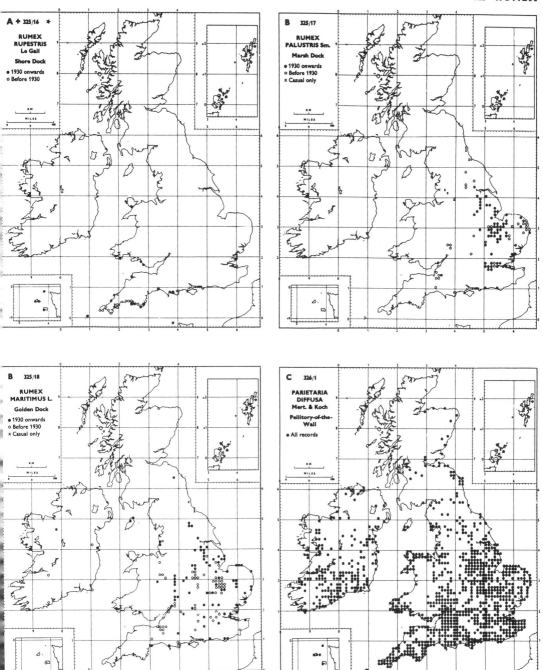

A + 325/16 *

RUMEX
RUPESTRIS
Le Gall

Shore Dock

• 1930 onwards
○ Before 1930

B 325/17

RUMEX
PALUSTRIS Sm.

Marsh Dock

• 1930 onwards
○ Before 1930
x Casual only

B 325/18

RUMEX
MARITIMUS L.

Golden Dock

• 1930 onwards
○ Before 1930
x Casual only

C 326/1

PARIETARIA
DIFFUSA
Mert. & Koch

Pellitory-of-the-
Wall

• All records

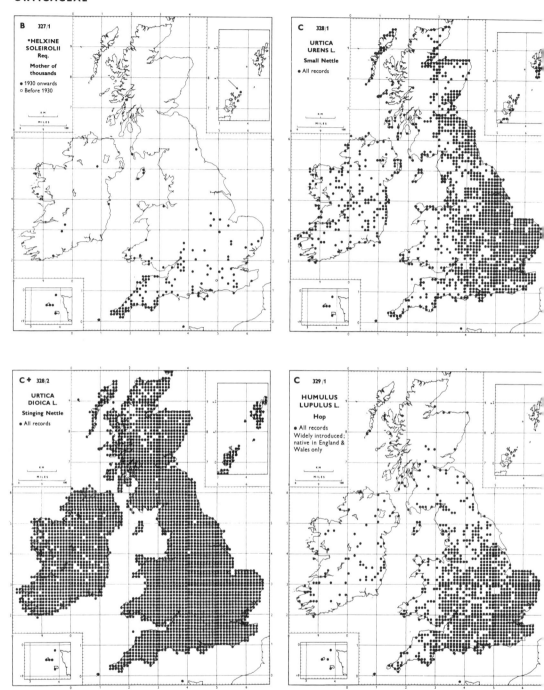

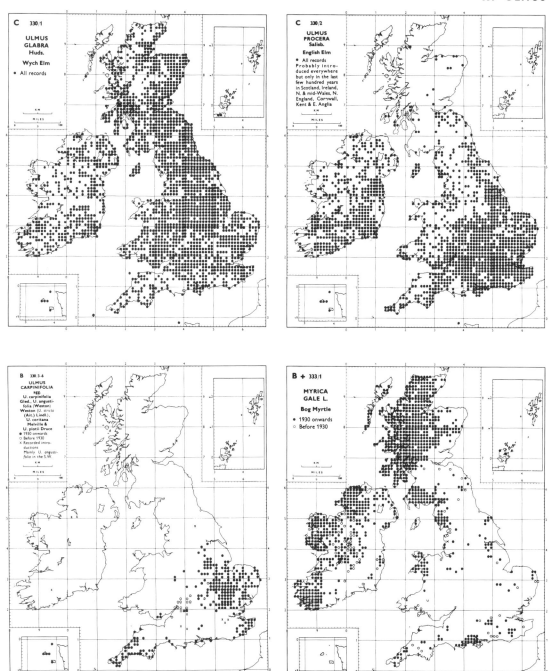

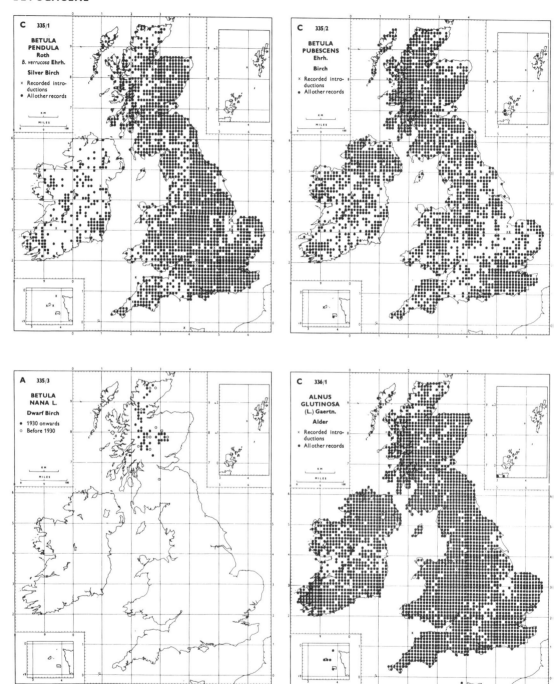

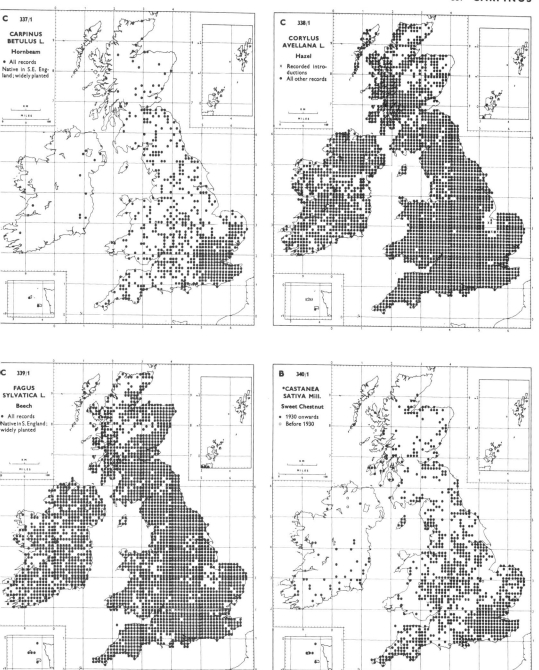

C 337/1

**CARPINUS
BETULUS L.**

Hornbeam

● All records
Native in S.E. Eng-
land; widely planted

C 338/1

**CORYLUS
AVELLANA L.**

Hazel

× Recorded intro-
ductions
● All other records

●C 339/1

**FAGUS
SYLVATICA L.**

Beech

● All records
Native in S. England;
widely planted

B 340/1

**°CASTANEA
SATIVA Mill.**

Sweet Chestnut

● 1930 onwards
○ Before 1930

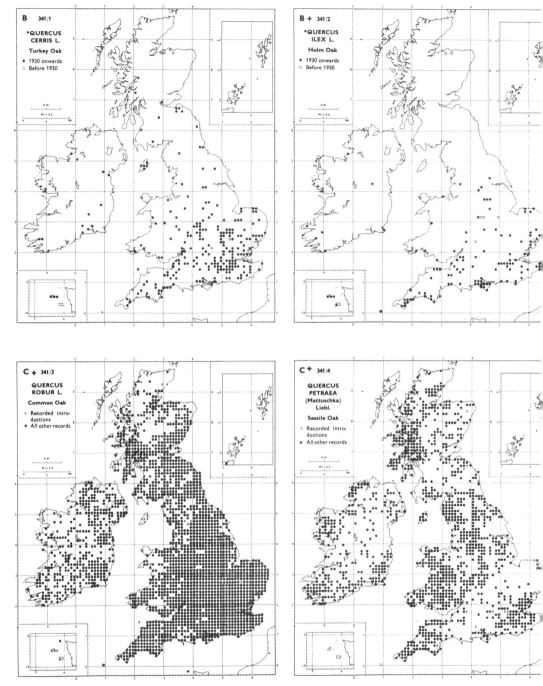

B 341/1

•QUERCUS
CERRIS L.

Turkey Oak

• 1930 onwards
○ Before 1930

B + 341/2

•QUERCUS
ILEX L.

Holm Oak

• 1930 onwards
○ Before 1930

C + 341/3

QUERCUS
ROBUR L.

Common Oak

× Recorded intro-
  ductions
• All other records

C + 341/4

QUERCUS
PETRAEA
(Mattuschka)
Liebl.

Sessile Oak

× Recorded intro-
  ductions
• All other records

186

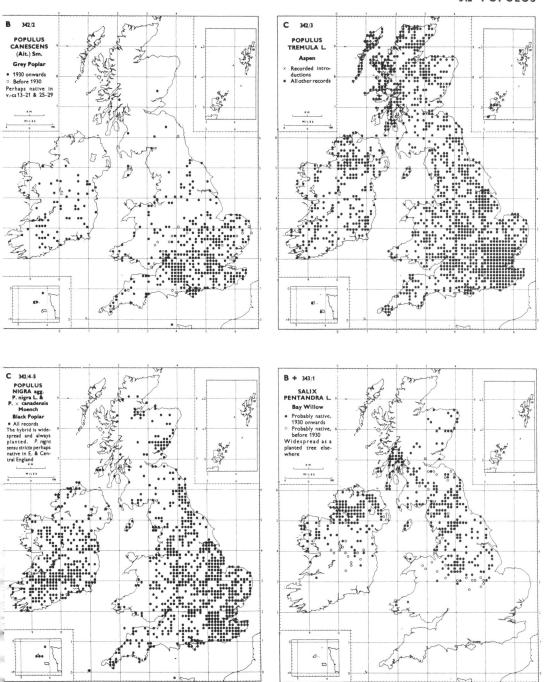

B 342/2

**POPULUS
CANESCENS**
(Ait.) Sm.

Grey Poplar

• 1930 onwards
○ Before 1930
Perhaps native in
v.-cs 13–21 & 25–29

C 342/3

**POPULUS
TREMULA L.**

Aspen

× Recorded intro-
ductions
• All other records

C 342/4–5

**POPULUS
NIGRA agg.**
P. nigra L. &
P. × canadensis
Moench

Black Poplar

• All records
The hybrid is wide-
spread and always
planted. *P. nigra*
sensu stricto perhaps
native in E. & Cen-
tral England

B + 343/1

**SALIX
PENTANDRA L.**

Bay Willow

• Probably native,
1930 onwards
○ Probably native,
before 1930
Widespread as a
planted tree else-
where

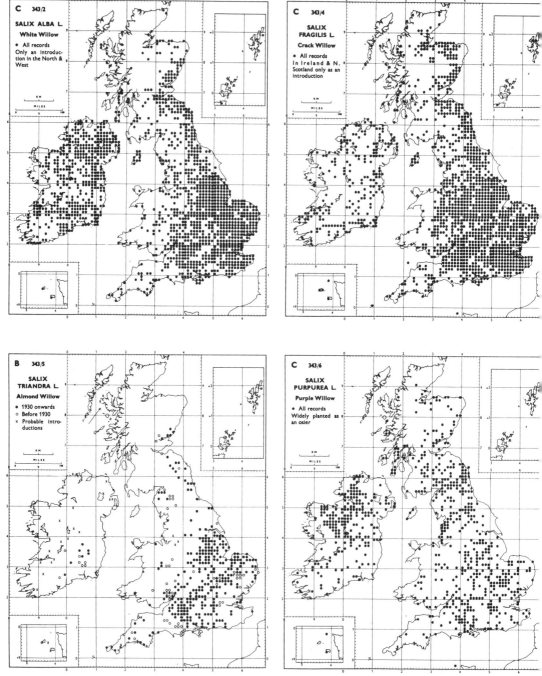

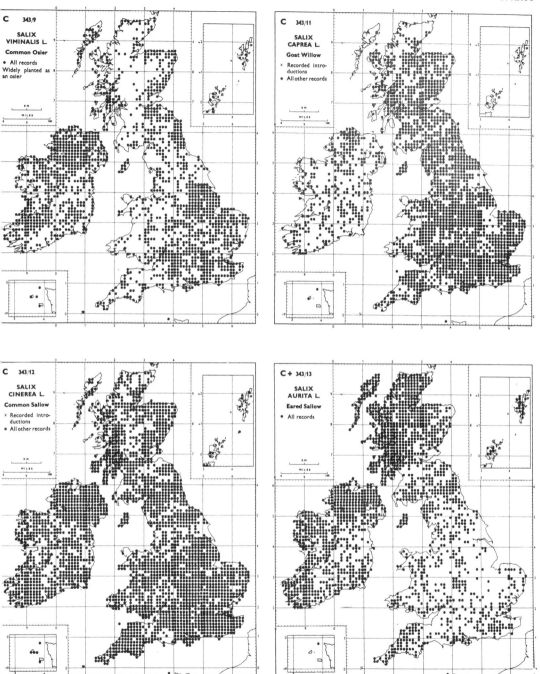

C 343/9

SALIX
VIMINALIS L.

Common Osier

● All records
Widely planted as
an osier

C 343/11

SALIX
CAPREA L.

Goat Willow

× Recorded intro-
ductions
● All other records

C 343/12

SALIX
CINEREA L.

Common Sallow

× Recorded intro-
ductions
● All other records

C + 343/13

SALIX
AURITA L.

Eared Sallow

● All records

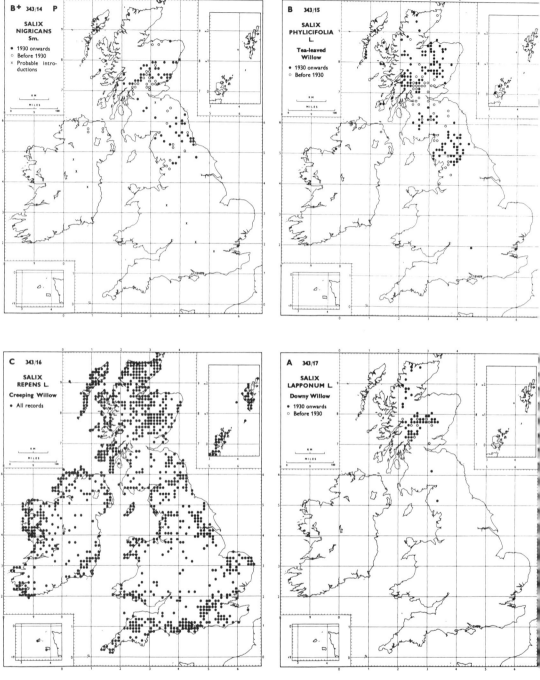

B+ 343/14 P
SALIX
NIGRICANS
Sm.
• 1930 onwards
○ Before 1930
× Probable Intro-
ductions

B 343/15
SALIX
PHYLICIFOLIA
L
Tea-leaved
Willow
• 1930 onwards
○ Before 1930

C 343/16
SALIX
REPENS L.
Creeping Willow
• All records

A 343/17
SALIX
LAPPONUM L.
Downy Willow
• 1930 onwards
○ Before 1930

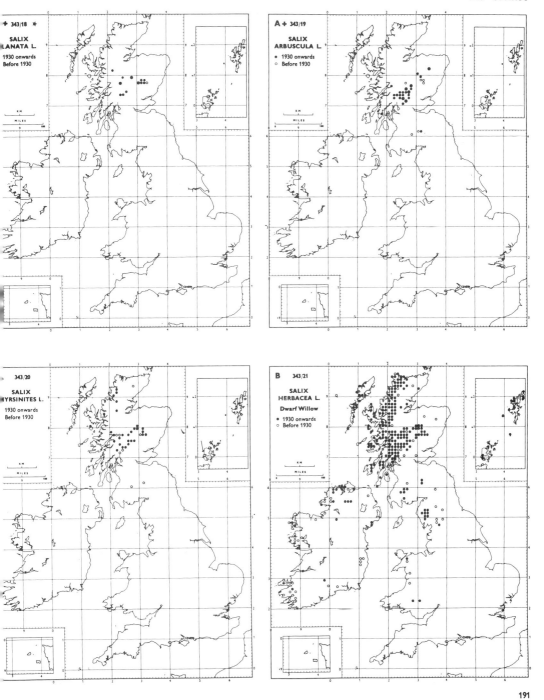

+ 343/18 +

SALIX
LANATA L.

1930 onwards
Before 1930

A + 343/19

SALIX
ARBUSCULA L.

• 1930 onwards
○ Before 1930

343/20

SALIX
MYRSINITES L.

1930 onwards
Before 1930

B   343/21

SALIX
HERBACEA L.

Dwarf Willow

• 1930 onwards
○ Before 1930

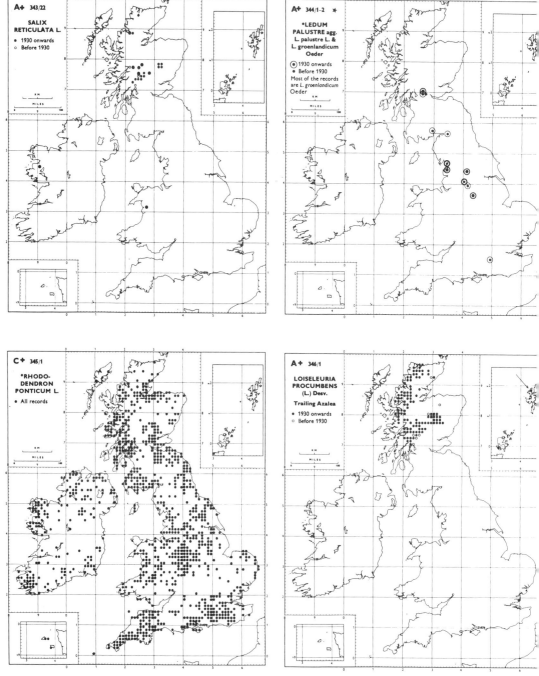

A+ 343/22

SALIX
RETICULATA L.

• 1930 onwards
○ Before 1930

A+ 344/1-2 ✱

*LEDUM
PALUSTRE agg.
L. palustre L. &
L. groenlandicum
Oeder

⊙ 1930 onwards
● Before 1930
Most of the records
are L. groenlandicum
Oeder

C+ 345/1

*RHODO-
DENDRON
PONTICUM L.

• All records

A+ 346/1

LOISELEURIA
PROCUMBENS
(L.) Desv.

Trailing Azalea

• 1930 onwards
○ Before 1930

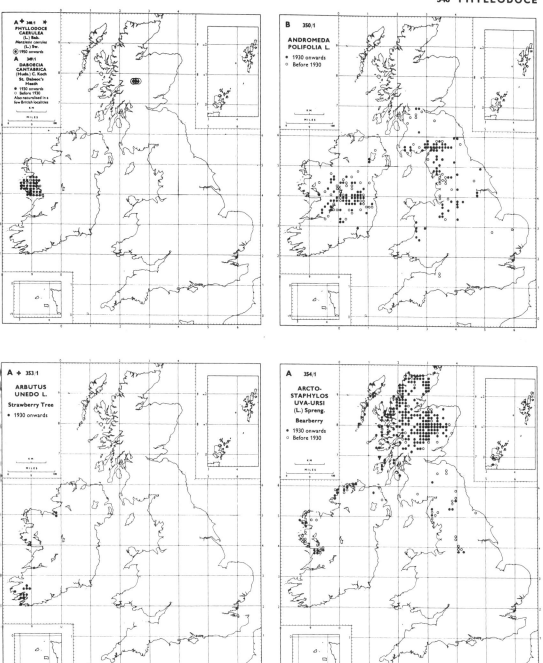

A 348/1
PHYLLODOCE CAERULEA
(L.) Bab.
Menziesia caerulea (L.) Sw.
● 1950 onwards
A 349/1
DABOECIA CANTABRICA
(Huds.) C. Koch
St. Daboec's Heath
● 1930 onwards
○ Before 1930
Also naturalised in a few British localities

B 350/1
ANDROMEDA POLIFOLIA L.
● 1930 onwards
○ Before 1930

A 353/1
ARBUTUS UNEDO L.
Strawberry Tree
● 1930 onwards

A 354/1
ARCTO-STAPHYLOS UVA-URSI
(L.) Spreng.
Bearberry
● 1930 onwards
○ Before 1930

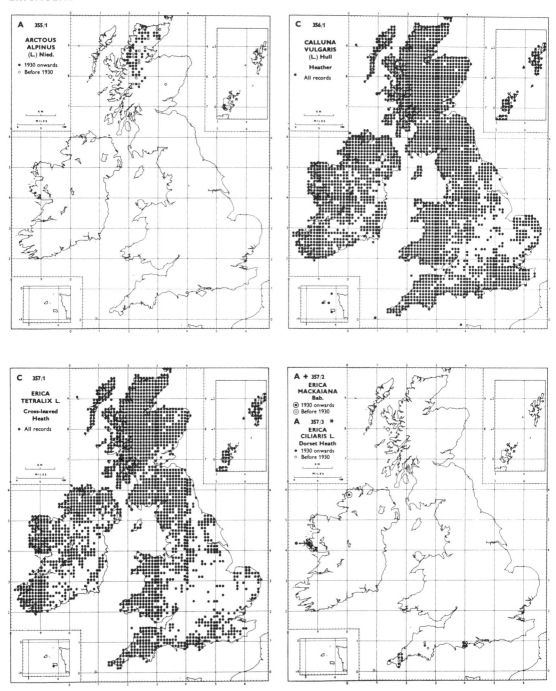

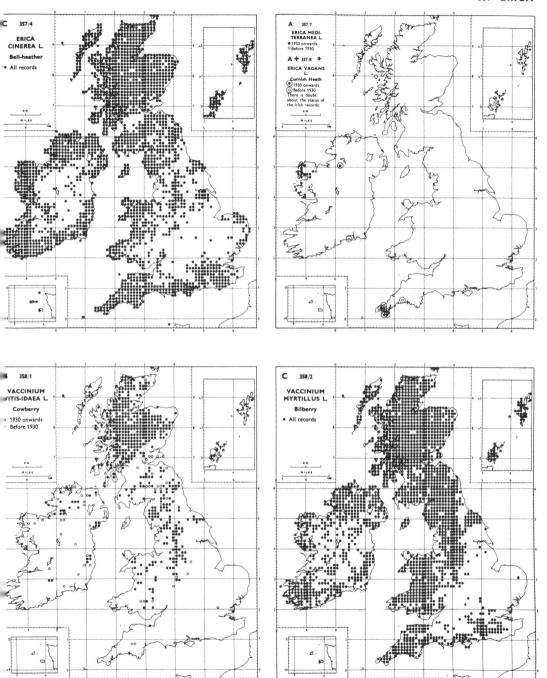

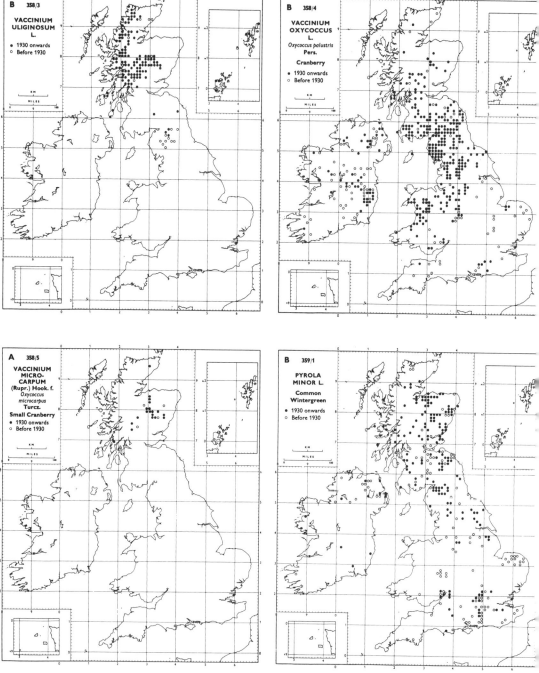

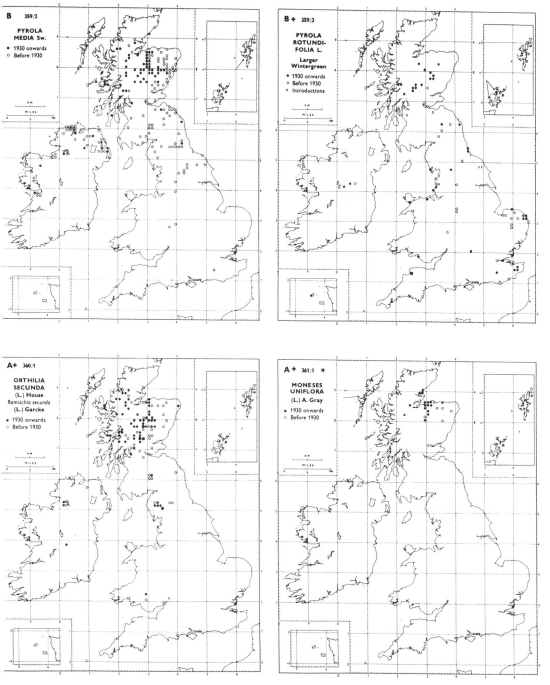

B 359/2

PYROLA
MEDIA Sw.

• 1930 onwards
○ Before 1930

B + 359/3

PYROLA
ROTUNDI-
FOLIA L.

Larger
Wintergreen

• 1930 onwards
○ Before 1930
× Introductions

A+ 360/1

ORTHILIA
SECUNDA
(L.) House
Ramischia secunda
(L.) Garcke

• 1930 onwards
○ Before 1930

A + 361/1 ✳

MONESES
UNIFLORA
(L.) A. Gray

• 1930 onwards
○ Before 1930

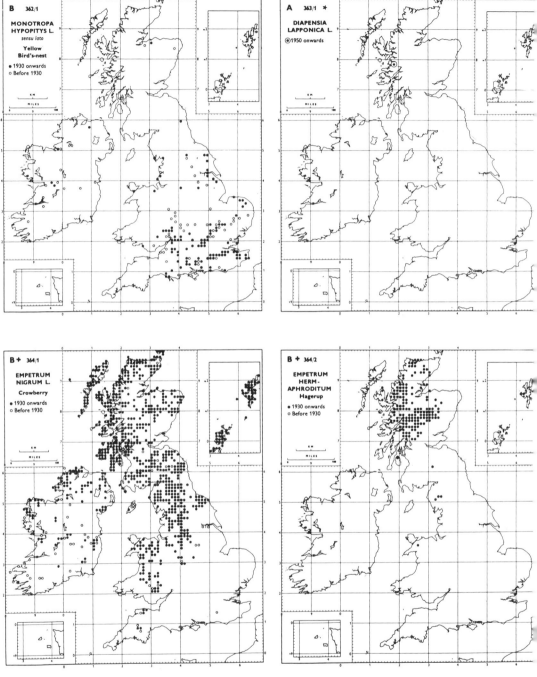

B 362/1

MONOTROPA HYPOPITYS L.
sensu lato
Yellow Bird's-nest
● 1930 onwards
○ Before 1930

A 363/1 ✱

DIAPENSIA LAPPONICA L.
◉ 1950 onwards

B + 364/1

EMPETRUM NIGRUM L.
Crowberry
● 1930 onwards
○ Before 1930

B + 364/2

EMPETRUM HERM-APHRODITUM
Hagerup
● 1930 onwards
○ Before 1930

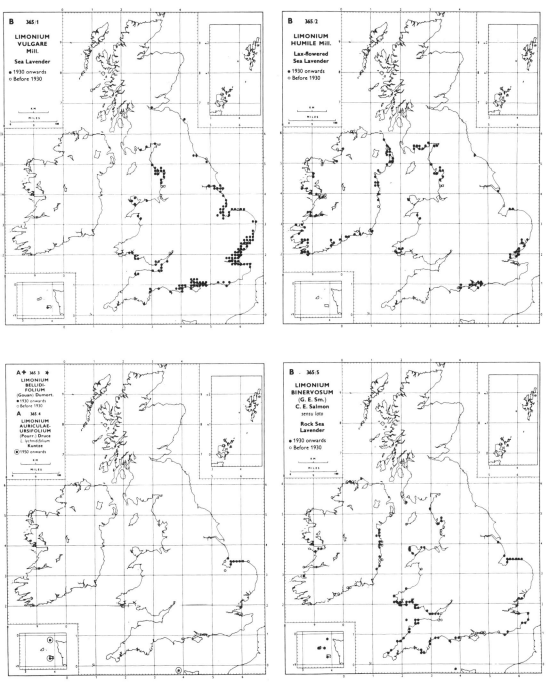

B   365/1
LIMONIUM
VULGARE
Mill.

Sea Lavender

• 1930 onwards
○ Before 1930

B   365/2
LIMONIUM
HUMILE Mill.

Lax-flowered
Sea Lavender

• 1930 onwards
○ Before 1930

A✛ 365 3 ✳
LIMONIUM
BELLIDI-
FOLIUM
(Gouan) Dumort.
• 1930 onwards
○ Before 1930

A
LIMONIUM
AURICULAE-
URSIFOLIUM
(Pourr.) Druce
L. lychnidifolium
Kuntze
⊙ 1950 onwards

B   365/5
LIMONIUM
BINERVOSUM
(G. E. Sm.)
C. E. Salmon
sensu lato

Rock Sea
Lavender

• 1930 onwards
○ Before 1930

199

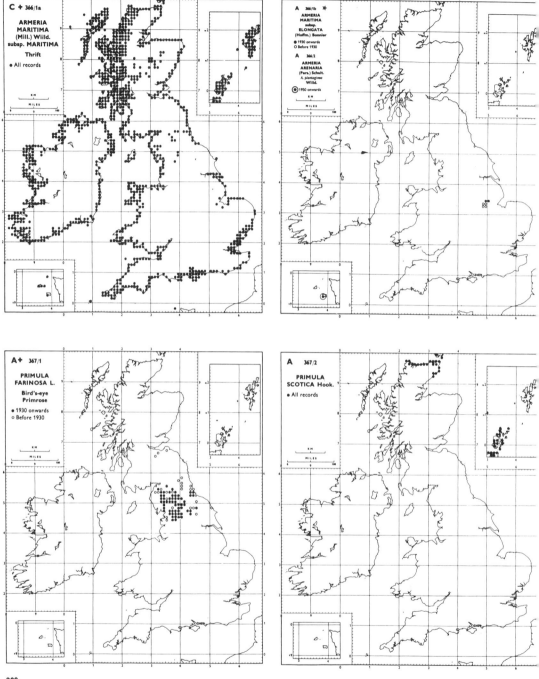

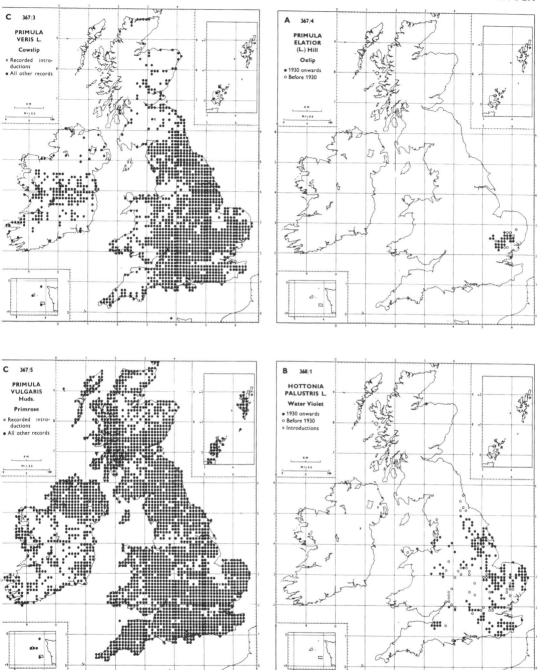

C 367/3

PRIMULA
VERIS L.

Cowslip

x Recorded intro-
ductions
• All other records

A 367/4

PRIMULA
ELATIOR
(L.) Hill

Oxlip

• 1930 onwards
○ Before 1930

C 367/5

PRIMULA
VULGARIS
Huds.

Primrose

x Recorded intro-
ductions
• All other records

B 368/1

HOTTONIA
PALUSTRIS L.

Water Violet

• 1930 onwards
○ Before 1930
x Introductions

PRIMULACEAE

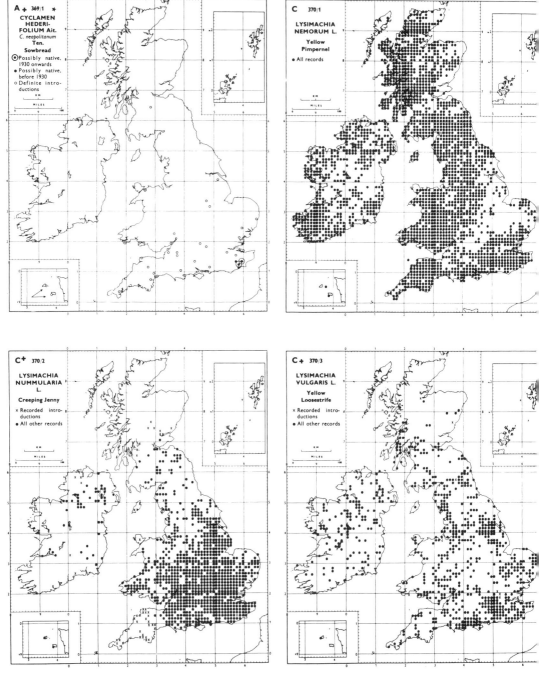

A ✚ 369/1 ✳
CYCLAMEN
HEDERI-
FOLIUM Ait.
C. neapolitanum
Ten.
Sowbread
⊙ Possibly native,
1930 onwards
● Possibly native,
before 1930
○ Definite intro-
ductions

C 370/1
LYSIMACHIA
NEMORUM L.
Yellow
Pimpernel
● All records

C ✚ 370/2
LYSIMACHIA
NUMMULARIA
L.
Creeping Jenny
✕ Recorded intro-
ductions
● All other records

C ✚ 370/3
LYSIMACHIA
VULGARIS L.
Yellow
Loosestrife
✕ Recorded intro-
ductions
● All other records

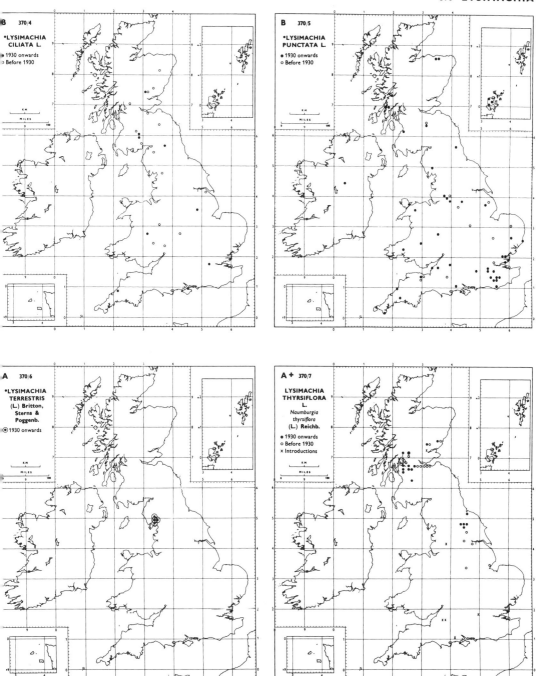

B 370/4

•LYSIMACHIA CILIATA L.

• 1930 onwards
○ Before 1930

B 370/5

•LYSIMACHIA PUNCTATA L.

• 1930 onwards
○ Before 1930

A 370/6

•LYSIMACHIA TERRESTRIS
(L) Britton, Sterns & Poggenb.

◉ 1930 onwards

A + 370/7

LYSIMACHIA THYRSIFLORA L.
Naumburgia thyrsiflora
(L) Reichb.

• 1930 onwards
○ Before 1930
x Introductions

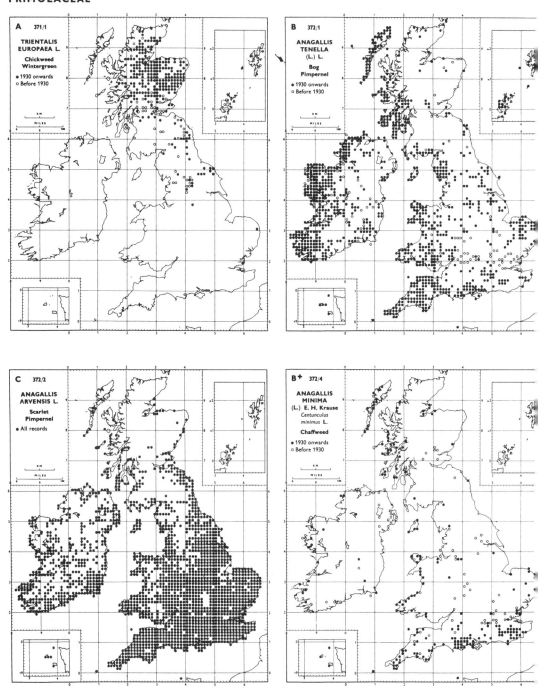

A 371/1

TRIENTALIS
EUROPAEA L.

Chickweed
Wintergreen

• 1930 onwards
○ Before 1930

KM
MILES

B 372/1

ANAGALLIS
TENELLA
(L.) L.

Bog
Pimpernel

• 1930 onwards
○ Before 1930

KM
MILES

C 372/2

ANAGALLIS
ARVENSIS L.

Scarlet
Pimpernel

• All records

KM
MILES

B⁺ 372/4

ANAGALLIS
MINIMA
(L.) E. H. Krause
Centunculus
minimus L.

Chaffweed

• 1930 onwards
○ Before 1930

KM
MILES

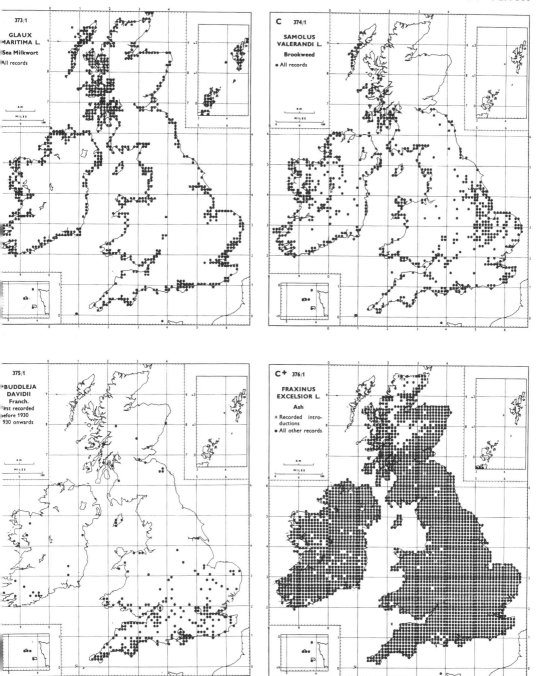

373/1
GLAUX
MARITIMA L.
Sea Milkwort
All records

C 374/1
SAMOLUS
VALERANDI L.
Brookweed
● All records

375/1
BUDDLEJA
DAVIDII
Franch.
First recorded
before 1930
1930 onwards

C⁺ 376/1
FRAXINUS
EXCELSIOR L.
Ash
x Recorded intro-
   ductions
● All other records

OLEACEAE

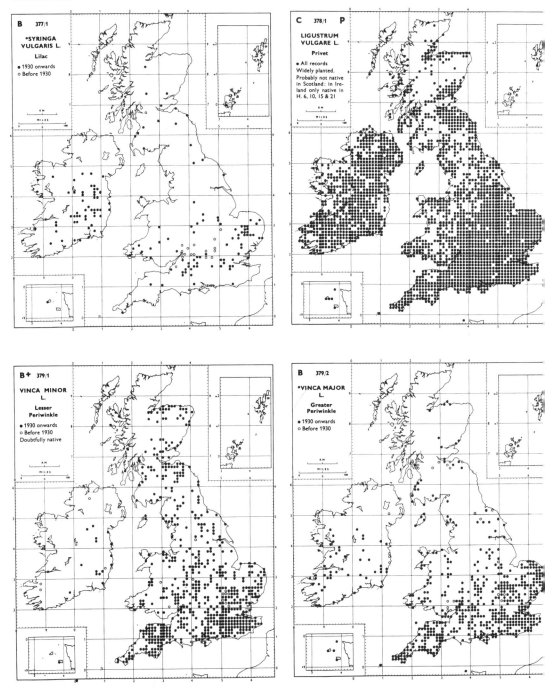

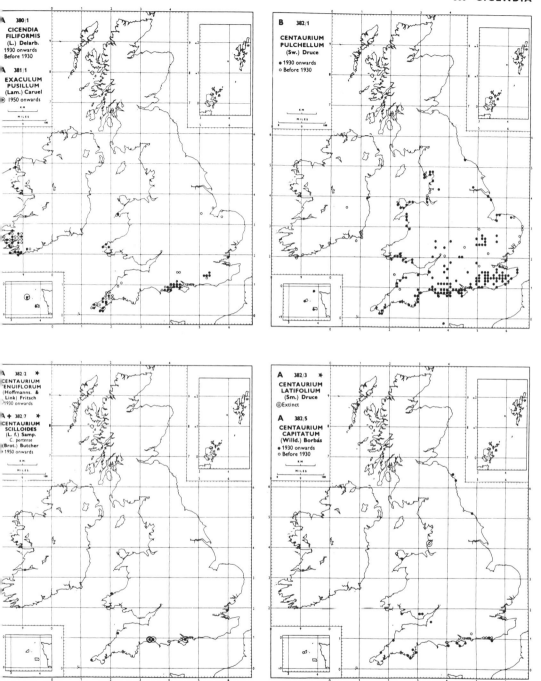

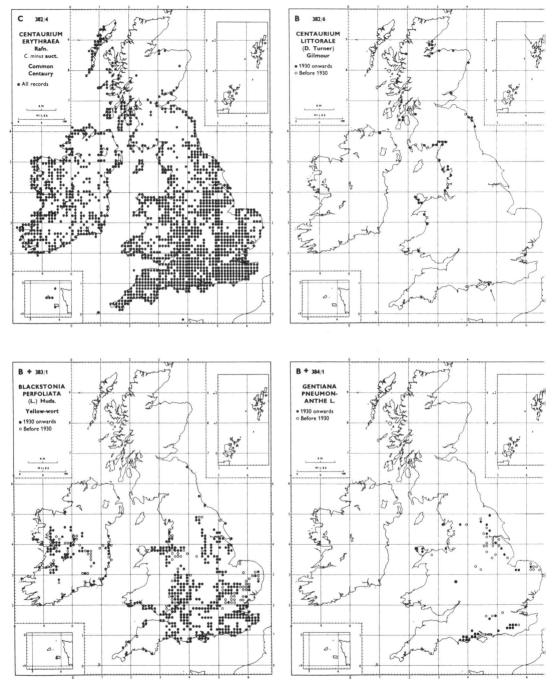

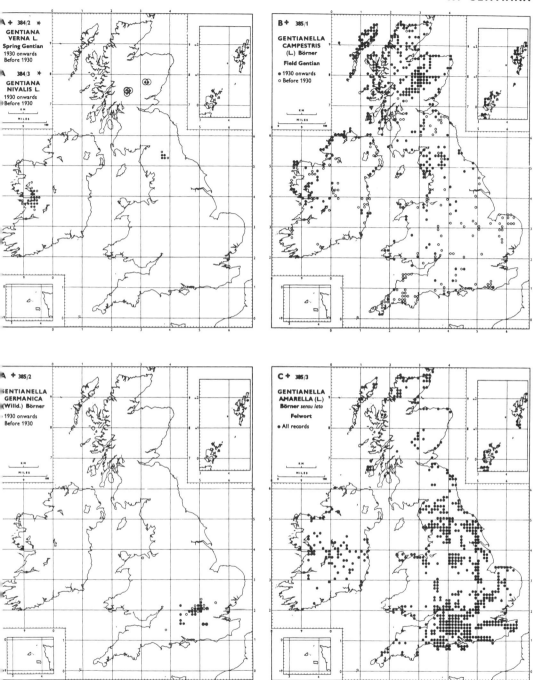

GENTIANACEAE

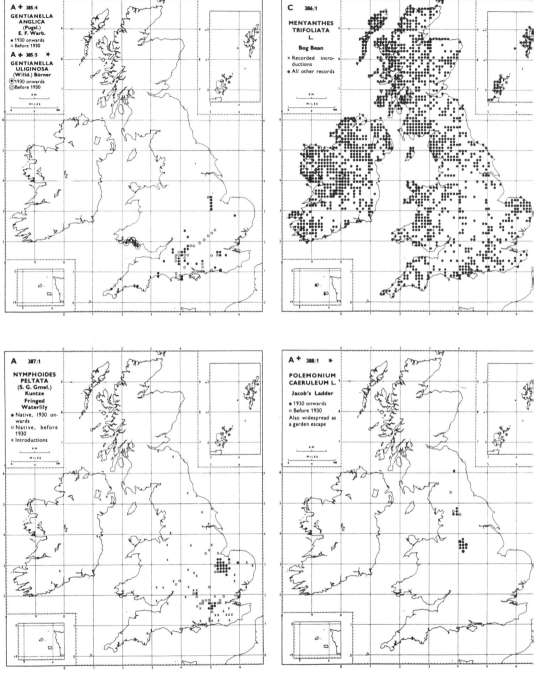

A + 385/4
GENTIANELLA ANGLICA
(Pugsl.)
E. F. Warb.
● 1930 onwards
○ Before 1930

A + 385/5 ✶
GENTIANELLA ULIGINOSA
(Willd.) Börner
◉ 1930 onwards
◎ Before 1930

C 386/1
MENYANTHES TRIFOLIATA L.

Bog Bean
× Recorded introductions
● All other records

A 387/1
NYMPHOIDES PELTATA
(S. G. Gmel.)
Kuntze
Fringed Waterlily
● Native, 1930 onwards
○ Native, before 1930
× Introductions

A + 388/1 ✶
POLEMONIUM CAERULEUM L.

Jacob's Ladder
● 1930 onwards
○ Before 1930
Also widespread as a garden escape

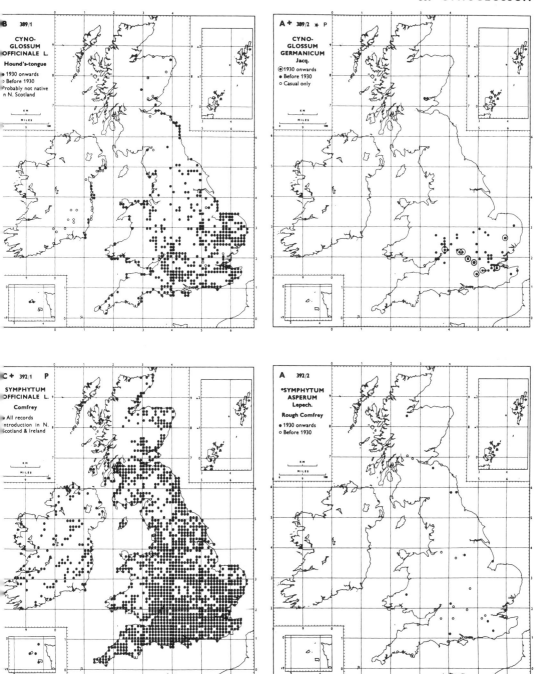

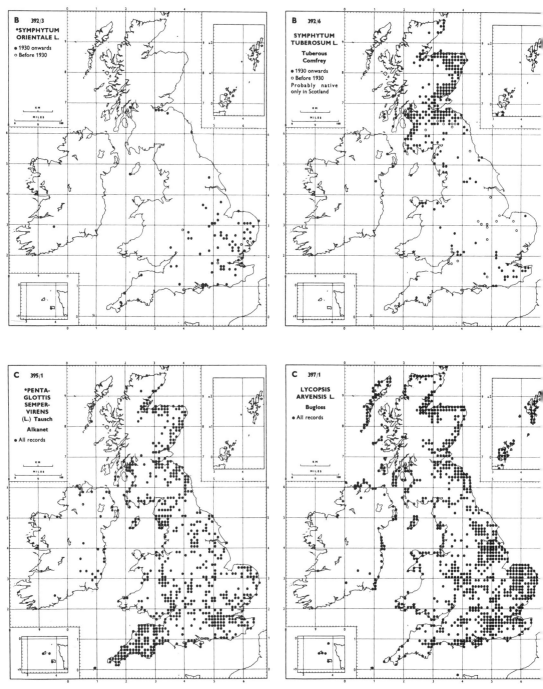

B 392/3
*SYMPHYTUM
ORIENTALE L.
● 1930 onwards
○ Before 1930

B 392/6
SYMPHYTUM
TUBEROSUM L.
**Tuberous
Comfrey**
● 1930 onwards
○ Before 1930
Probably native
only in Scotland

C 395/1
*PENTA-
GLOTTIS
SEMPER-
VIRENS
(L.) Tausch
Alkanet
● All records

C 397/1
LYCOPSIS
ARVENSIS L.
Bugloss
● All records

212

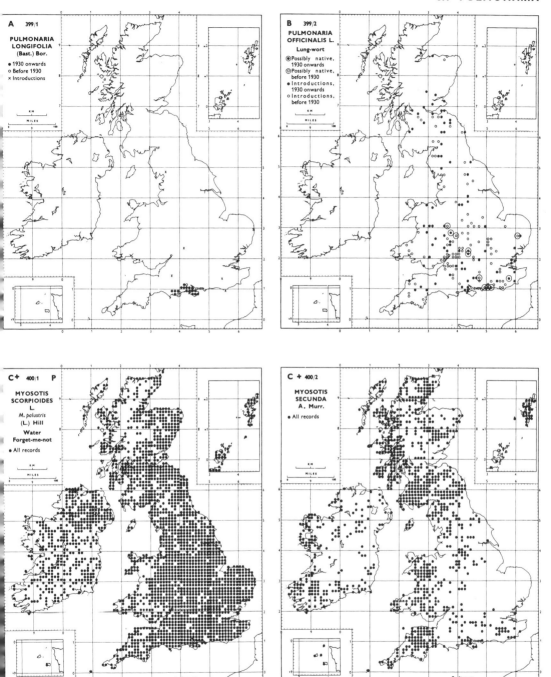

A 399/1

PULMONARIA
LONGIFOLIA
(Bast.) Bor.

● 1930 onwards
○ Before 1930
× Introductions

K M
MILES

B 399/2

PULMONARIA
OFFICINALIS L.
Lung-wort

◉ Possibly native,
   1930 onwards
◉ Possibly native,
   before 1930
● Introductions,
   1930 onwards
○ Introductions,
   before 1930

K M
MILES

C⁺ 400/1 P

MYOSOTIS
SCORPIOIDES
L.
M. palustris
(L.) Hill
Water
Forget-me-not

● All records

K M
MILES

C ✛ 400/2

MYOSOTIS
SECUNDA
A. Murr.

● All records

K M
MILES

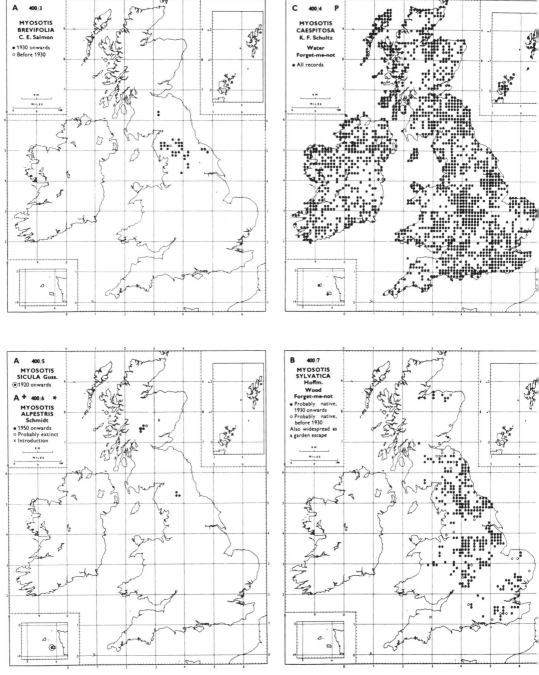

A 400/3
MYOSOTIS
BREVIFOLIA
C. E. Salmon
● 1930 onwards
○ Before 1930

C 400/4 P
MYOSOTIS
CAESPITOSA
K. F. Schultz
Water
Forget-me-not
● All records

A 400/5
MYOSOTIS
SICULA Guss.
⊙ 1920 onwards
A + 400/6 ✱
MYOSOTIS
ALPESTRIS
Schmidt
● 1950 onwards
○ Probably extinct
✕ Introduction

B 400/7
MYOSOTIS
SYLVATICA
Hoffm.
Wood
Forget-me-not
● Probably native,
1930 onwards
○ Probably native,
before 1930
Also widespread as
a garden escape

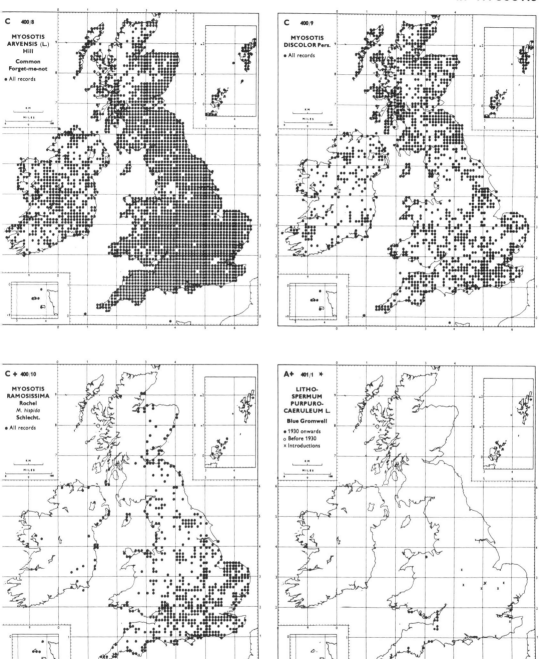

C 400/8

MYOSOTIS
ARVENSIS (L.)
Hill

Common
Forget-me-not

● All records

C 400/9

MYOSOTIS
DISCOLOR Pers.

● All records

C + 400/10

MYOSOTIS
RAMOSISSIMA
Rochel
M. hispida
Schlecht.

● All records

A+ 401/1 ✳

LITHO-
SPERMUM
PURPURO-
CAERULEUM L.

Blue Gromwell

● 1930 onwards
○ Before 1930
× Introductions

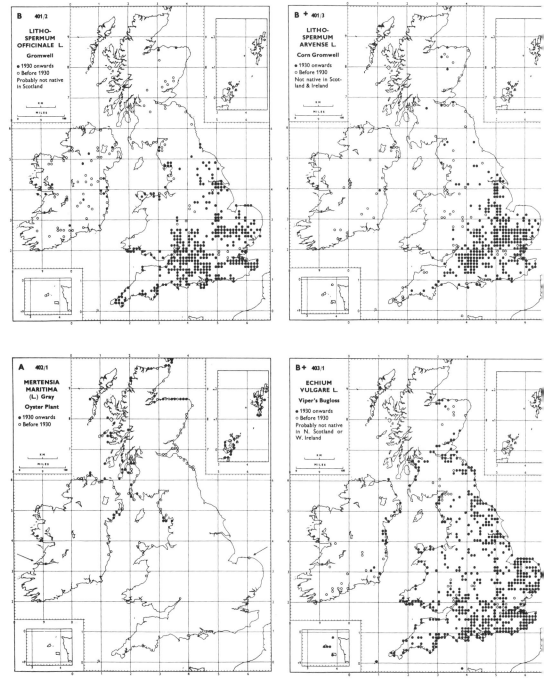

B 401/2

LITHO-SPERMUM OFFICINALE L.

Gromwell

● 1930 onwards
○ Before 1930
Probably not native in Scotland

B + 401/3

LITHO-SPERMUM ARVENSE L.

Corn Gromwell

● 1930 onwards
○ Before 1930
Not native in Scotland & Ireland

A 402/1

MERTENSIA MARITIMA (L.) Gray

Oyster Plant

● 1930 onwards
○ Before 1930

B + 403/1

ECHIUM VULGARE L.

Viper's Bugloss

● 1930 onwards
○ Before 1930
Probably not native in N. Scotland or W. Ireland

216

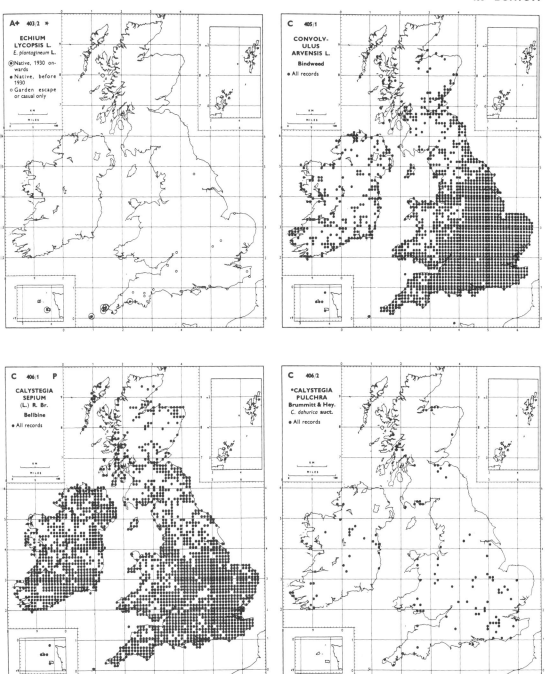

A+ 403/2 ✻

ECHIUM
LYCOPSIS L.
E. plantagineum L.

⊙ Native, 1930 on-
wards
● Native, before
1930
○ Garden escape
or casual only

C 405/1

CONVOLV-
ULUS
ARVENSIS L.
Bindweed

● All records

C 406/1 P

CALYSTEGIA
SEPIUM
(L.) R. Br.
Bellbine

● All records

C 406/2

*CALYSTEGIA
PULCHRA
Brummitt & Hey.
C. dahurica auct.

● All records

217

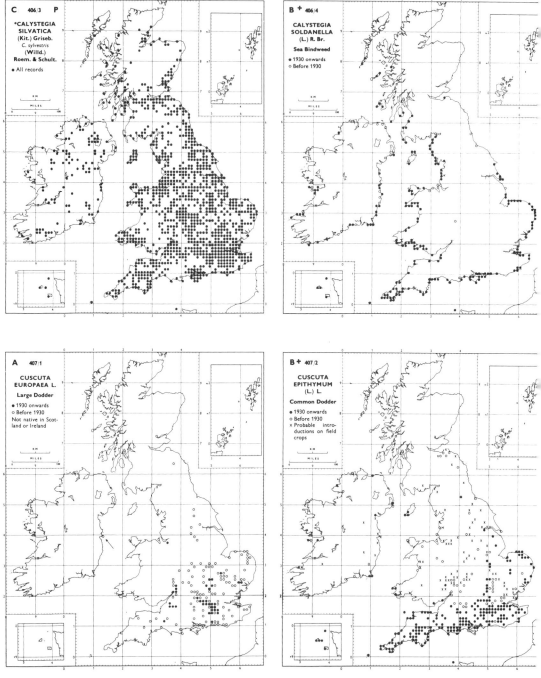

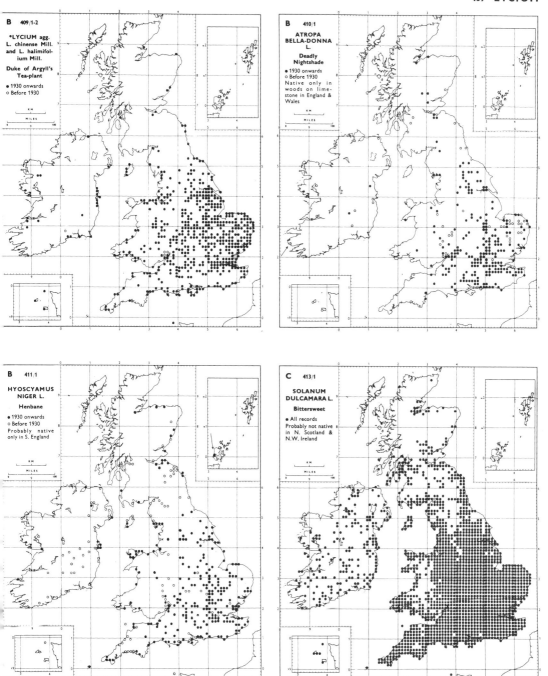

B 409/1-2

*LYCIUM agg.
L. chinense Mill.
and L. halimifol-
ium Mill.

Duke of Argyll's
Tea-plant

● 1930 onwards
○ Before 1930

B 410/1

ATROPA
BELLA-DONNA
L.

Deadly
Nightshade

● 1930 onwards
○ Before 1930
Native only in
woods on lime-
stone in England &
Wales

B 411/1

HYOSCYAMUS
NIGER L.

Henbane

● 1930 onwards
○ Before 1930
Probably native
only in S. England

C 413/1

SOLANUM
DULCAMARA L.

Bittersweet

● All records
Probably not native
in N. Scotland &
N.W. Ireland

SOLANACEAE

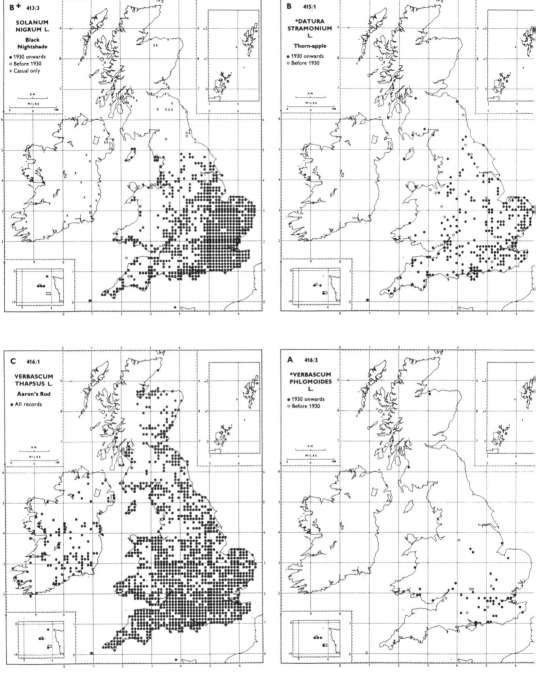

B + 413/3
SOLANUM NIGRUM L.
Black Nightshade
● 1930 onwards
○ Before 1930
× Casual only

B 415/1
°DATURA STRAMONIUM L.
Thorn-apple
● 1930 onwards
○ Before 1930

C 416/1
VERBASCUM THAPSUS L.
Aaron's Rod
● All records

A 416/3
°VERBASCUM PHLOMOIDES L.
● 1930 onwards
○ Before 1930

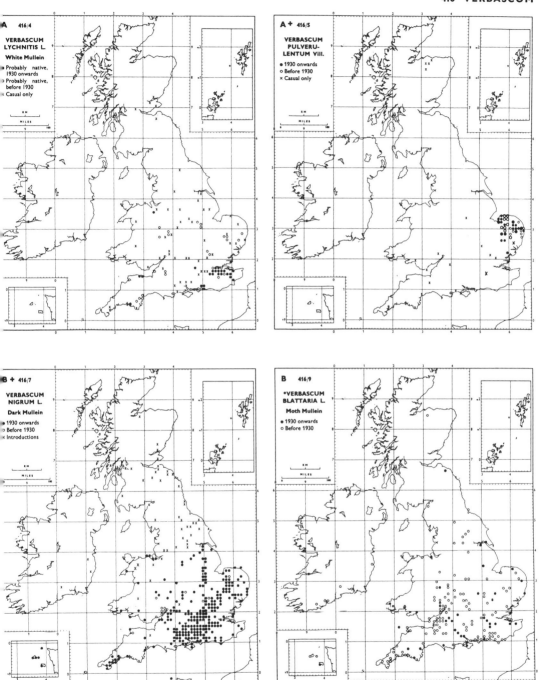

A 416/4

**VERBASCUM
LYCHNITIS L.**

White Mullein

● Probably native,
1930 onwards
○ Probably native,
before 1930
× Casual only

A + 416/5

**VERBASCUM
PULVERU-
LENTUM Vill.**

● 1930 onwards
○ Before 1930
× Casual only

B + 416/7

**VERBASCUM
NIGRUM L.**

Dark Mullein

● 1930 onwards
○ Before 1930
× Introductions

B 416/9

*VERBASCUM
BLATTARIA L.

Moth Mullein

● 1930 onwards
○ Before 1930

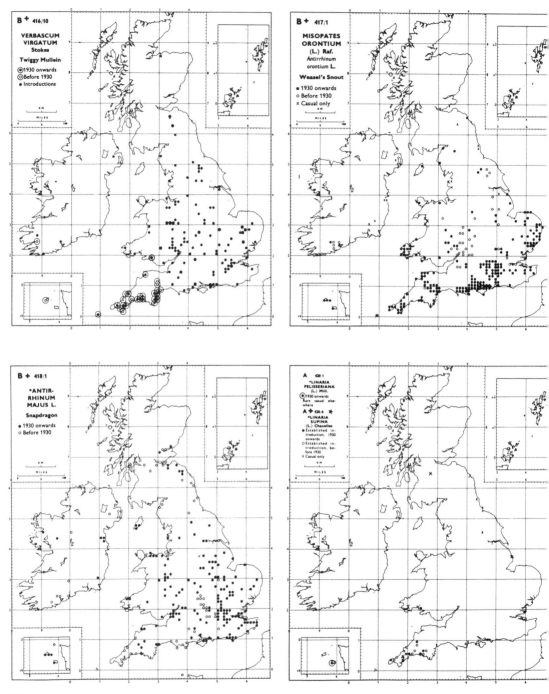

B + 416/10

VERBASCUM
VIRGATUM
Stokes

Twiggy Mullein

⦿ 1930 onwards
⊙ Before 1930
● Introductions

B + 417/1

MISOPATES
ORONTIUM
(L.) Raf.
Antirrhinum
orontium L.

Weasel's Snout

● 1930 onwards
○ Before 1930
✕ Casual only

B + 418/1

*ANTIR-
RHINUM
MAJUS L.

Snapdragon

● 1930 onwards
○ Before 1930

A 420/1
*LINARIA
PELISSERIANA
(L.) Mill.
⦿ 1930 onwards
Rare casual else-
where

A ✳ 420/6
*LINARIA
SUPINA
(L.) Chazelles
● Established in-
troduction, 1930
onwards
○ Established in-
troduction, be-
fore 1930
✕ Casual only

222

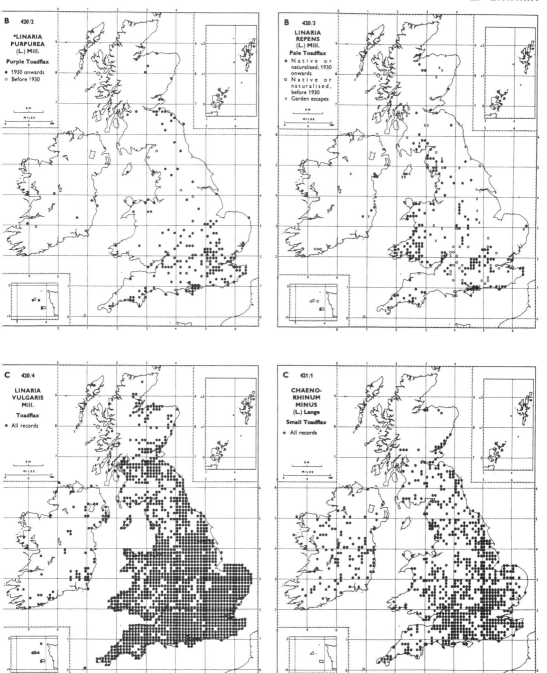

B 420/2

*LINARIA
PURPUREA
(L.) Mill.

Purple Toadflax

● 1930 onwards
○ Before 1930

B 420/3

LINARIA
REPENS
(L.) Mill.

Pale Toadflax

● Native or
naturalised, 1930
onwards
○ Native or
naturalised,
before 1930
x Garden escapes

C 420/4

LINARIA
VULGARIS
Mill.

Toadflax

● All records

C 421/1

CHAENO-
RHINUM
MINUS
(L.) Lange

Small Toadflax

● All records

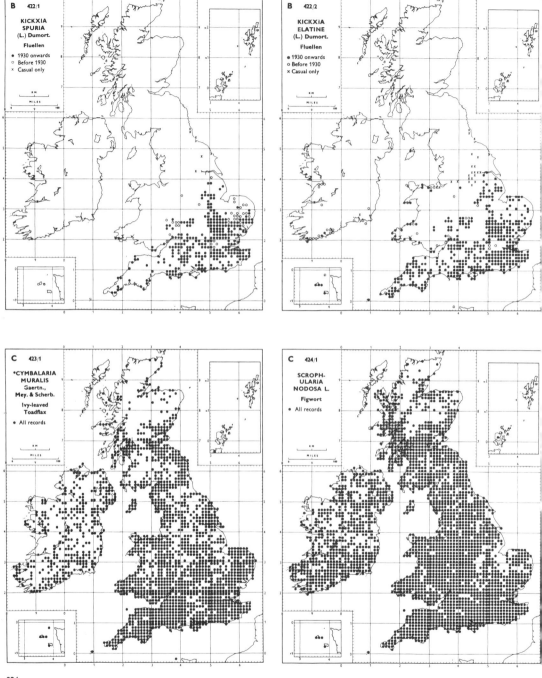

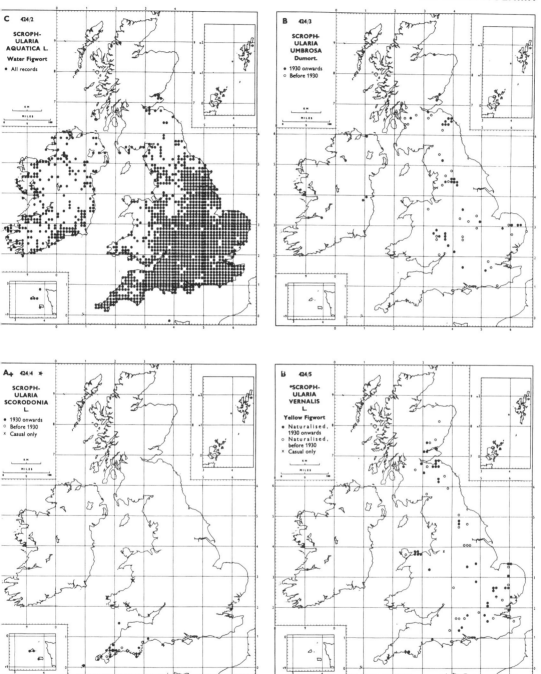

C 424/2

SCROPH-
ULARIA
AQUATICA L.

Water Figwort

● All records

B 424/3

SCROPH-
ULARIA
UMBROSA
Dumort.

● 1930 onwards
○ Before 1930

A╋ 424/4 ✲

SCROPH-
ULARIA
SCORODONIA
L.

● 1930 onwards
○ Before 1930
✕ Casual only

B 424/5

*SCROPH-
ULARIA
VERNALIS
L.

Yellow Figwort

● Naturalised,
  1930 onwards
○ Naturalised,
  before 1930
✕ Casual only

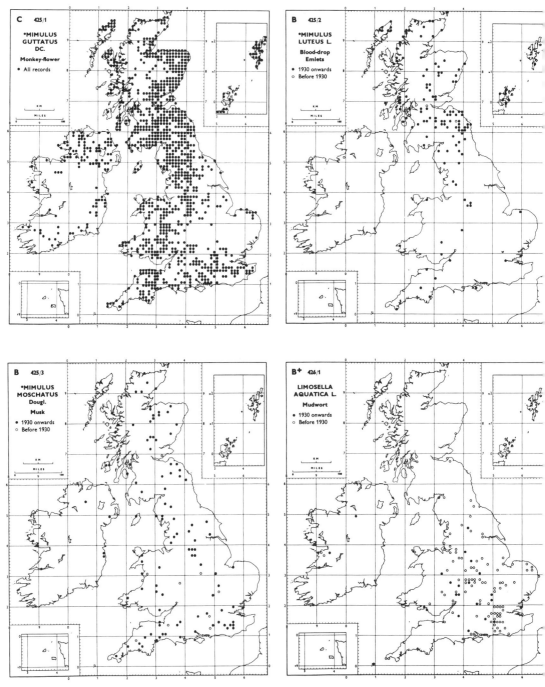

C 425/1

*MIMULUS
GUTTATUS
DC.

Monkey-flower

• All records

KM
MILES

B 425/2

*MIMULUS LUTEUS L.

Blood-drop
Emlets

• 1930 onwards
○ Before 1930

KM
MILES

B 425/3

*MIMULUS
MOSCHATUS
Dougl.

Musk

• 1930 onwards
○ Before 1930

KM
MILES

B⁺ 426/1

LIMOSELLA
AQUATICA L.

Mudwort

• 1930 onwards
○ Before 1930

KM
MILES

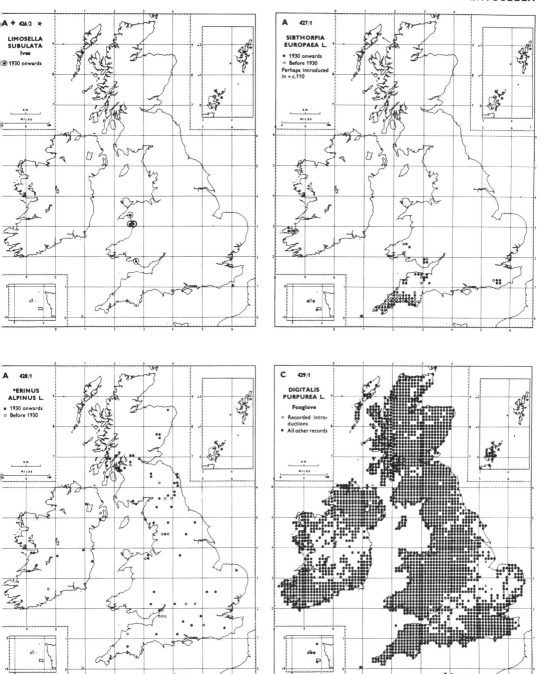

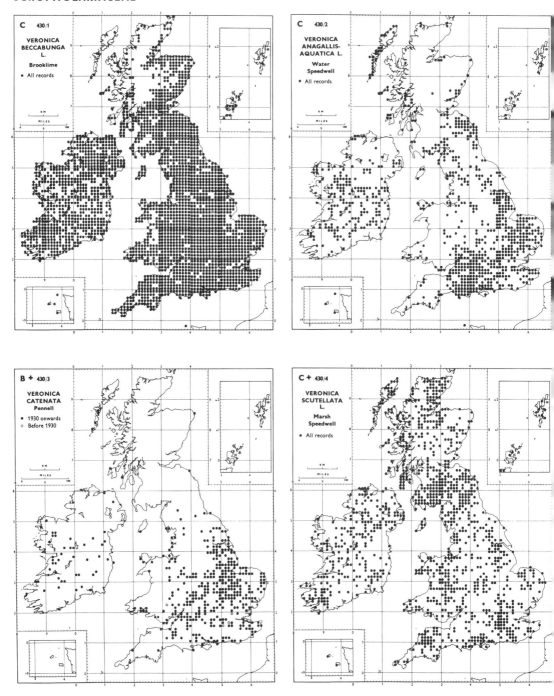

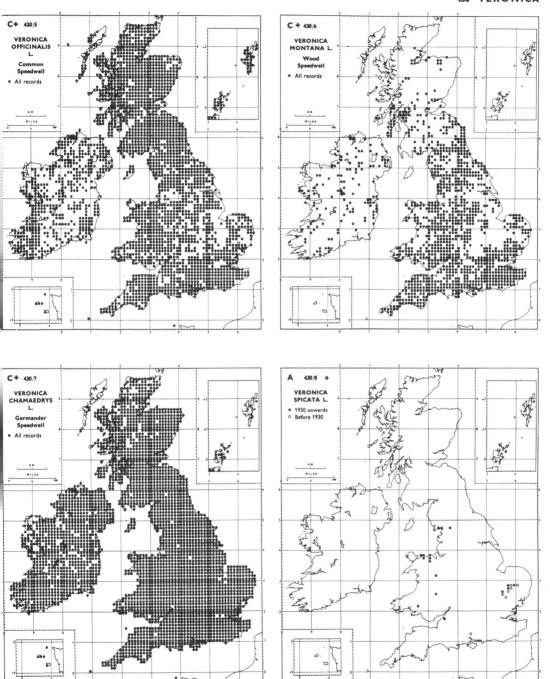

C+ 430/5

VERONICA
OFFICINALIS
L.

Common
Speedwell

• All records

C + 430/6

VERONICA
MONTANA L.

Wood
Speedwell

• All records

C+ 430/7

VERONICA
CHAMAEDRYS
L.

Germander
Speedwell

• All records

A 430/8 ✷

VERONICA
SPICATA L.

• 1930 onwards
○ Before 1930

229

SCROPHULARIACEAE

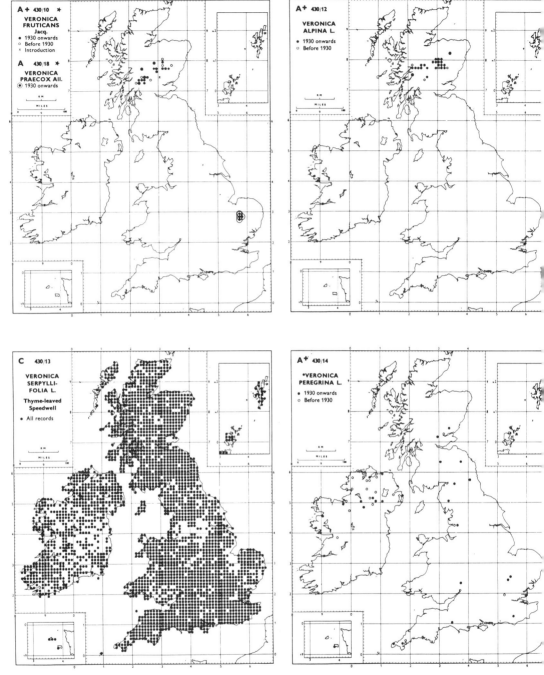

A + 430/10 ✱
VERONICA
FRUTICANS
Jacq.
● 1930 onwards
○ Before 1930
× Introduction

A 430/18 ✱
VERONICA
PRAECOX All.
◉ 1930 onwards

A⁺ 430/12
VERONICA
ALPINA L.
● 1930 onwards
○ Before 1930

C 430/13
VERONICA
SERPYLLI-
FOLIA L.
Thyme-leaved
Speedwell
● All records

A⁺ 430/14
*VERONICA PEREGRINA L.
● 1930 onwards
○ Before 1930

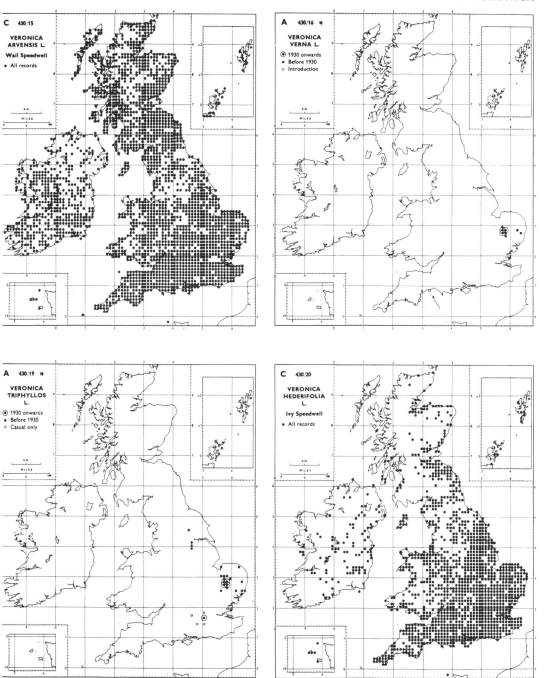

C 430/15

VERONICA
ARVENSIS L.
Wall Speedwell
• All records

A 430/16 ✴

VERONICA
VERNA L.
⊙ 1930 onwards
• Before 1930
○ Introduction

A 430/19 ✴

VERONICA
TRIPHYLLOS
L.
⊙ 1930 onwards
• Before 1930
○ Casual only

C 430/20

VERONICA
HEDERIFOLIA
L.
Ivy Speedwell
• All records

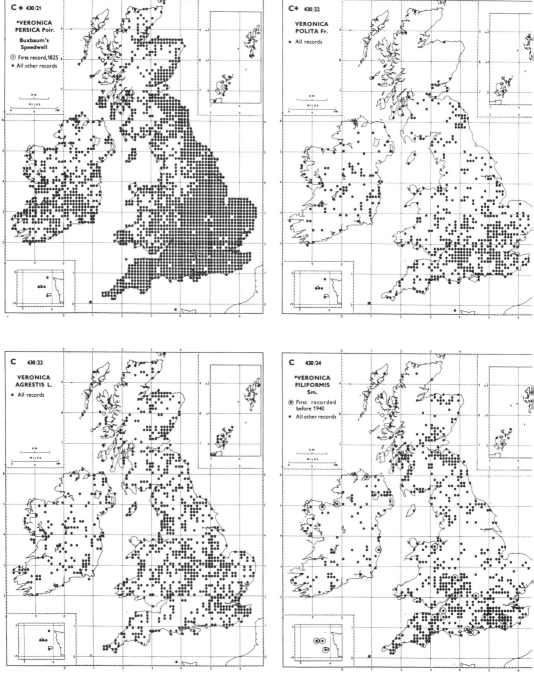

C ✛ 430/21

*VERONICA PERSICA Poir.

Buxbaum's Speedwell

⊙ First record, 1825
• All other records

C✛ 430/22

VERONICA POLITA Fr.

• All records

C 430/23

VERONICA AGRESTIS L.

• All records

C 430/24

*VERONICA FILIFORMIS Sm.

⊙ First recorded before 1940
• All other records

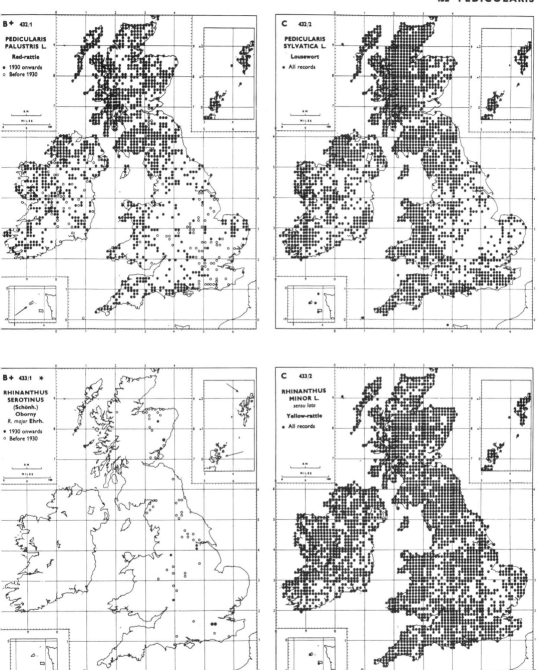

SCROPHULARIACEAE

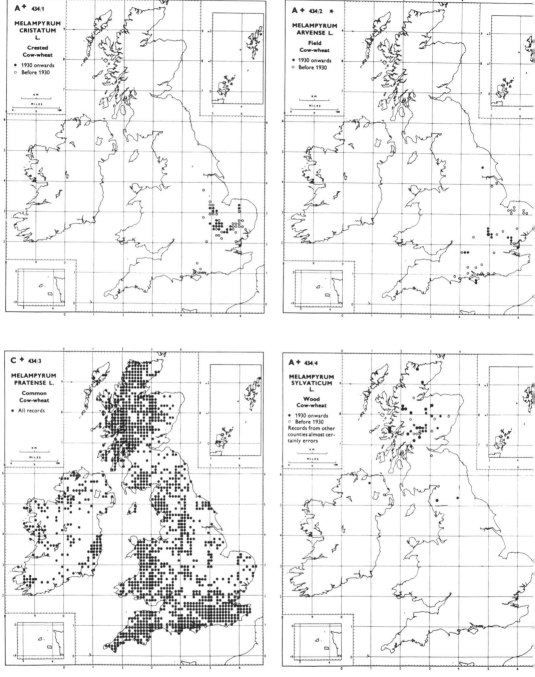

A + 434/1

MELAMPYRUM
CRISTATUM
L.

Crested
Cow-wheat

● 1930 onwards
○ Before 1930

KM
MILES

A + 434/2 ★

MELAMPYRUM
ARVENSE L.

Field
Cow-wheat

● 1930 onwards
○ Before 1930

KM
MILES

C + 434/3

MELAMPYRUM
PRATENSE L.

Common
Cow-wheat

● All records

KM
MILES

A + 434/4

MELAMPYRUM
SYLVATICUM
L.

Wood
Cow-wheat

● 1930 onwards
○ Before 1930
Records from other
counties almost cer-
tainly errors

KM
MILES

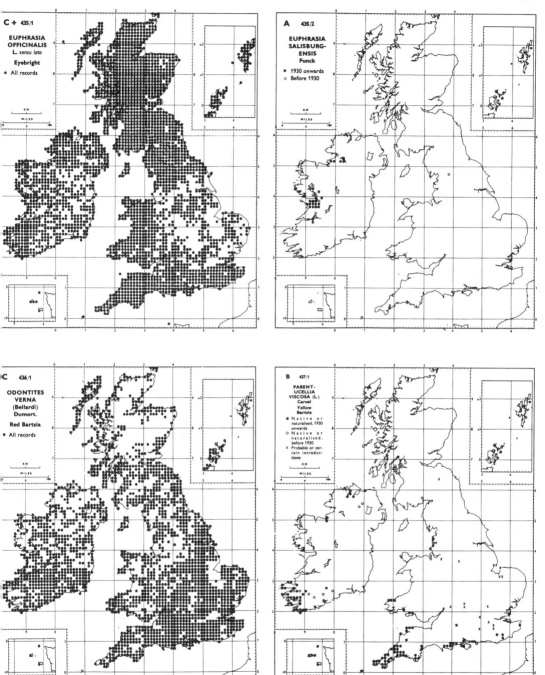

C + 435/1

EUPHRASIA
OFFICINALIS
L. sensu lato

Eyebright

● All records

A 435/2

EUPHRASIA
SALISBURG-
ENSIS
Funck

● 1930 onwards
○ Before 1930

C 436/1

ODONTITES
VERNA
(Bellardi)
Dumort.

Red Bartsia

● All records

B 437/1

PARENT-
UCELLIA
VISCOSA (L.)
Caruel
Yellow
Bartsia

● Native or
naturalised, 1930
onwards
○ Native or
naturalised,
before 1930
× Probable or cer-
tain introduc-
tions

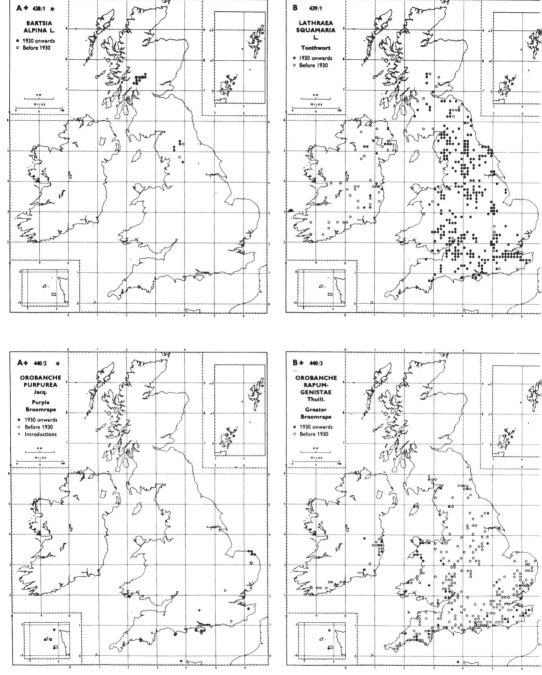

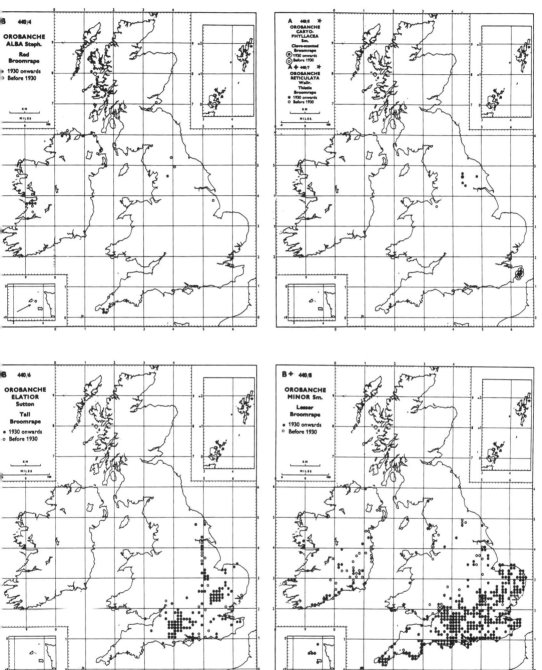

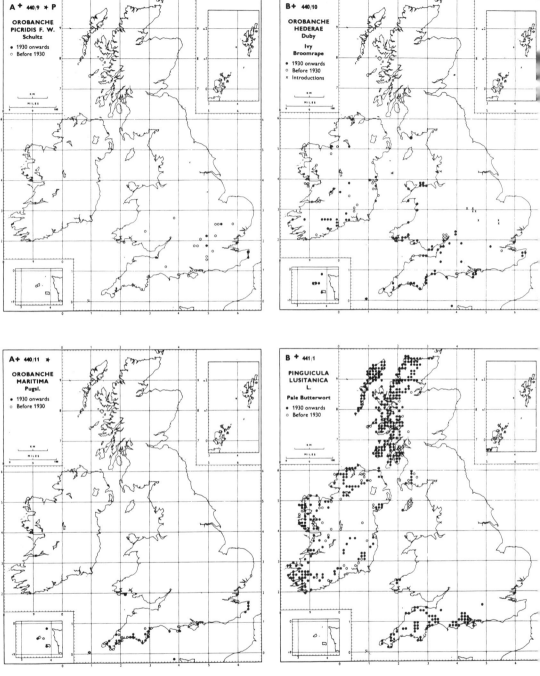

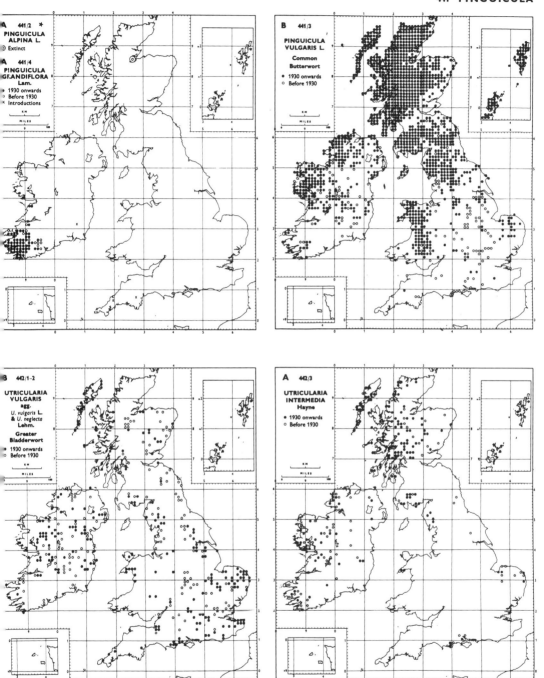

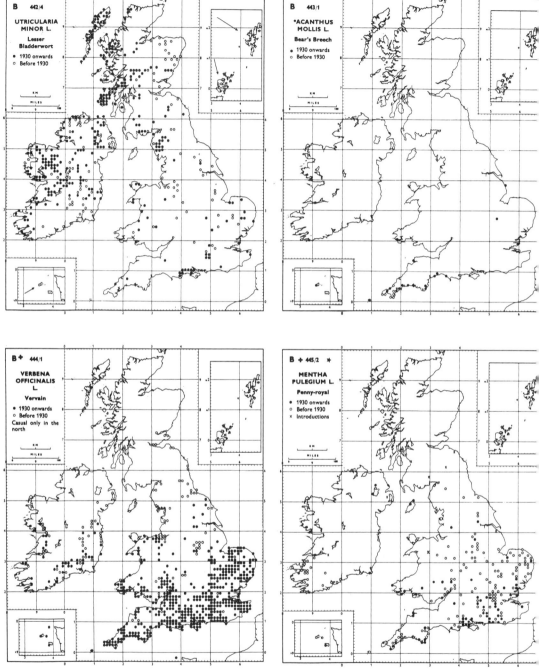

B 442/4

UTRICULARIA
MINOR L.

Lesser
Bladderwort

● 1930 onwards
○ Before 1930

B 443/1

*ACANTHUS
MOLLIS L.

Bear's Breech

● 1930 onwards
○ Before 1930

B ✦ 444/1

VERBENA
OFFICINALIS
L.

Vervain

● 1930 onwards
○ Before 1930
Casual only in the
north

B ✦ 445/2 ✱

MENTHA
PULEGIUM L.

Penny-royal

● 1930 onwards
○ Before 1930
✕ Introductions

240

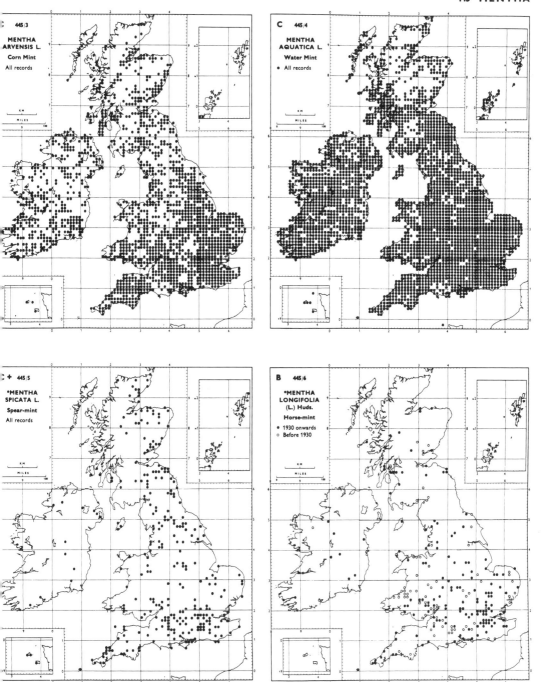

445/3

MENTHA
ARVENSIS L.
Corn Mint
All records

445/4

MENTHA
AQUATICA L.
Water Mint
• All records

445/5

°MENTHA
SPICATA L.
Spear-mint
All records

445/6

°MENTHA
LONGIFOLIA
(L.) Huds.
Horse-mint
• 1930 onwards
○ Before 1930

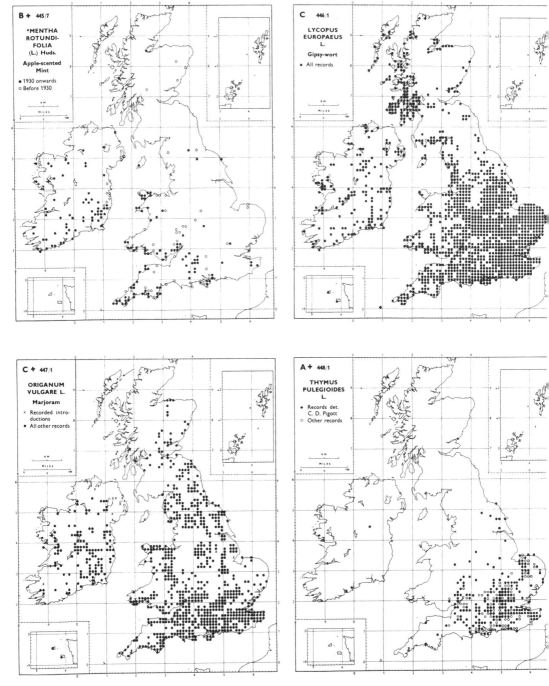

B + 445/7

*MENTHA
ROTUNDI-
FOLIA
(L.) Huds.

Apple-scented
Mint

● 1930 onwards
○ Before 1930

C 446/1

LYCOPUS
EUROPAEUS
L.

Gipsy-wort

● All records

C + 447/1

ORIGANUM
VULGARE L.

Marjoram

× Recorded intro-
ductions
● All other records

A + 448/1

THYMUS
PULEGIOIDES
L.

● Records det.
C. D. Pigott
○ Other records

242

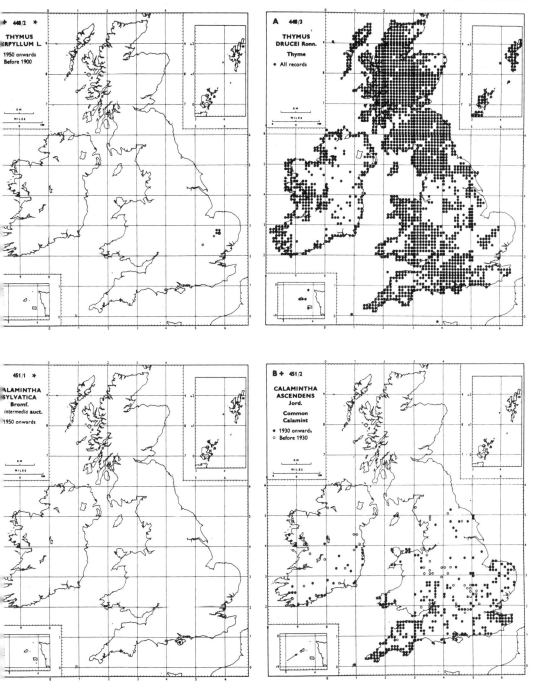

* 448/2 *

THYMUS
ERPYLLUM L.

1950 onwards
Before 1900

A 448/3

THYMUS
DRUCEI Ronn.

Thyme

● All records

451/1 *

ALAMINTHA
SYLVATICA
Bromf.
intermedia auct.

1950 onwards

B + 451/2

CALAMINTHA
ASCENDENS
Jord.

Common
Calamint

● 1930 onwards
○ Before 1930

LABIATAE

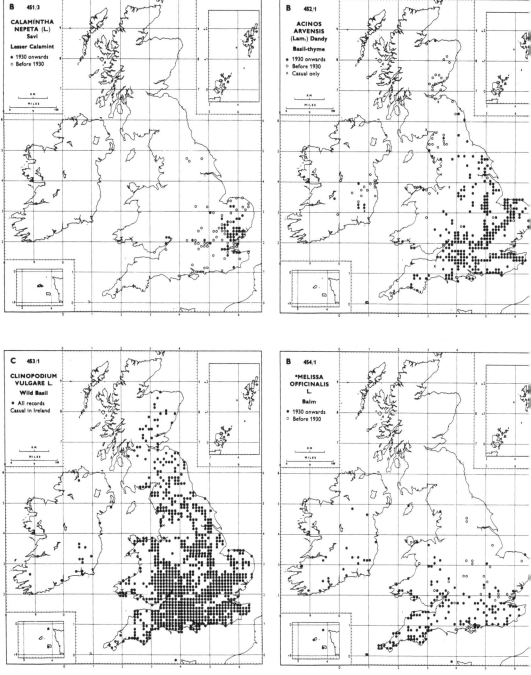

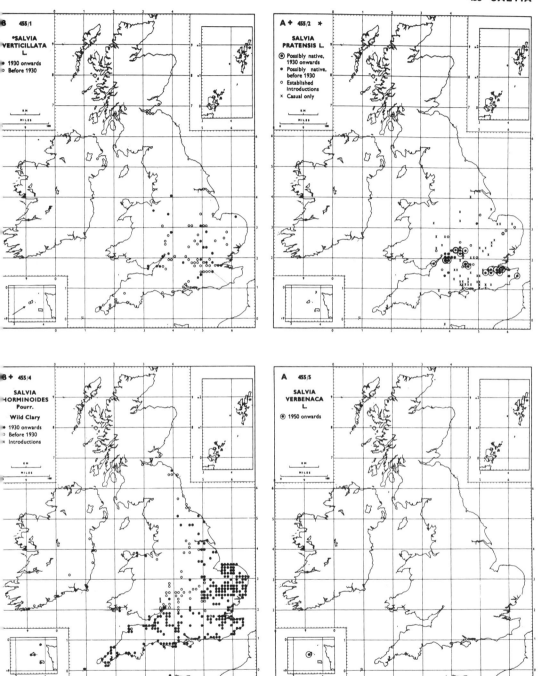

LABIATAE

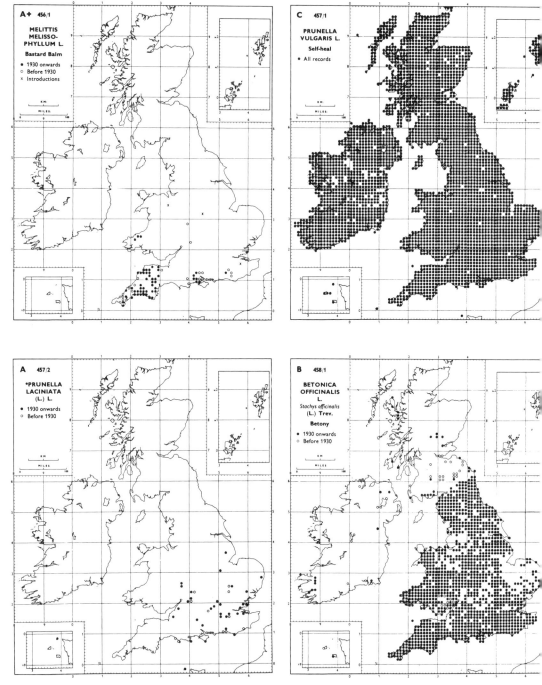

A+ 456/1

MELITTIS
MELISSO-
PHYLLUM L.

Bastard Balm

● 1930 onwards
○ Before 1930
× Introductions

C 457/1

PRUNELLA
VULGARIS L.

Self-heal

● All records

A 457/2

*PRUNELLA
LACINIATA
(L.) L.

● 1930 onwards
○ Before 1930

B 458/1

BETONICA
OFFICINALIS
L.

Stachys officinalis
(L.) Trev.

Betony

● 1930 onwards
○ Before 1930

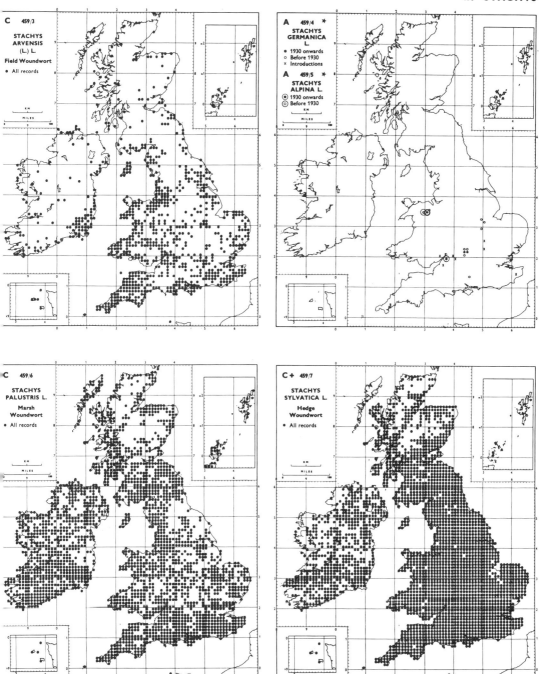

C 459/3
STACHYS
ARVENSIS
(L.) L.
Field Woundwort
● All records

A 459/4 ✱
STACHYS
GERMANICA
L.
● 1930 onwards
○ Before 1930
✕ Introductions
A 459/5 ✱
STACHYS
ALPINA L.
⦿ 1930 onwards
⊙ Before 1930

C 459/6
STACHYS
PALUSTRIS L.
Marsh
Woundwort
● All records

C + 459/7
STACHYS
SYLVATICA L.
Hedge
Woundwort
● All records

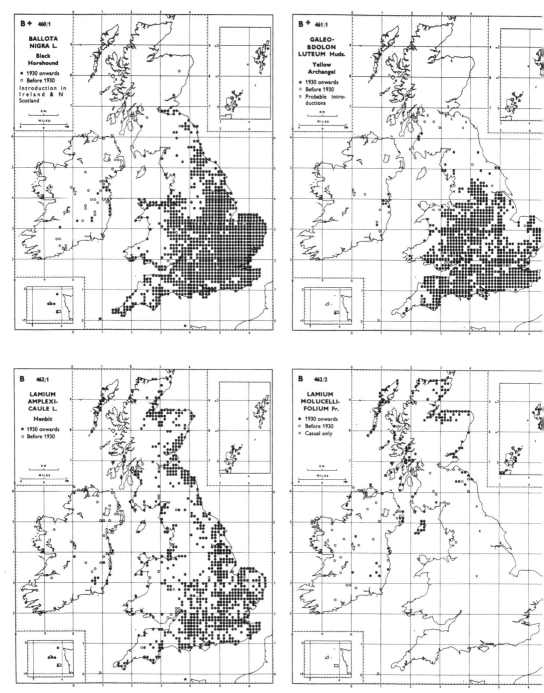

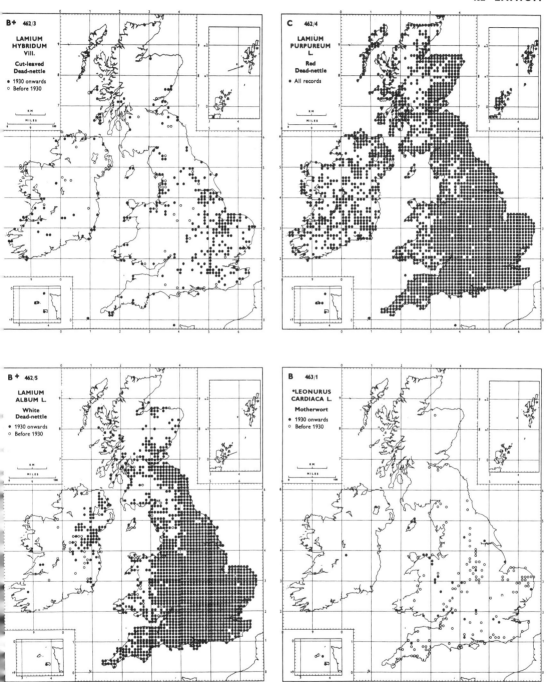

B+ 462/3

LAMIUM
HYBRIDUM
Vill.

Cut-leaved
Dead-nettle

● 1930 onwards
○ Before 1930

C 462/4

LAMIUM
PURPUREUM
L

Red
Dead-nettle

● All records

B+ 462/5

LAMIUM
ALBUM L.

White
Dead-nettle

● 1930 onwards
○ Before 1930

B 463/1

*LEONURUS
CARDIACA L.

Motherwort

● 1930 onwards
○ Before 1930

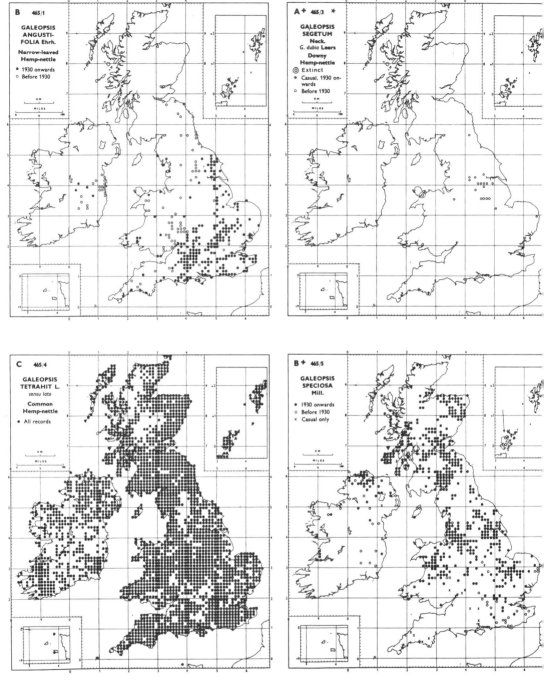

B   465/1

GALEOPSIS
ANGUSTI-
FOLIA Ehrh.

Narrow-leaved
Hemp-nettle

• 1930 onwards
○ Before 1930

A + 465/3   *

GALEOPSIS
SEGETUM
Neck.
G. dubia Leers
Downy
Hemp-nettle

⊚ Extinct
• Casual, 1930 on-
    wards
○ Before 1930

C   465/4

GALEOPSIS
TETRAHIT L.
sensu lato

Common
Hemp-nettle

• All records

B + 465/5

GALEOPSIS
SPECIOSA
Mill.

• 1930 onwards
○ Before 1930
× Casual only

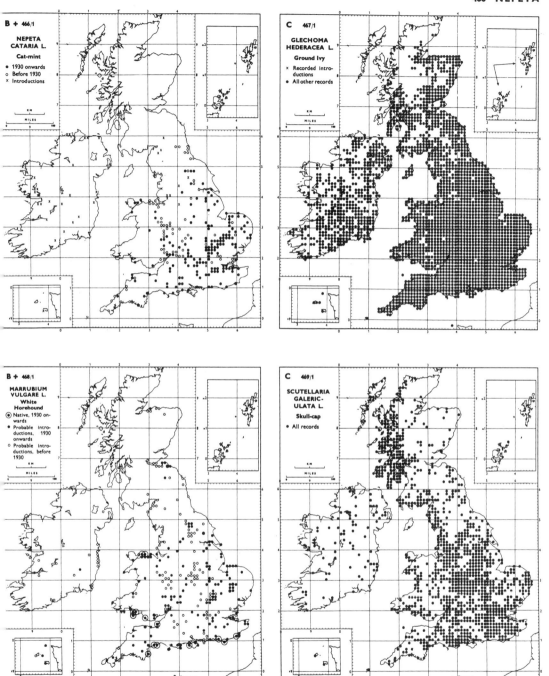

B ✛ 466/1

NEPETA
CATARIA L.
Cat-mint

• 1930 onwards
○ Before 1930
× Introductions

C   467/1

GLECHOMA
HEDERACEA L.
Ground Ivy

× Recorded intro-
   ductions
• All other records

B ✛ 468/1

MARRUBIUM
VULGARE L.
White
Horehound

⊙ Native, 1930 on-
   wards
• Probable intro-
   ductions,   1930
   onwards
○ Probable Intro-
   ductions, before
   1930

C   469/1

SCUTELLARIA
GALERIC-
ULATA L.
Skull-cap

• All records

LABIATAE

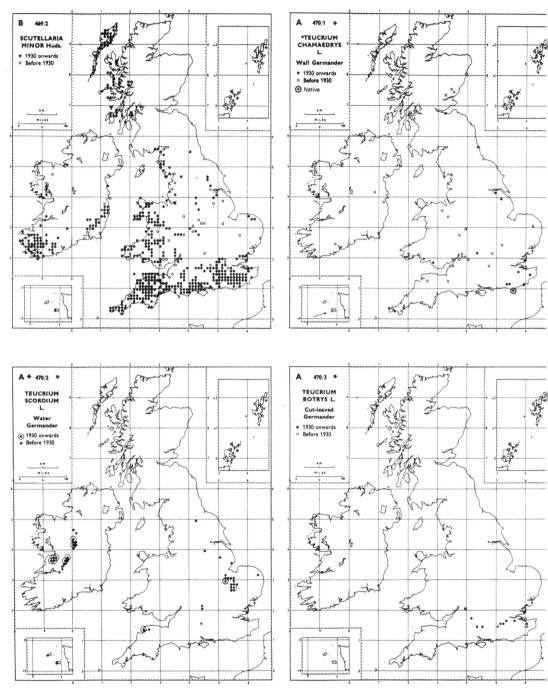

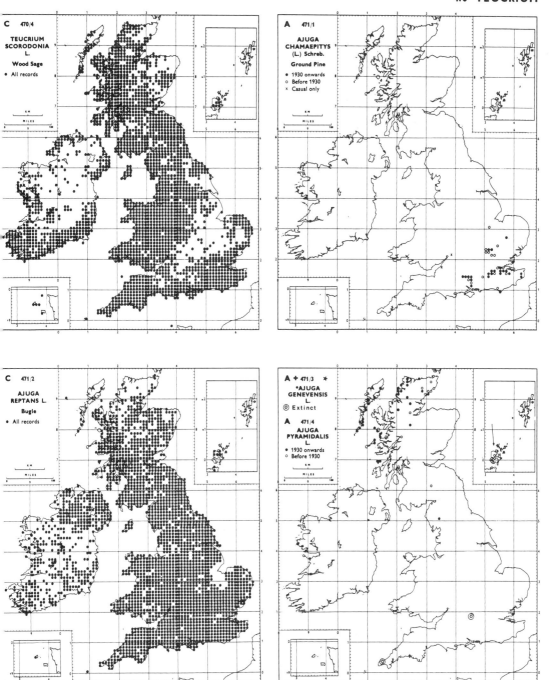

C 470/4
TEUCRIUM
SCORODONIA
L.
Wood Sage
● All records

A 471/1
AJUGA
CHAMAEPITYS
(L.) Schreb.
Ground Pine
● 1930 onwards
○ Before 1930
× Casual only

C 471/2
AJUGA
REPTANS L.
Bugle
● All records

A + 471/3 ✳
*AJUGA
GENEVENSIS
L.
◎ Extinct

A 471/4
AJUGA
PYRAMIDALIS
L.
● 1930 onwards
○ Before 1930

PLANTAGINACEAE

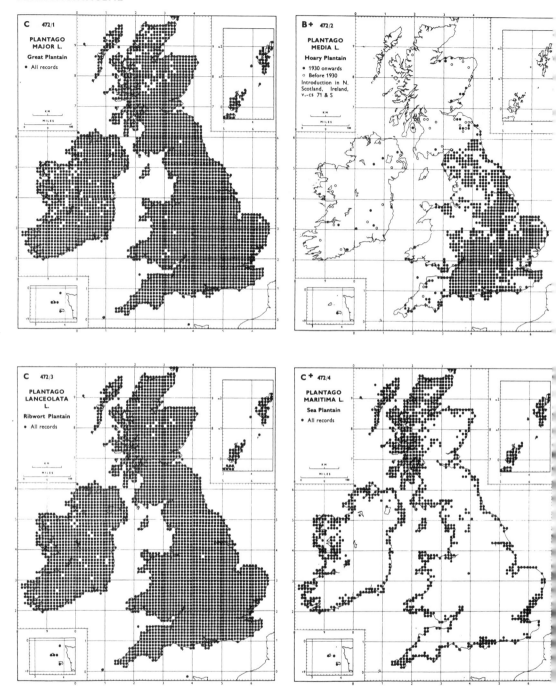

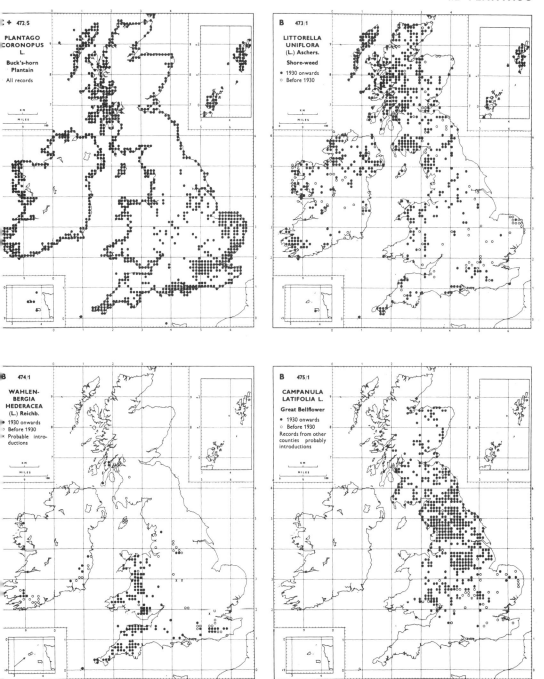

C + 472/5

PLANTAGO
CORONOPUS
L.

Buck's-horn
Plantain

All records

B 473/1

LITTORELLA
UNIFLORA
(L.) Aschers.

Shore-weed

• 1930 onwards
○ Before 1930

B 474/1

WAHLEN-
BERGIA
HEDERACEA
(L.) Reichb.

■ 1930 onwards
○ Before 1930
✕ Probable intro-
ductions

B 475/1

CAMPANULA
LATIFOLIA L.

Great Bellflower

• 1930 onwards
○ Before 1930
Records from other
counties probably
introductions

CAMPANULACEAE

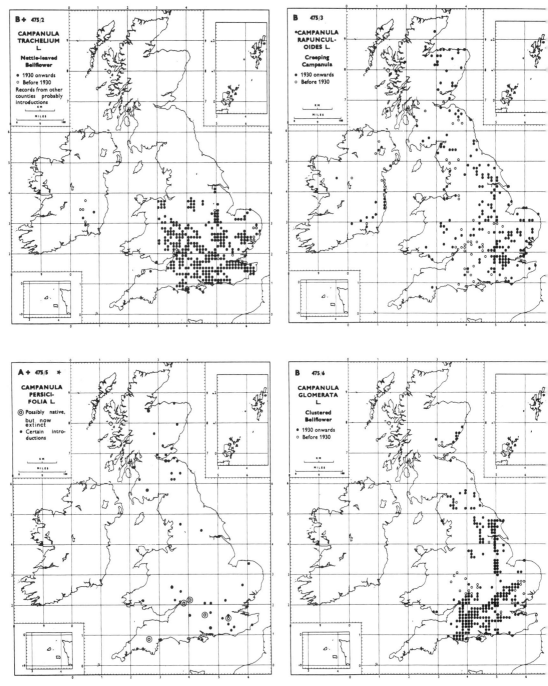

B+ 475/2

CAMPANULA
TRACHELIUM
L.

**Nettle-leaved
Bellflower**

● 1930 onwards
○ Before 1930
Records from other
counties probably
introductions

B 475/3

●CAMPANULA
RAPUNCUL-
OIDES L.

**Creeping
Campanula**

● 1930 onwards
○ Before 1930

A+ 475/5 *

CAMPANULA
PERSICI-
FOLIA L.

◎ Possibly native,
but now
extinct
● Certain intro-
ductions

B 475/6

CAMPANULA
GLOMERATA
L.

**Clustered
Bellflower**

● 1930 onwards
○ Before 1930

256

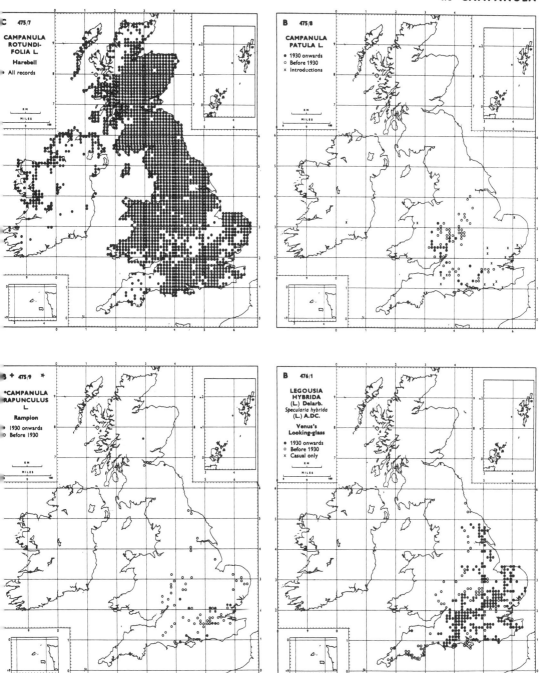

C 475/7

CAMPANULA
ROTUNDI-
FOLIA L.

Harebell

All records

B 475/8

CAMPANULA
PATULA L

• 1930 onwards
○ Before 1930
× Introductions

+ 475/9 *

*CAMPANULA
RAPUNCULUS
L

Rampion

• 1930 onwards
○ Before 1930

B 476/1

LEGOUSIA
HYBRIDA
(L.) Delarb.
Specularia hybrida
(L.) A.DC.

Venus's
Looking-glass

• 1930 onwards
○ Before 1930
× Casual only

CAMPANULACEAE

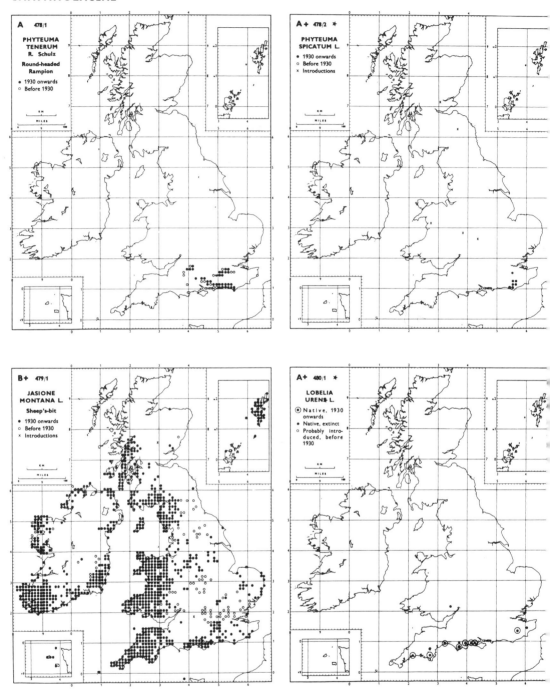

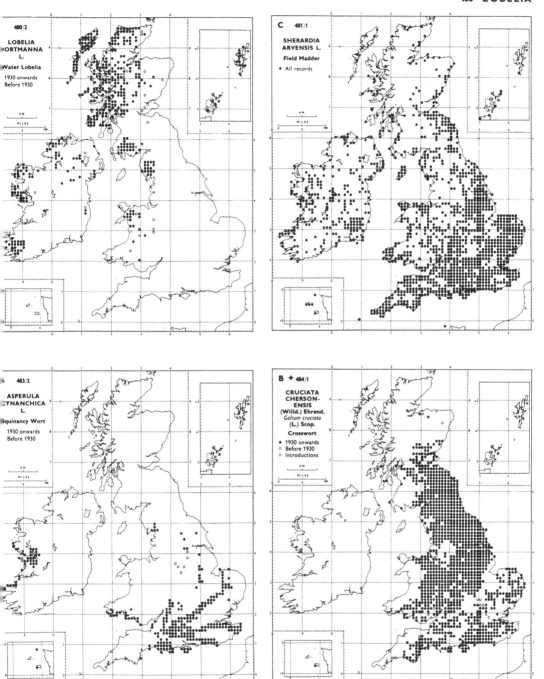

480/2

LOBELIA
DORTMANNA
L.

Water Lobelia

● 1930 onwards
○ Before 1930

C 481/1

SHERARDIA
ARVENSIS L.

Field Madder

● All records

483/2

ASPERULA
CYNANCHICA
L.

Squinancy Wort

● 1930 onwards
○ Before 1930

B + 484/1

CRUCIATA
CHERSON-
ENSIS
(Willd.) Ehrend.
Galium cruciata
(L.) Scop.

Crosswort

● 1930 onwards
○ Before 1930
× Introductions

259

RUBIACEAE

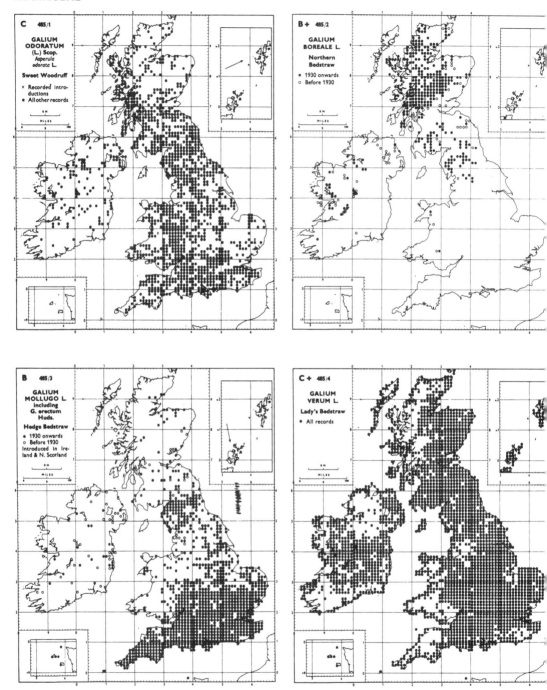

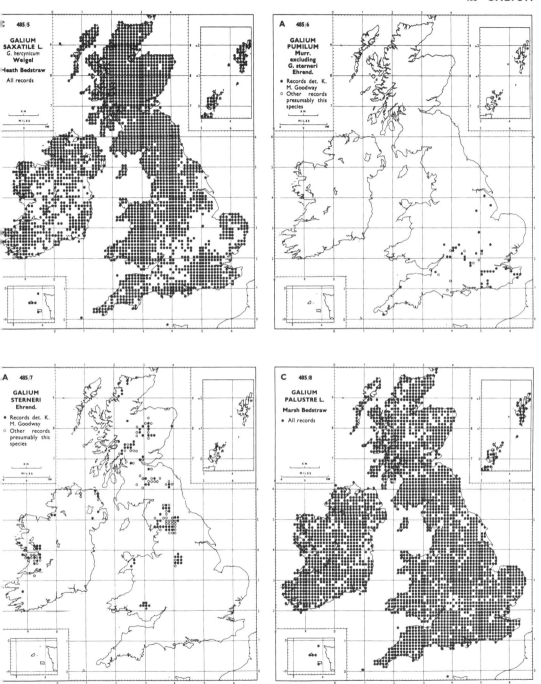

C 485/5

**GALIUM
SAXATILE L.**
G. hercynicum
Weigel
Heath Bedstraw
All records

KM
MILES

A 485/6

**GALIUM
PUMILUM
Murr.**
excluding
G. sterneri
Ehrend.
● Records det. K.
M. Goodway
○ Other records
presumably this
species

KM
MILES

A 485/7

**GALIUM
STERNERI**
Ehrend.
● Records det. K.
M. Goodway
○ Other records
presumably this
species

KM
MILES

C 485/8

**GALIUM
PALUSTRE L.**
Marsh Bedstraw
● All records

KM
MILES

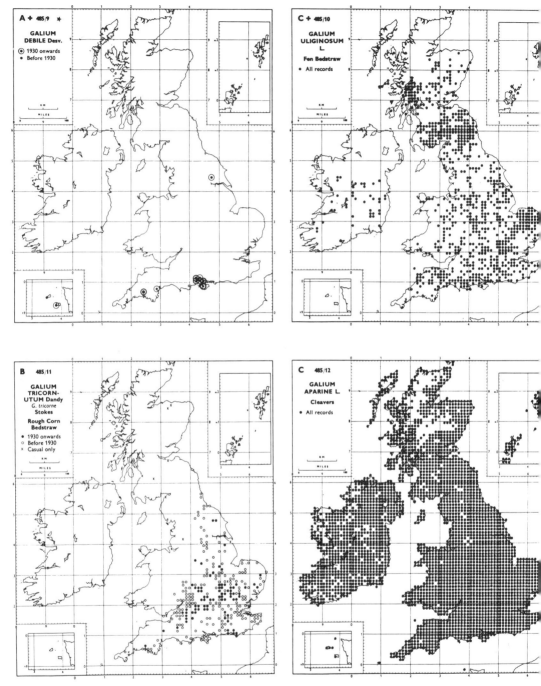

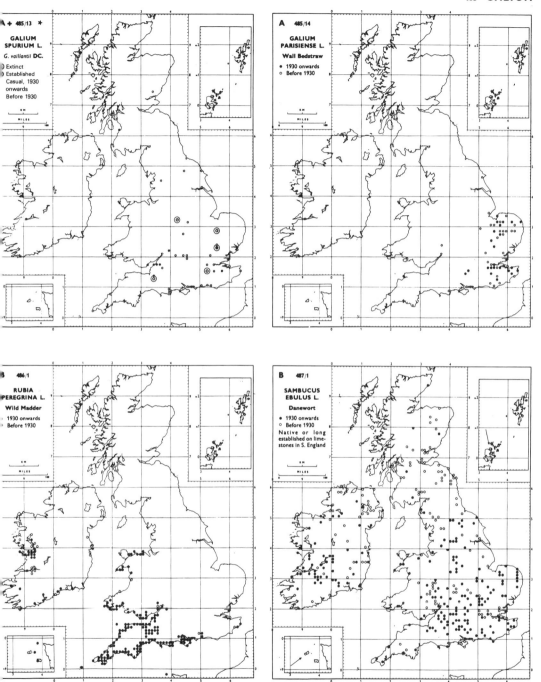

A + 485/13 ✶

**GALIUM
SPURIUM L.**

G. vaillantii DC.

⊙ Extinct
◐ Established
● Casual, 1930
onwards
○ Before 1930

K M
MILES

A 485/14

**GALIUM
PARISIENSE L.**

Wall Bedstraw

● 1930 onwards
○ Before 1930

K M
MILES

B 486/1

**RUBIA
PEREGRINA L.**

Wild Madder

● 1930 onwards
◑ Before 1930

K M
MILES

B 487/1

**SAMBUCUS
EBULUS L.**

Danewort

● 1930 onwards
○ Before 1930
Native or long
established on lime-
stones in S. England

K M
MILES

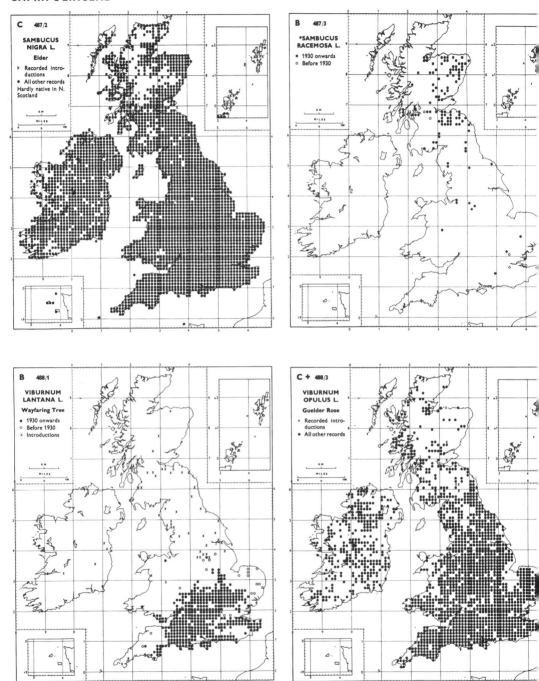

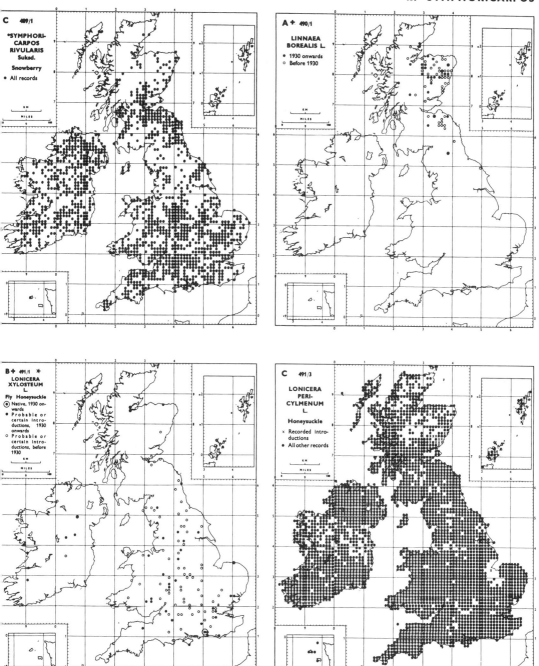

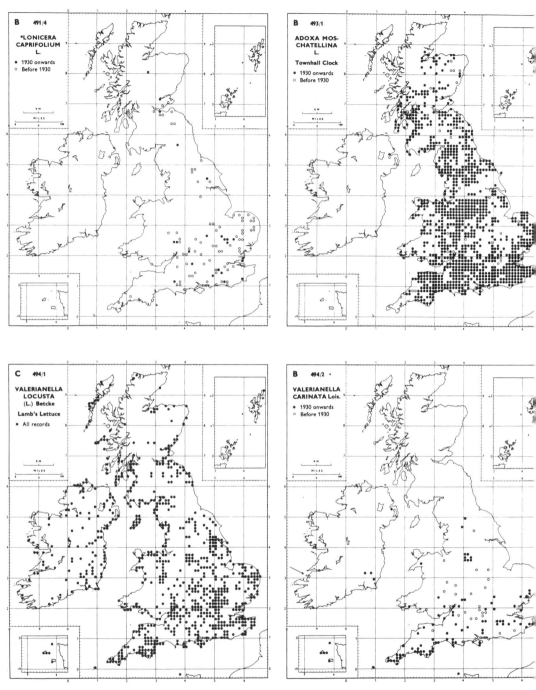

B 491/4

*LONICERA CAPRIFOLIUM L.

- 1930 onwards
- Before 1930

KM
MILES

B 493/1

ADOXA MOS-CHATELLINA L.

Townhall Clock

- 1930 onwards
- Before 1930

KM
MILES

C 494/1

VALERIANELLA LOCUSTA (L.) Betcke

Lamb's Lettuce

- All records

KM
MILES

B 494/2

VALERIANELLA CARINATA Lois.

- 1930 onwards
- Before 1930

KM
MILES

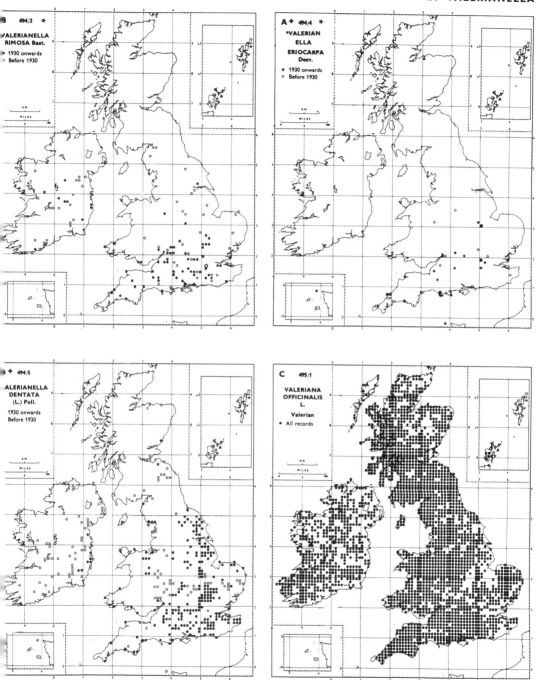

B 494/3 ⋆

**VALERIANELLA
RIMOSA Bast.**

● 1930 onwards
○ Before 1930

A + 494/4 ⋆

**VALERIAN
ELLA
ERIOCARPA
Desv.**

● 1930 onwards
○ Before 1930

+ 494/5

**VALERIANELLA
DENTATA
(L.) Poll.**

1930 onwards
Before 1930

C 495/1

**VALERIANA
OFFICINALIS
L.**

Valerian

● All records

267

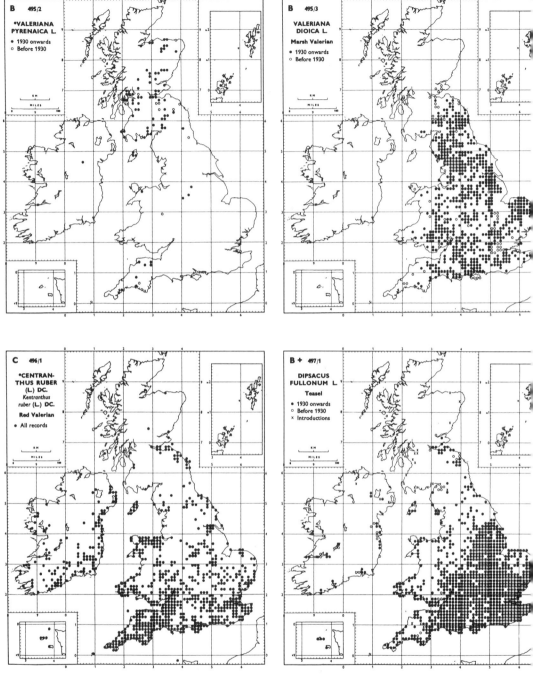

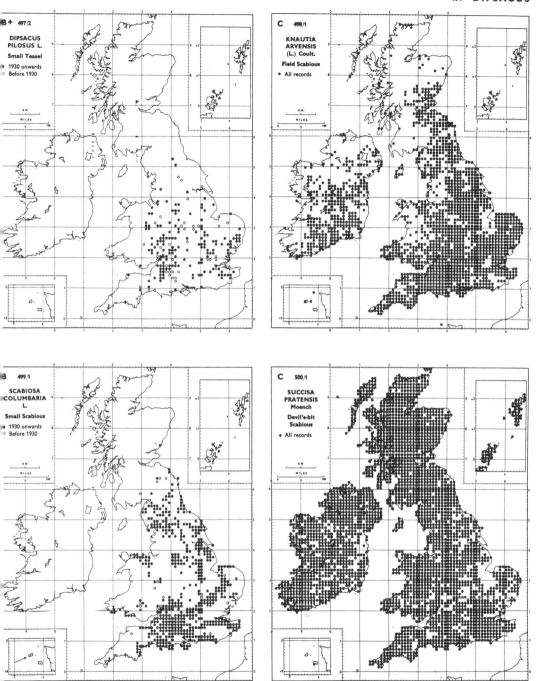

B + 497/2

DIPSACUS PILOSUS L.

Small Teasel

● 1930 onwards
○ Before 1930

C 498/1

KNAUTIA ARVENSIS (L.) Coult.

Field Scabious

● All records

B 499/1

SCABIOSA COLUMBARIA L.

Small Scabious

● 1930 onwards
○ Before 1930

C 500/1

SUCCISA PRATENSIS Moench

Devil's-bit Scabious

● All records

COMPOSITAE

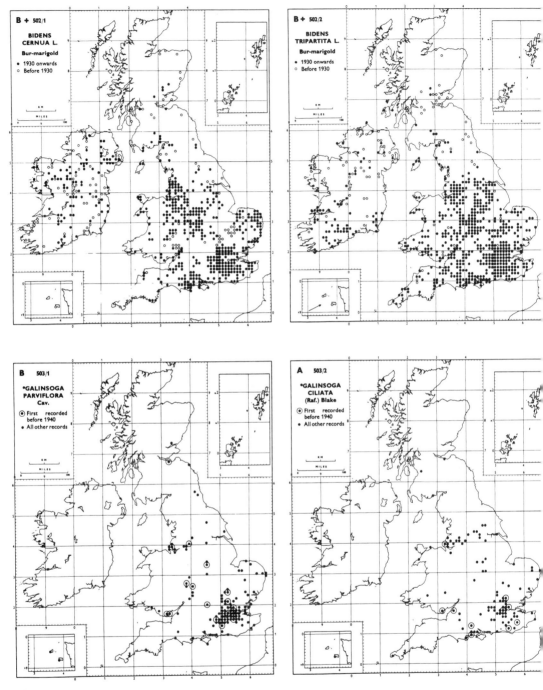

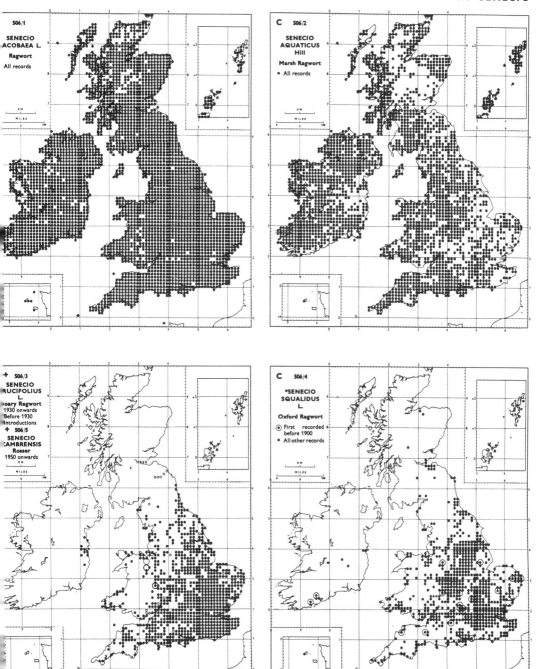

506/1
SENECIO
JACOBAEA L.
Ragwort
All records

C 506/2
SENECIO
AQUATICUS
Hill
Marsh Ragwort
• All records

506/3
SENECIO
ERUCIFOLIUS
L.
Hoary Ragwort
1930 onwards
Before 1930
Introductions
506/5
SENECIO
CAMBRENSIS
Rosser
1950 onwards

C 506/4
*SENECIO
SQUALIDUS
L.
Oxford Ragwort
⊙ First recorded
  before 1900
• All other records

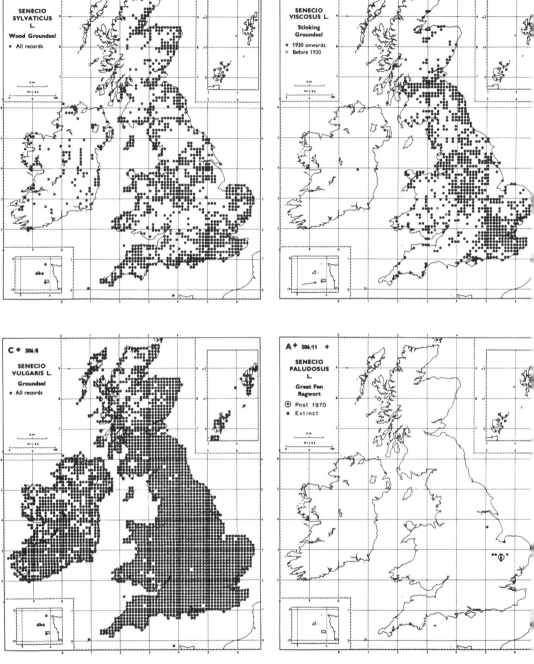

C 506/6

SENECIO
SYLVATICUS
L.

Wood Groundsel

• All records

B 506/7

SENECIO
VISCOSUS L.

Stinking
Groundsel

• 1930 onwards
○ Before 1930

C ✛ 506/8

SENECIO
VULGARIS L.

Groundsel

• All records

A ✛ 506/11 ✻

SENECIO
PALUDOSUS
L.

Great Fen
Ragwort

⊙ Post 1970
• Extinct

272

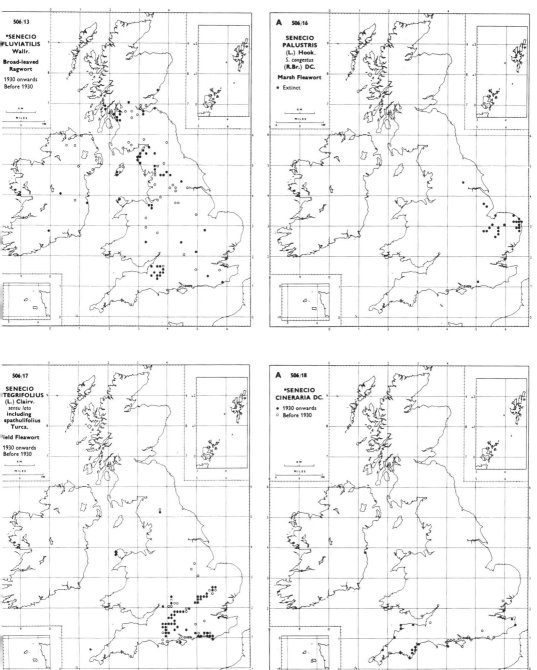

506/13

*SENECIO
FLUVIATILIS
Wallr.

Broad-leaved
Ragwort

● 1930 onwards
○ Before 1930

A 506/16

SENECIO
PALUSTRIS
(L.) Hook.
S. congestus
(R.Br.) DC.

Marsh Fleawort

● Extinct

506/17

SENECIO
TEGRIFOLIUS
(L.) Clairv.
sensu lato
including
spathulifolius
Turcz.

Field Fleawort

● 1930 onwards
○ Before 1930

A 506/18

*SENECIO
CINERARIA DC.

● 1930 onwards
○ Before 1930

273

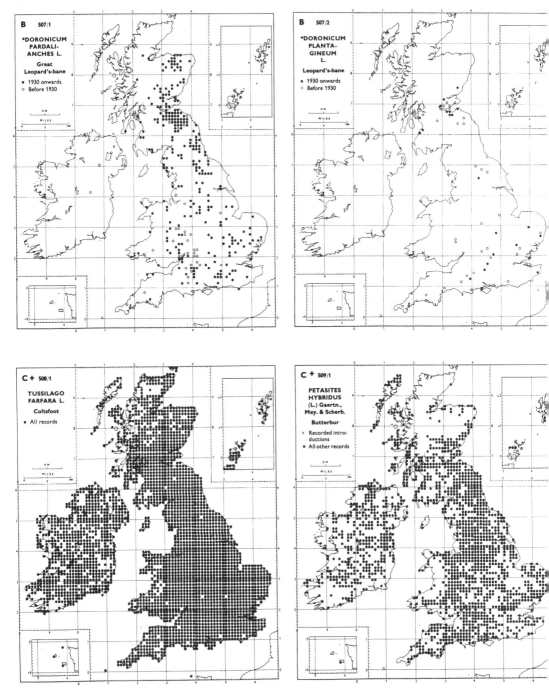

B 507/1

*DORONICUM
PARDALI-
ANCHES L.

Great
Leopard's-bane

• 1930 onwards
○ Before 1930

B 507/2

*DORONICUM
PLANTA-
GINEUM
L.

Leopard's-bane

• 1930 onwards
○ Before 1930

C + 508/1

TUSSILAGO
FARFARA L.

Coltsfoot

• All records

C + 509/1

PETASITES
HYBRIDUS
(L.) Gaertn.,
Mey. & Scherb.

Butterbur

× Recorded intro-
ductions
• All other records

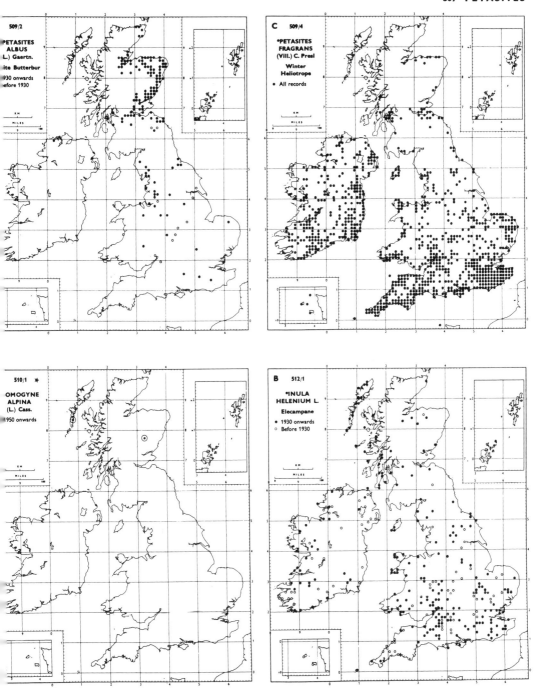

509/2

PETASITES
ALBUS
(L.) Gaertn.

ite Butterbur

930 onwards
efore 1930

C   509/4

PETASITES
FRAGRANS
(Vill.) C. Presl

Winter
Heliotrope

● All records

510/1   ★

OMOGYNE
ALPINA
(L.) Cass.

950 onwards

B   512/1

INULA
HELENIUM L.

Elecampane

● 1930 onwards
○ Before 1930

275

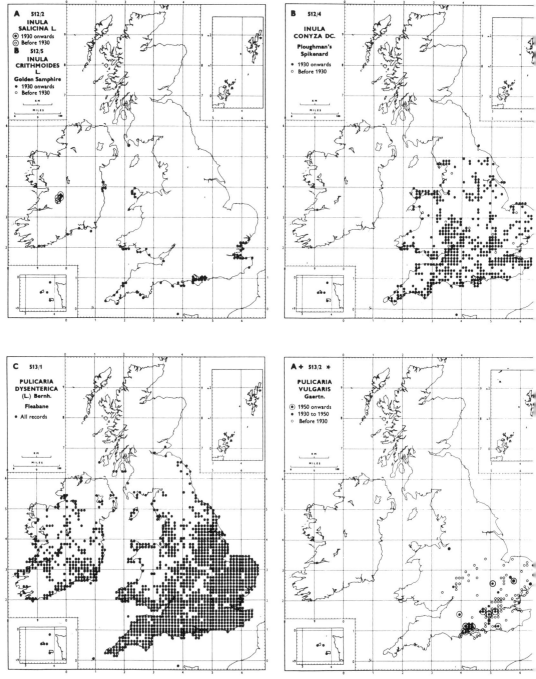

A 512/2
INULA
SALICINA L.
⊙ 1930 onwards
◉ Before 1930
B 512/5
INULA
CRITHMOIDES
L.
Golden Samphire
• 1930 onwards
○ Before 1930

B 512/4
INULA
CONYZA DC.
Ploughman's
Spikenard
• 1930 onwards
○ Before 1930

C 513/1
PULICARIA
DYSENTERICA
(L.) Bernh.
Fleabane
• All records

A + 513/2 ✱
PULICARIA
VULGARIS
Gaertn.
⊙ 1950 onwards
• 1930 to 1950
○ Before 1930

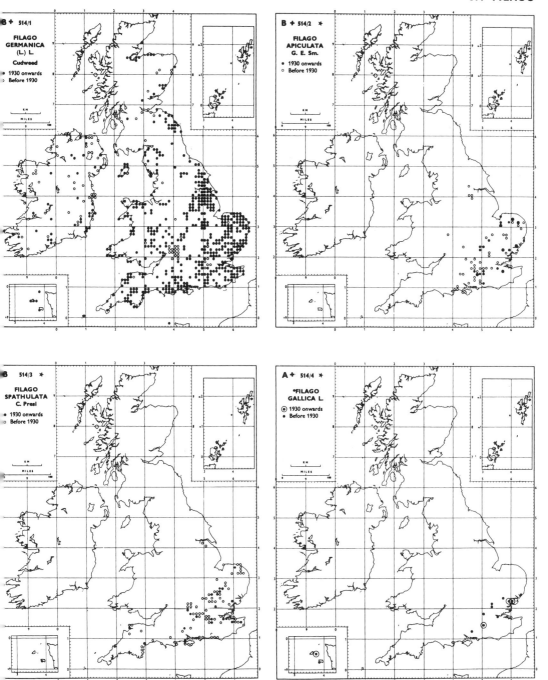

COMPOSITAE

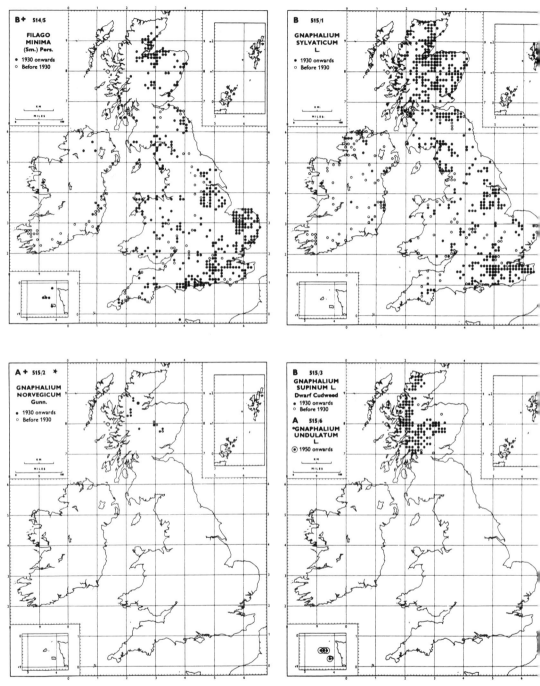

B+ 514/5

FILAGO
MINIMA
(Sm.) Pers.

- 1930 onwards
- Before 1930

B 515/1

GNAPHALIUM
SYLVATICUM
L.

- 1930 onwards
- Before 1930

A+ 515/2 *

GNAPHALIUM
NORVEGICUM
Gunn.

- 1930 onwards
- Before 1930

B 515/3

GNAPHALIUM
SUPINUM L.
Dwarf Cudweed
- 1930 onwards
- Before 1930

A 515/6
*GNAPHALIUM
UNDULATUM
L.
- 1950 onwards

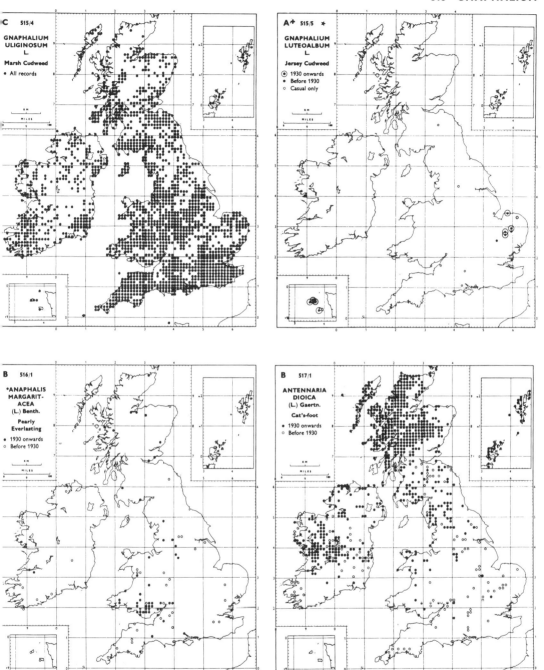

C 515/4

GNAPHALIUM
ULIGINOSUM
L.

Marsh Cudweed

• All records

A+ 515/5 ✱

GNAPHALIUM
LUTEOALBUM
L.

Jersey Cudweed

⊙ 1930 onwards
• Before 1930
○ Casual only

B 516/1

*ANAPHALIS
MARGARIT-
ACEA
(L.) Benth.

Pearly
Everlasting

• 1930 onwards
○ Before 1930

B 517/1

ANTENNARIA
DIOICA
(L.) Gaertn.

Cat's-foot

• 1930 onwards
○ Before 1930

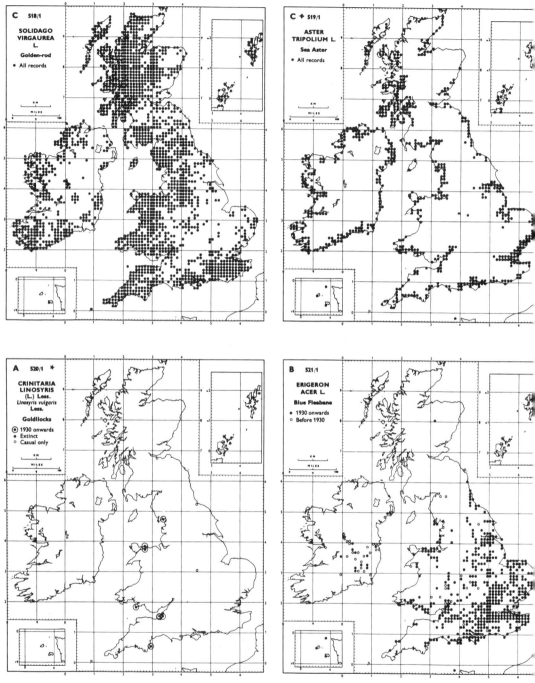

C 518/1

SOLIDAGO
VIRGAUREA
L.

Golden-rod

• All records

C + 519/1

ASTER
TRIPOLIUM L.

Sea Aster

• All records

A 520/1 *

CRINITARIA
LINOSYRIS
(L.) Less.
Linosyris vulgaris
Less.

Goldilocks

⊙ 1930 onwards
• Extinct
○ Casual only

B 521/1

ERIGERON
ACER L.

Blue Fleabane

• 1930 onwards
○ Before 1930

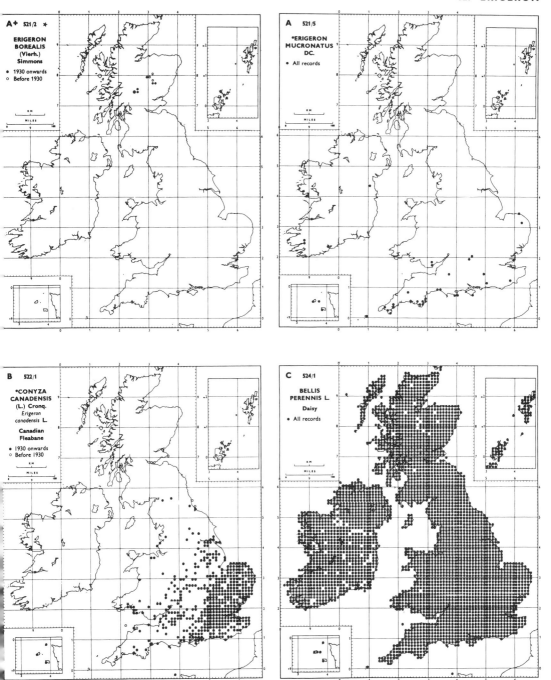

A+ 521/2

ERIGERON
BOREALIS
(Vierh.)
Simmons

• 1930 onwards
○ Before 1930

A 521/5

*ERIGERON
MUCRONATUS
DC.

• All records

B 522/1

*CONYZA
CANADENSIS
(L.) Cronq.
Erigeron
canadensis L.

Canadian
Fleabane

• 1930 onwards
○ Before 1930

C 524/1

BELLIS
PERENNIS L.

Daisy

• All records

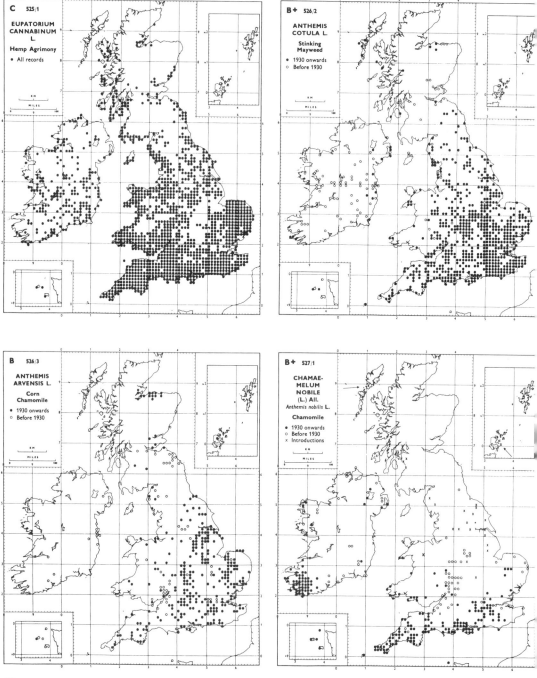

C 525/1

EUPATORIUM
CANNABINUM
L.

Hemp Agrimony

• All records

B+ 526/2

ANTHEMIS
COTULA L.

Stinking
Mayweed

• 1930 onwards
○ Before 1930

B 526/3

ANTHEMIS
ARVENSIS L.

Corn
Chamomile

• 1930 onwards
○ Before 1930

B+ 527/1

CHAMAE-
MELUM
NOBILE
(L.) All.
Anthemis nobilis L.

Chamomile

• 1930 onwards
○ Before 1930
× Introductions

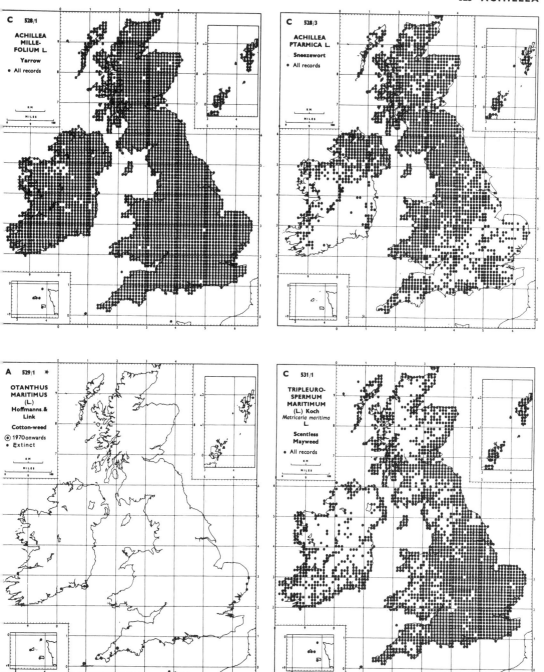

C 528/1
ACHILLEA
MILLE-
FOLIUM L.
Yarrow
• All records

C 528/3
ACHILLEA
PTARMICA L.
Sneezewort
• All records

A 529/1 ✳
OTANTHUS
MARITIMUS
(L.)
Hoffmanns. &
Link
Cotton-weed
⊙ 1970 onwards
• Extinct

C 531/1
TRIPLEURO-
SPERMUM
MARITIMUM
(L.) Koch
Matricaria maritima
L.
Scentless
Mayweed
• All records

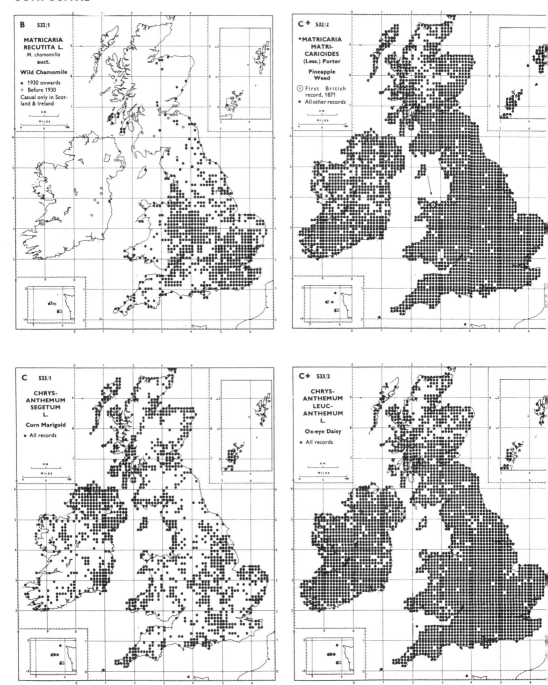

B 532/1

**MATRICARIA
RECUTITA L.**
M. chamomilla
auct.

Wild Chamomile

- 1930 onwards
- Before 1930
Casual only in Scot-
land & Ireland

C+ 532/2

•**MATRICARIA
MATRI-
CARIOIDES**
(Less.) Porter

**Pineapple
Weed**

⊗ First British
record, 1871
• All other records

C 533/1

**CHRYS-
ANTHEMUM
SEGETUM
L.**

Corn Marigold

• All records

C+ 533/2

**CHRYS-
ANTHEMUM
LEUC-
ANTHEMUM
L.**

Ox-eye Daisy

• All records

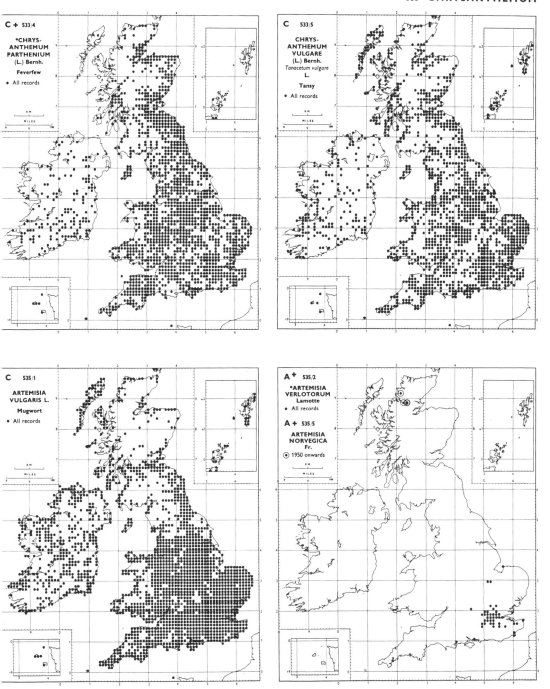

C + 533/4
*CHRYS-
ANTHEMUM
PARTHENIUM
(L.) Bernh.
Feverfew
● All records

C 533/5
CHRYS-
ANTHEMUM
VULGARE
(L.) Bernh.
Tanacetum vulgare
L.
Tansy
● All records

C 535/1
ARTEMISIA
VULGARIS L.
Mugwort
● All records

A + 535/2
*ARTEMISIA
VERLOTORUM
Lamotte
● All records

A + 535/5
ARTEMISIA
NORVEGICA
Fr.
⊙ 1950 onwards

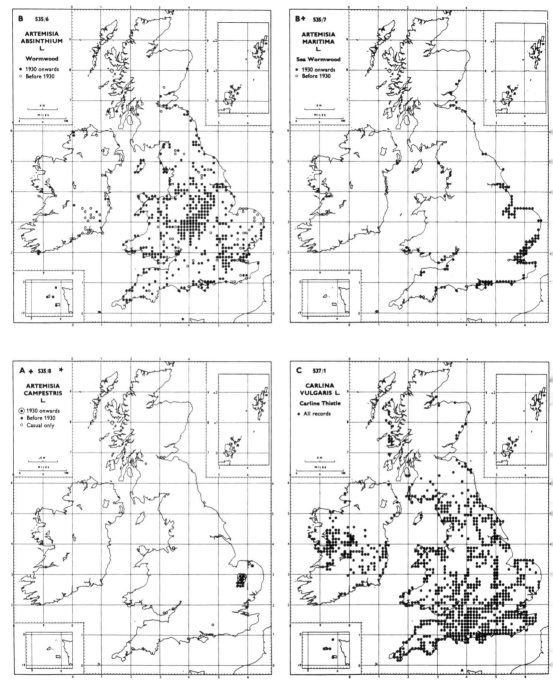

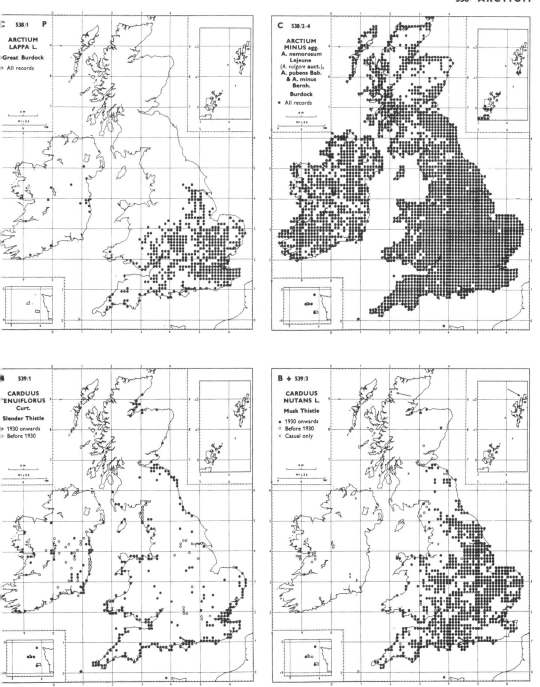

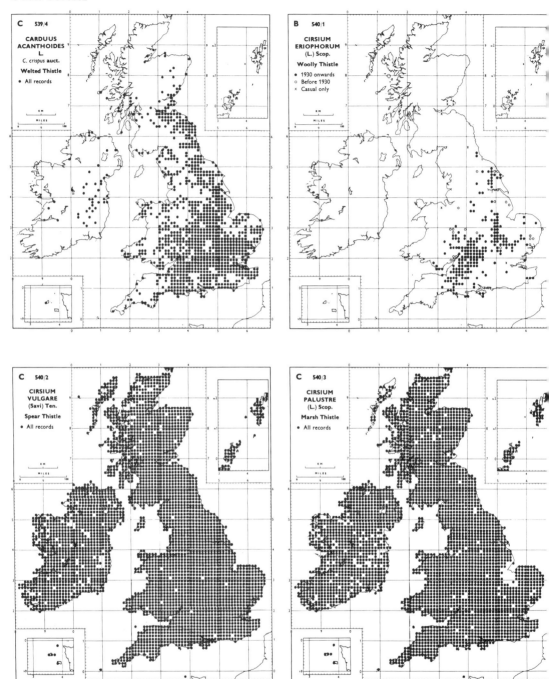

C 539/4

CARDUUS
ACANTHOIDES
L.
C. crispus auct.

Welted Thistle

• All records

K M
MILES

B 540/1

CIRSIUM
ERIOPHORUM
(L.) Scop.

Woolly Thistle

• 1930 onwards
○ Before 1930
× Casual only

K M
MILES

C 540/2

CIRSIUM
VULGARE
(Savi) Ten.

Spear Thistle

• All records

K M
MILES

C 540/3

CIRSIUM
PALUSTRE
(L.) Scop.

Marsh Thistle

• All records

K M
MILES

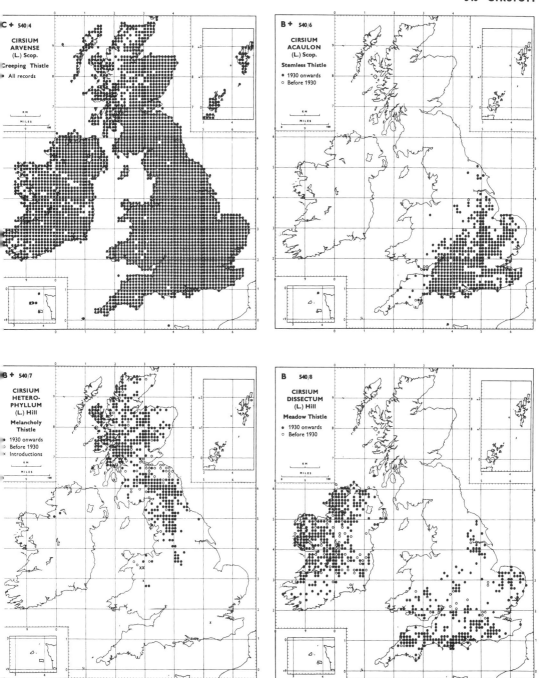

C + 540/4
CIRSIUM
ARVENSE
(L.) Scop.
Creeping Thistle
● All records

B + 540/6
CIRSIUM
ACAULON
(L.) Scop.
Stemless Thistle
● 1930 onwards
○ Before 1930

B + 540/7
CIRSIUM
HETERO-
PHYLLUM
(L.) Hill
Melancholy
Thistle
● 1930 onwards
○ Before 1930
× Introductions

B 540/8
CIRSIUM
DISSECTUM
(L.) Hill
Meadow Thistle
● 1930 onwards
○ Before 1930

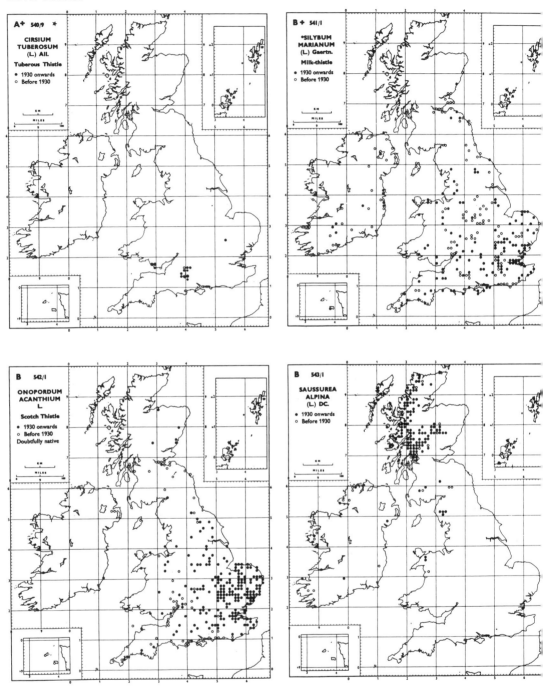

A+ 540/9 ✳
CIRSIUM
TUBEROSUM
(L.) All.
Tuberous Thistle
● 1930 onwards
○ Before 1930

B+ 541/1
*SILYBUM
MARIANUM
(L.) Gaertn.
Milk-thistle
● 1930 onwards
○ Before 1930

B 542/1
ONOPORDUM
ACANTHIUM
L.
Scotch Thistle
● 1930 onwards
○ Before 1930
Doubtfully native

B 543/1
SAUSSUREA
ALPINA
(L.) DC.
● 1930 onwards
○ Before 1930

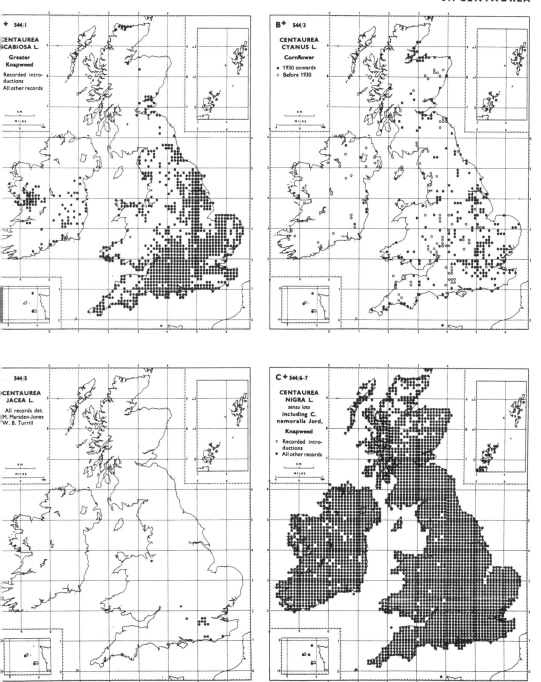

+ 544/1

CENTAUREA
SCABIOSA L.
Greater
Knapweed

Recorded intro-
ductions
All other records

B+ 544/3

CENTAUREA
CYANUS L.
Cornflower

● 1930 onwards
○ Before 1930

544/5

CENTAUREA
JACEA L.
All records det.
M. Marsden-Jones
W. B. Turrill

C+ 544/6–7

CENTAUREA
NIGRA L.
sensu lato
including C.
nemoralis Jord.

Knapweed

× Recorded intro-
ductions
● All other records

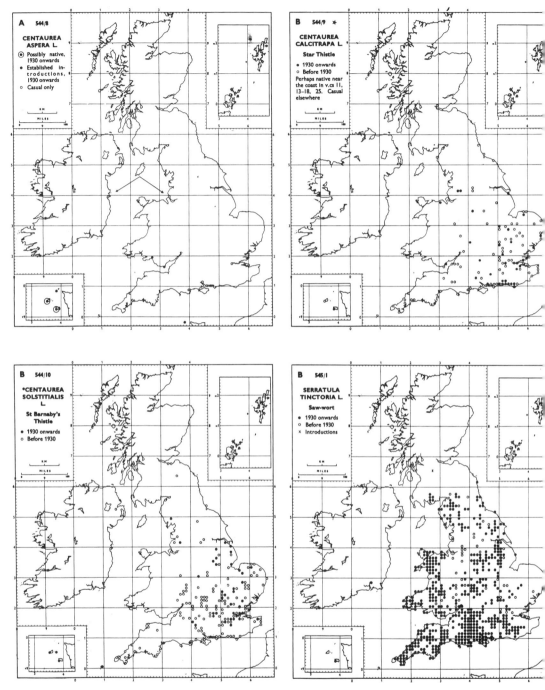

A 544/8

CENTAUREA ASPERA L.

⊙ Possibly native, 1930 onwards
● Established introductions, 1930 onwards
○ Casual only

B 544/9 ✶

CENTAUREA CALCITRAPA L.

Star Thistle

● 1930 onwards
○ Before 1930
Perhaps native near the coast in v.cs 11, 13-18, 25. Casual elsewhere

B 544/10

***CENTAUREA SOLSTITIALIS L.**

St Barnaby's Thistle

● 1930 onwards
○ Before 1930

B 545/1

SERRATULA TINCTORIA L.

Saw-wort

● 1930 onwards
○ Before 1930
× Introductions

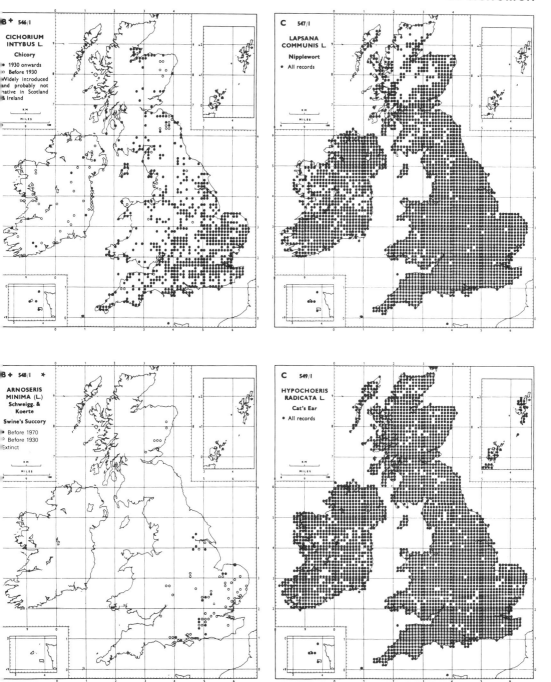

B + 546/1

CICHORIUM
INTYBUS L.

Chicory

● 1930 onwards
○ Before 1930
Widely introduced
and probably not
native in Scotland
& Ireland

C 547/1

LAPSANA
COMMUNIS L.

Nipplewort

● All records

B + 548/1 ✱

ARNOSERIS
MINIMA (L.)
Schweigg. &
Koerte

Swine's Succory

● Before 1970
○ Before 1930
Extinct

C 549/1

HYPOCHOERIS
RADICATA L.

Cat's Ear

● All records

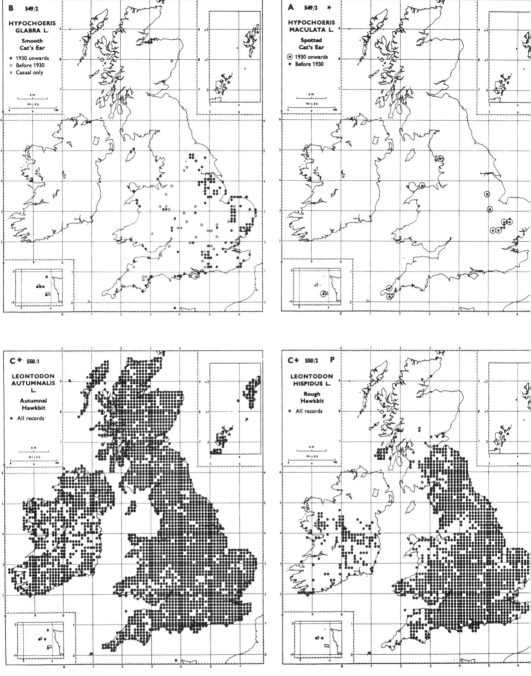

B 549/2

HYPOCHOERIS
GLABRA L.

Smooth
Cat's Ear

● 1930 onwards
○ Before 1930
× Casual only

A 549/3 ★

HYPOCHOERIS
MACULATA L.

Spotted
Cat's Ear

⊙ 1930 onwards
● Before 1930

C+ 550/1

LEONTODON
AUTUMNALIS
L.

Autumnal
Hawkbit

● All records

C+ 550/2 P

LEONTODON
HISPIDUS L.

Rough
Hawkbit

● All records

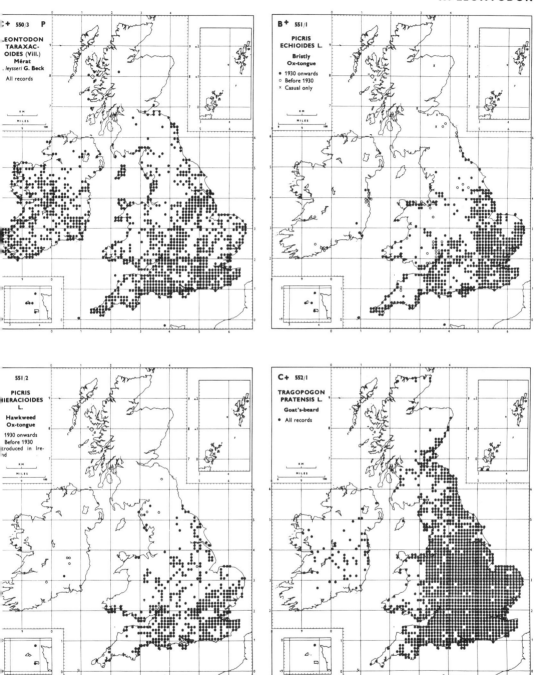

C + 550/3 P

LEONTODON
TARAXAC-
OIDES (Vill.)
Mérat
. leysseri G. Beck

All records

KM
MILES

B + 551/1

PICRIS
ECHIOIDES L.

Bristly
Ox-tongue

• 1930 onwards
○ Before 1930
× Casual only

KM
MILES

551/2

PICRIS
HIERACIOIDES
L.

Hawkweed
Ox-tongue

1930 onwards
Before 1930
troduced in Ire-
nd

KM
MILES

C + 552/1

TRAGOPOGON
PRATENSIS L.

Goat's-beard
• All records

KM
MILES

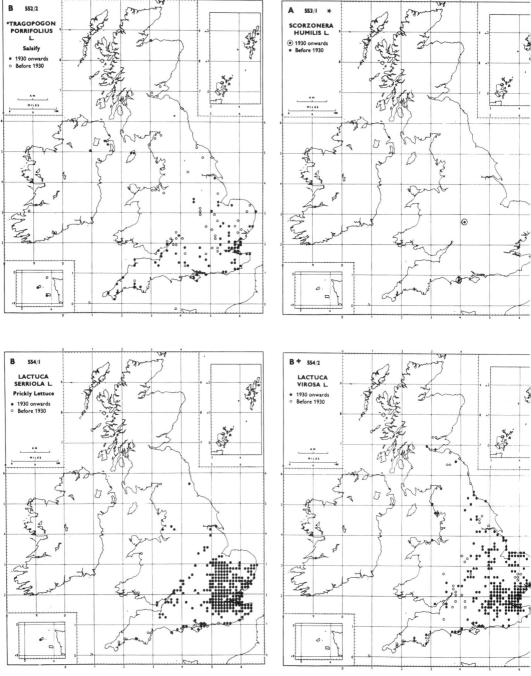

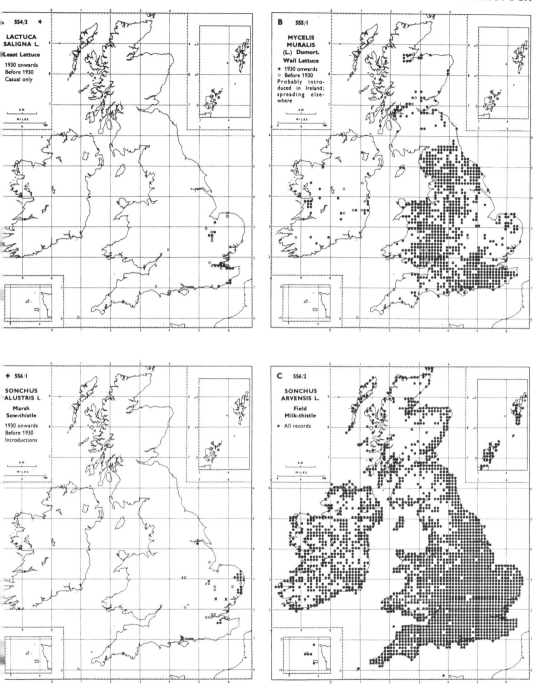

554/3 ✲

LACTUCA
SALIGNA L.
Least Lettuce

■ 1930 onwards
□ Before 1930
× Casual only

KM
MILES

B 555/I

MYCELIS
MURALIS
(L.) Dumort.
Wall Lettuce

● 1930 onwards
○ Before 1930
Probably intro-
duced in Ireland;
spreading else-
where

KM
MILES

✚ 556/I

SONCHUS
ALUSTRIS L.
Marsh
Sow-thistle

● 1930 onwards
○ Before 1930
× Introductions

KM
MILES

C 556/2

SONCHUS
ARVENSIS L.
Field
Milk-thistle

● All records

KM
MILES

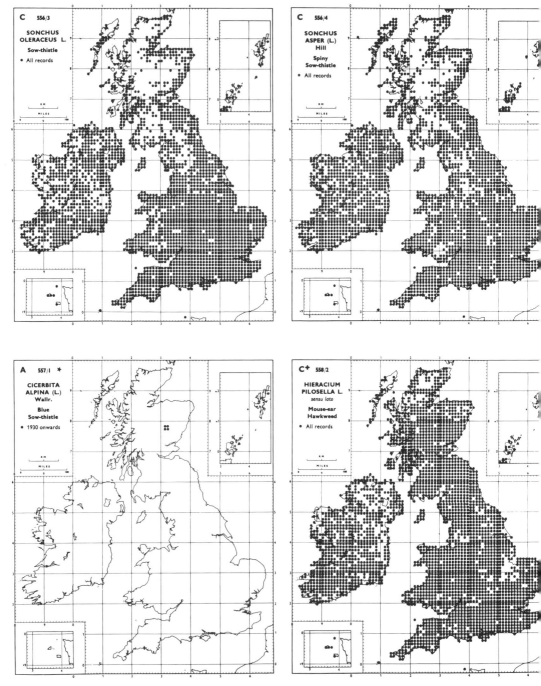

C 556/3
SONCHUS
OLERACEUS L.
Sow-thistle
● All records

C 556/4
SONCHUS
ASPER (L.)
Hill
Spiny
Sow-thistle
● All records

A 557/1 ★
CICERBITA
ALPINA (L.)
Wallr.
Blue
Sow-thistle
● 1930 onwards

C⁺ 558/2
HIERACIUM
PILOSELLA L.
sensu lato
Mouse-ear
Hawkweed
● All records

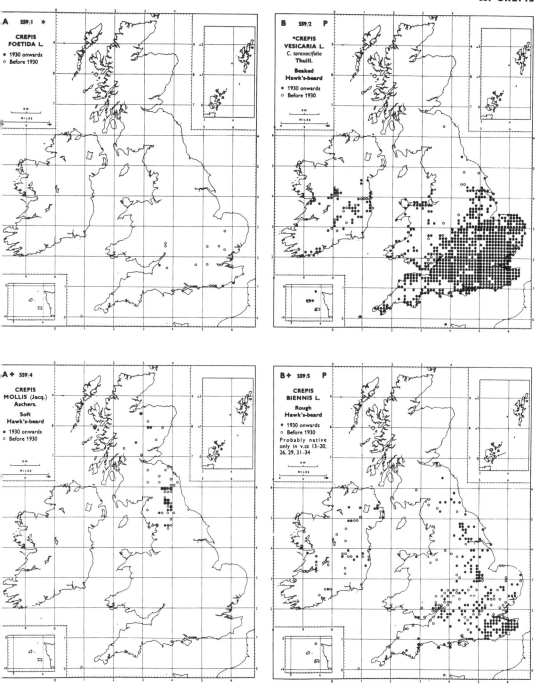

A 559/1 ✳

CREPIS
FOETIDA L.

● 1930 onwards
○ Before 1930

KM
MILES

B 559/2 P

°CREPIS
VESICARIA L.
C. taraxacifolia
Thuill.

Beaked
Hawk's-beard

● 1930 onwards
○ Before 1930

KM
MILES

A ✛ 559/4

CREPIS
MOLLIS (Jacq.)
Aschers.

Soft
Hawk's-beard

● 1930 onwards
○ Before 1930

KM
MILES

B ✛ 559/5 P

CREPIS
BIENNIS L.

Rough
Hawk's-beard

● 1930 onwards
○ Before 1930
Probably native
only in v.cs 13–20,
26, 29, 31–34

KM
MILES

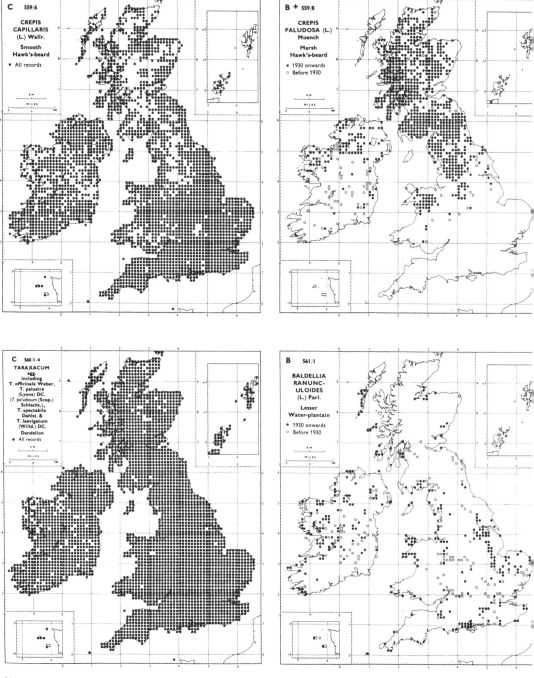

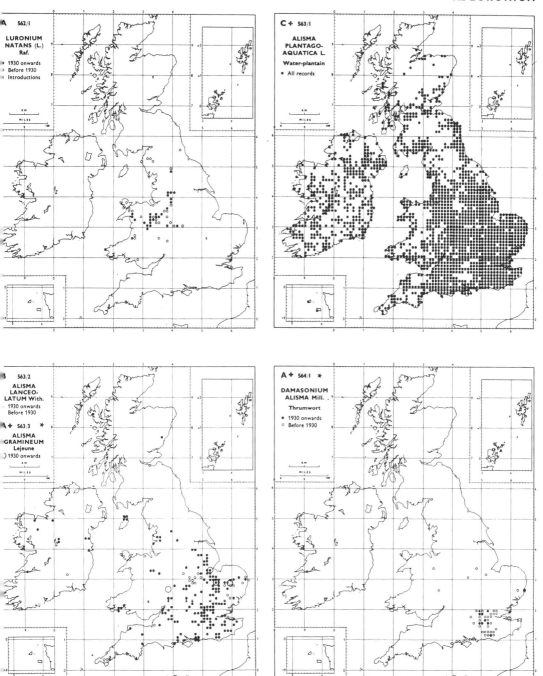

A 562/1

LURONIUM NATANS (L.) Raf.
- 1930 onwards
- Before 1930
- Introductions

C + 563/1

ALISMA PLANTAGO-AQUATICA L.
Water-plantain
- All records

B 563/2

ALISMA LANCEOLATUM With.
- 1930 onwards
- Before 1930

A + 563/3

ALISMA GRAMINEUM Lejeune
- 1930 onwards

A + 564/1

DAMASONIUM ALISMA Mill.
Thrumwort
- 1930 onwards
- Before 1930

ALISMATACEAE

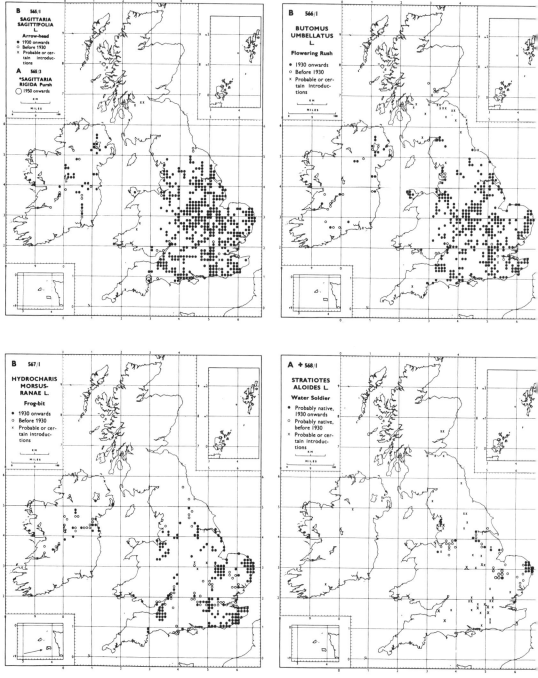

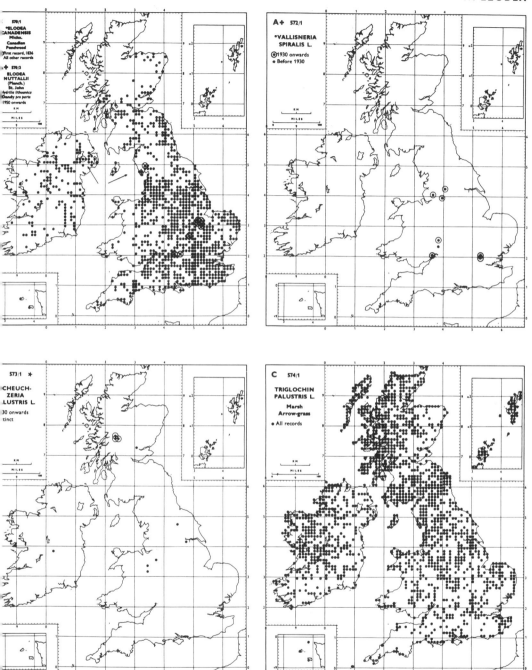

570/1
**ELODEA
CANADENSIS**
Michx.
Canadian
Pondweed
First record, 1836
All other records

570/3
**ELODEA
NUTTALLII**
(Planch.)
St. John
hydrilla lithuanica
Dandy pro parte
1950 onwards

A+ 572/1
**VALLISNERIA
SPIRALIS L.**
1930 onwards
Before 1930

573/1
**SCHEUCH-
ZERIA
PALUSTRIS L.**
1930 onwards
Extinct

C 574/1
**TRIGLOCHIN
PALUSTRIS L.**
Marsh
Arrow-grass
All records

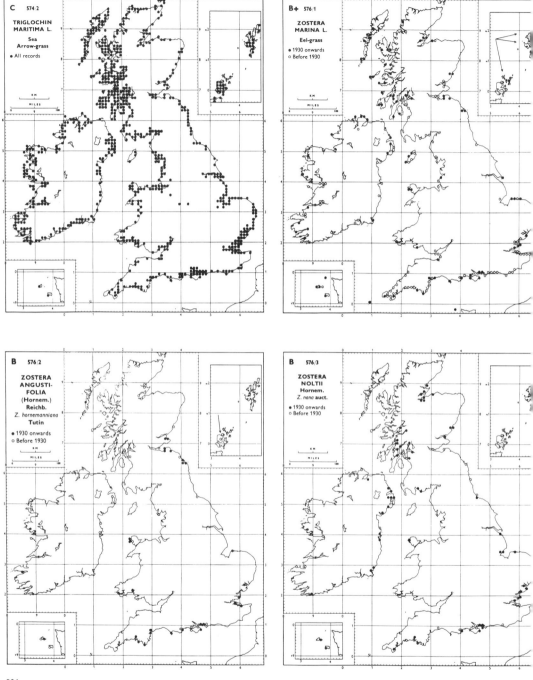

C 574/2

TRIGLOCHIN
MARITIMA L.

Sea
Arrow-grass

● All records

B+ 576/1

ZOSTERA
MARINA L.

Eel-grass

● 1930 onwards
○ Before 1930

B 576/2

ZOSTERA
ANGUSTI-
FOLIA
(Hornem.)
Reichb.

Z. hornemanniana
Tutin

● 1930 onwards
○ Before 1930

B 576/3

ZOSTERA
NOLTII
Hornem.
Z. nana auct.

● 1930 onwards
○ Before 1930

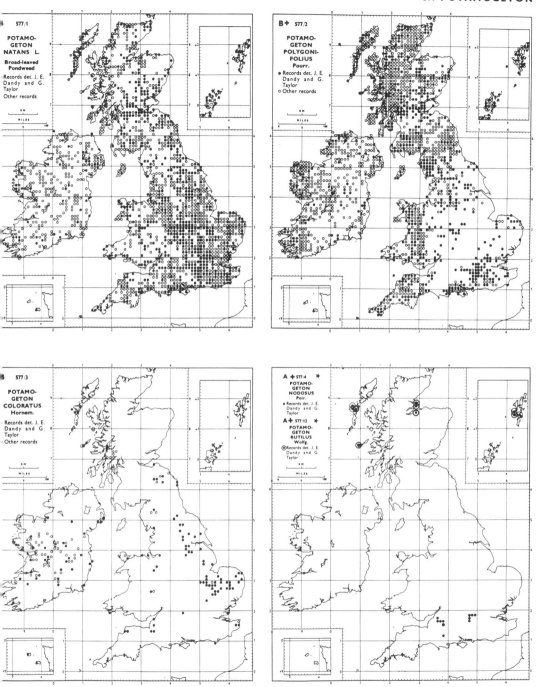

577/1

POTAMO-
GETON
NATANS L.
Broad-leaved
Pondweed

● Records det. J. E.
Dandy and G.
Taylor
○ Other records

B+ 577/2

POTAMO-
GETON
POLYGONI-
FOLIUS
Pourr.

● Records det. J. E.
Dandy and G.
Taylor
○ Other records

577/3

POTAMO-
GETON
COLORATUS
Hornem.

● Records det. J. E.
Dandy and G.
Taylor
○ Other records

A+ 577/4

POTAMO-
GETON
NODOSUS
Poir.

● Records det. J. E.
Dandy and G.
Taylor

A+ 577/12

POTAMO-
GETON
RUTILUS
Wolfg.

⊙ Records det. J. E.
Dandy and G.
Taylor

POTAMOGETONACEAE

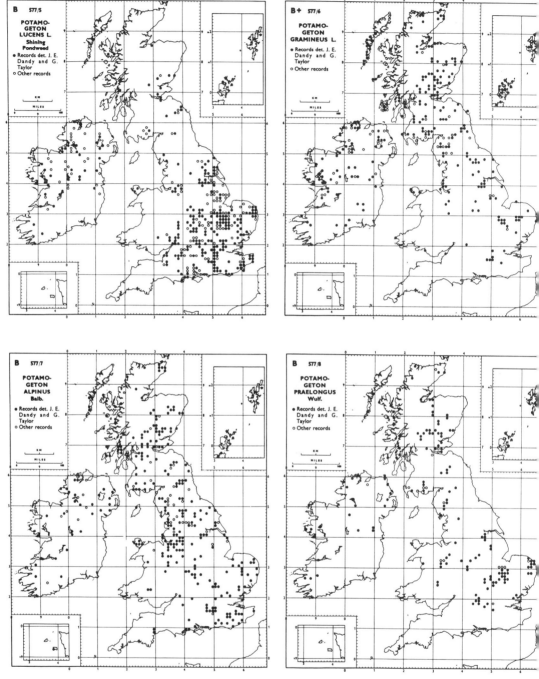

B   577/5
POTAMO-
GETON
LUCENS L.
Shining
Pondweed
● Records det. J. E.
  Dandy and G.
  Taylor
○ Other records

B✛   577/6
POTAMO-
GETON
GRAMINEUS L.
● Records det. J. E.
  Dandy and G.
  Taylor
○ Other records

B   577/7
POTAMO-
GETON
ALPINUS
Balb.
● Records det. J. E.
  Dandy and G.
  Taylor
○ Other records

B   577/8
POTAMO-
GETON
PRAELONGUS
Wulf.
● Records det. J. E.
  Dandy and G.
  Taylor
○ Other records

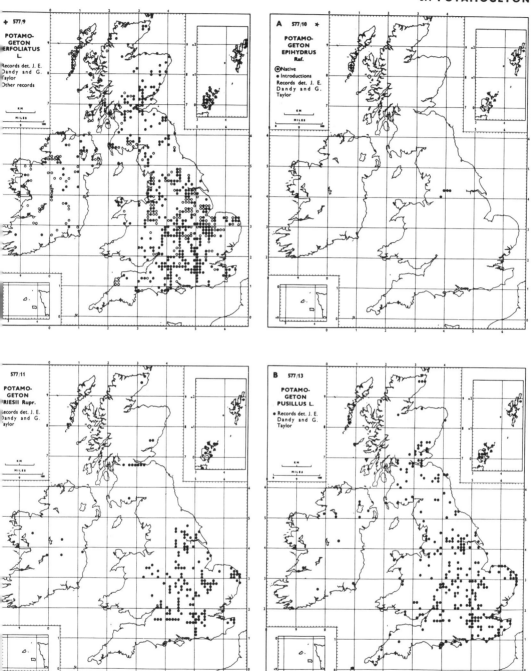

+ 577/9

POTAMO-
GETON
ERFOLIATUS
L

Records det. J. E.
Dandy and G.
Taylor
Other records

A 577/10 ★

POTAMO-
GETON
EPIHYDRUS
Raf.

Native
Introductions
Records det. J. E.
Dandy and G.
Taylor

577/11

POTAMO-
GETON
RIESII Rupr.

Records det. J. E.
Dandy and G.
Taylor

B 577/13

POTAMO-
GETON
PUSILLUS L.

Records det. J. E.
Dandy and G.
Taylor

POTAMOGETONACEAE

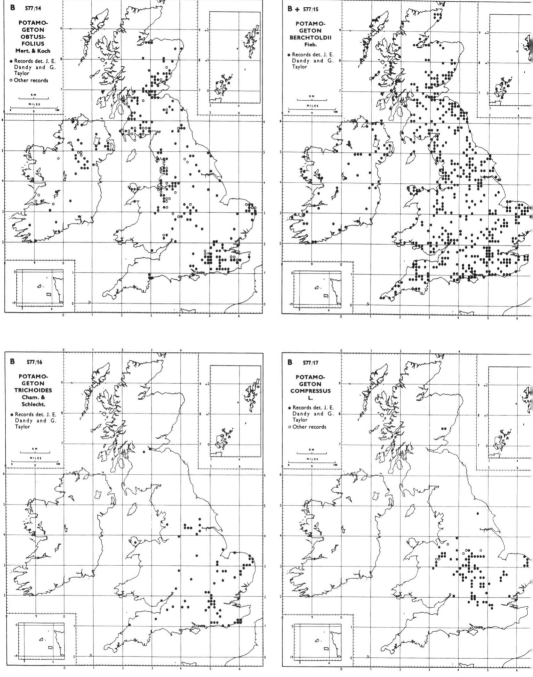

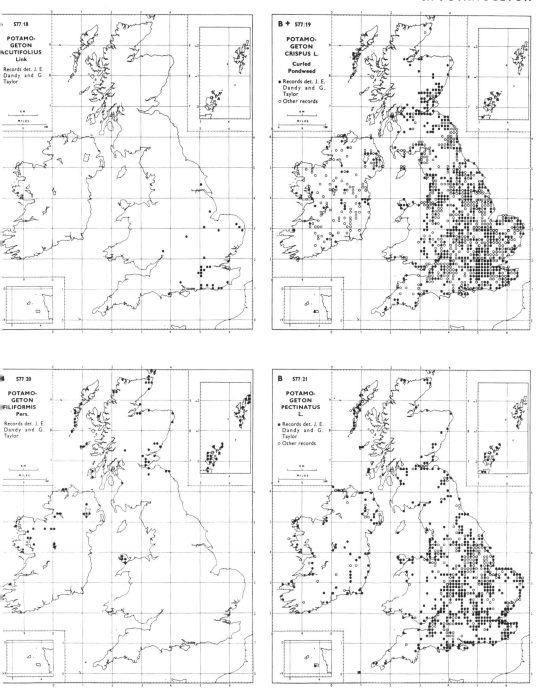

577/18

POTAMO-
GETON
ACUTIFOLIUS
Link

Records det. J. E.
Dandy and G.
Taylor

KM
MILES

B ✛ 577/19

POTAMO-
GETON
CRISPUS L.

Curled
Pondweed

● Records det. J. E.
Dandy and G.
Taylor
○ Other records

KM
MILES

577/20

POTAMO-
GETON
FILIFORMIS
Pers.

Records det. J. E.
Dandy and G.
Taylor

KM
MILES

B   577/21

POTAMO-
GETON
PECTINATUS
L.

● Records det. J. E.
Dandy and G.
Taylor
○ Other records

309

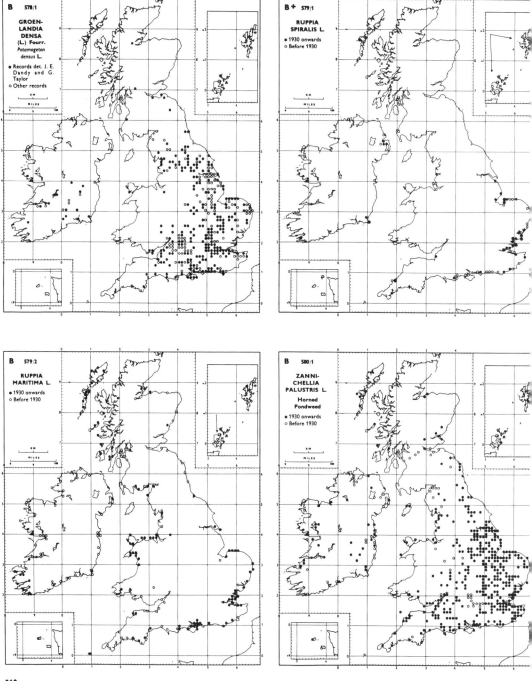

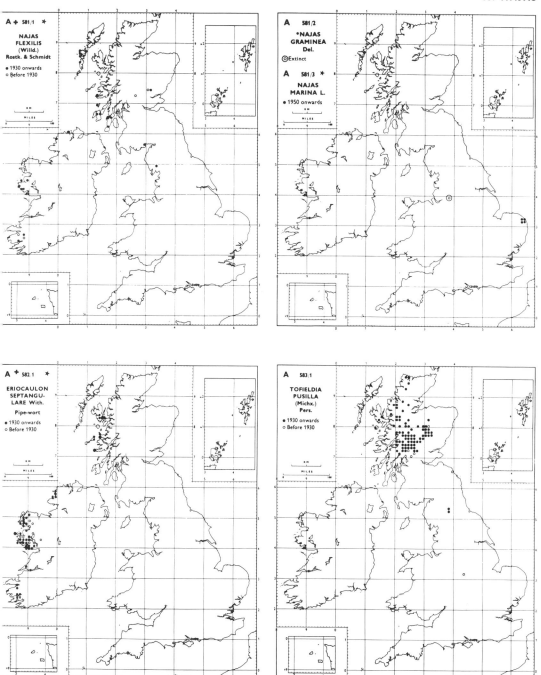

A + 581/1 ✳

NAJAS
FLEXILIS
(Willd.)
Rostk. & Schmidt
● 1930 onwards
○ Before 1930

A   581/2
•NAJAS
GRAMINEA
Del.
⊚Extinct

A   581/3 ✳
NAJAS
MARINA L.
● 1950 onwards

A + 582/1 ✳

ERIOCAULON
SEPTANGU-
LARE With.
Pipe-wort
● 1930 onwards
○ Before 1930

A   583/1

TOFIELDIA
PUSILLA
(Michx.)
Pers.
● 1930 onwards
○ Before 1930

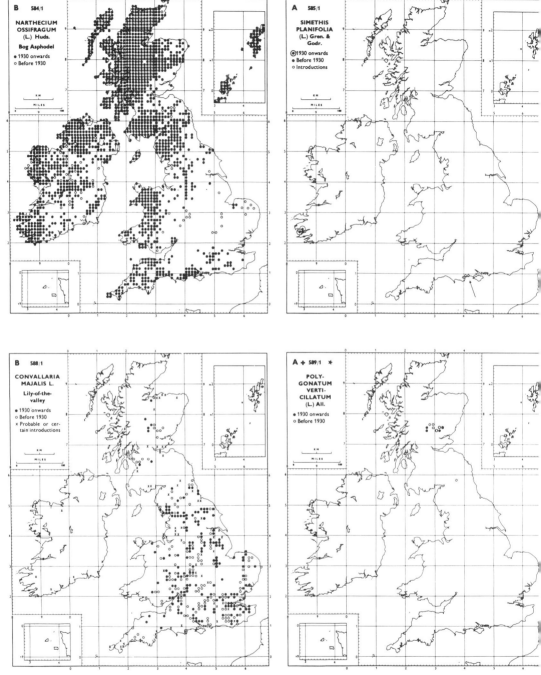

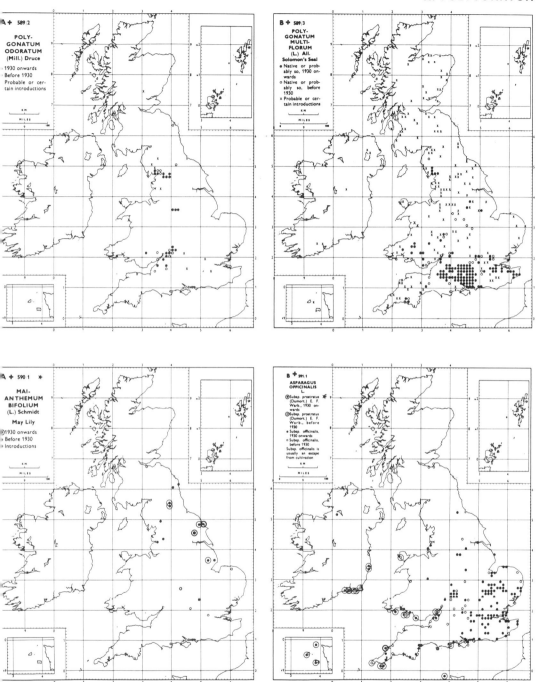

A ✛ 589/2

POLY-GONATUM ODORATUM (Mill.) Druce
- 1930 onwards
- Before 1930
- Probable or certain introductions

B ✛ 589/3

POLY-GONATUM MULTI-FLORUM (L.) All. Solomon's Seal
- Native or probably so, 1930 onwards
- Native or probably so, before 1930
- × Probable or certain introductions

A ✛ 590/1 ✳

MAI-ANTHEMUM BIFOLIUM (L.) Schmidt May Lily
- 1930 onwards
- Before 1930
- Introductions

B ✛ 591/1

ASPARAGUS OFFICINALIS L.
- Subsp. prostratus (Dumort.) E. F. Warb., 1930 onwards
- Subsp. prostratus (Dumort.) E. F. Warb., before 1930
- Subsp. officinalis, 1930 onwards
- Subsp. officinalis, before 1930
  Subsp. officinalis is usually an escape from cultivation

LILIACEAE

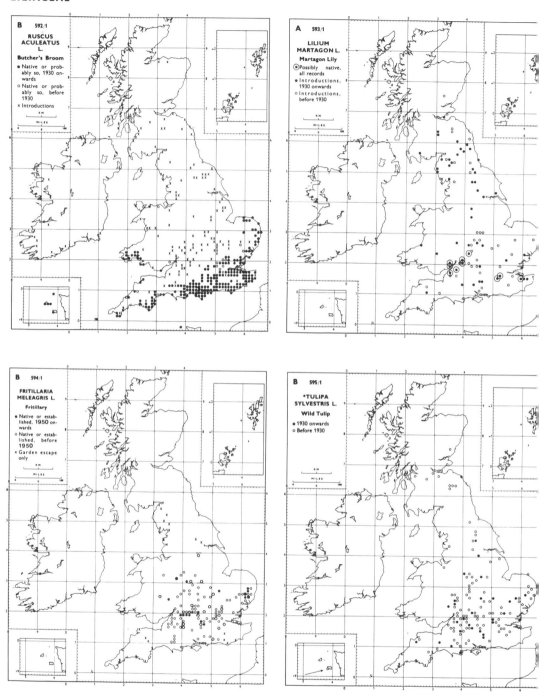

B 592/1
RUSCUS ACULEATUS L.
Butcher's Broom
- Native or probably so, 1930 onwards
- Native or probably so, before 1930
- x Introductions

A 593/1
LILIUM MARTAGON L.
Martagon Lily
- Possibly native, all records
- Introductions, 1930 onwards
- Introductions, before 1930

B 594/1
FRITILLARIA MELEAGRIS L.
Fritillary
- Native or established, 1950 onwards
- Native or established, before 1950
- x Garden escape only

B 595/1
TULIPA SYLVESTRIS L.
Wild Tulip
- 1930 onwards
- Before 1930

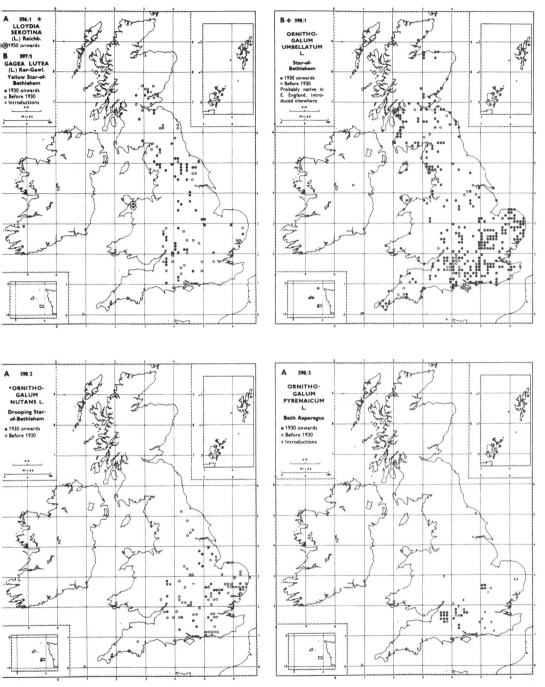

A 596/1 ✶
LLOYDIA
SEROTINA
(L.) Reichb.
⊙1950 onwards

B 597/1
GAGEA LUTEA
(L.) Ker-Gawl.
Yellow Star-of-
Bethlehem

● 1930 onwards
o Before 1930
× Introductions

B ✛ 598/1
ORNITHO-
GALUM
UMBELLATUM
L
Star-of-
Bethlehem

● 1930 onwards
o Before 1930
Probably native in
E. England. Intro-
duced elsewhere

A 598/2
*ORNITHO-
GALUM
NUTANS L.
Drooping Star-
of-Bethlehem

● 1930 onwards
o Before 1930

A 598/3
ORNITHO-
GALUM
PYRENAICUM
L
Bath Asparagus

● 1930 onwards
o Before 1930
× Introductions

LILIACEAE

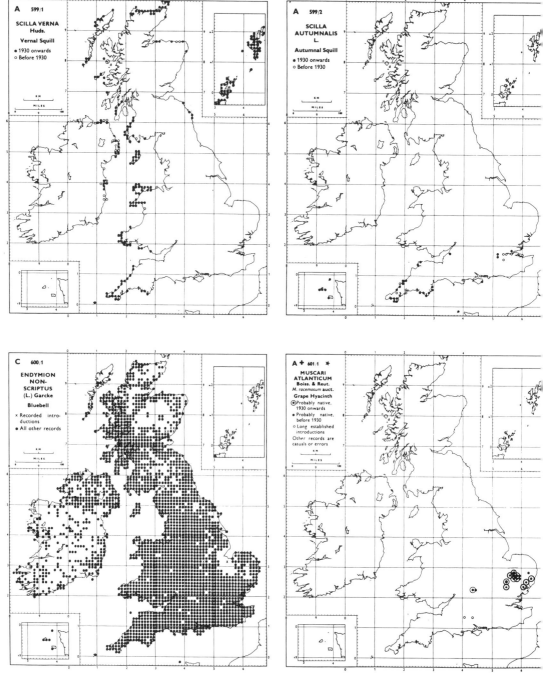

A 599/1
SCILLA VERNA
Huds.
Vernal Squill
● 1930 onwards
○ Before 1930

A 599/2
SCILLA AUTUMNALIS
L.
Autumnal Squill
● 1930 onwards
○ Before 1930

C 600/1
ENDYMION NON-SCRIPTUS
(L.) Garcke
Bluebell
× Recorded introductions
● All other records

A ✛ 601/1 ✱
MUSCARI ATLANTICUM
Boiss. & Reut.
M. racemosum auct.
Grape Hyacinth
⊕ Probably native, 1930 onwards
● Probably native, before 1930
○ Long established introductions
Other records are casuals or errors

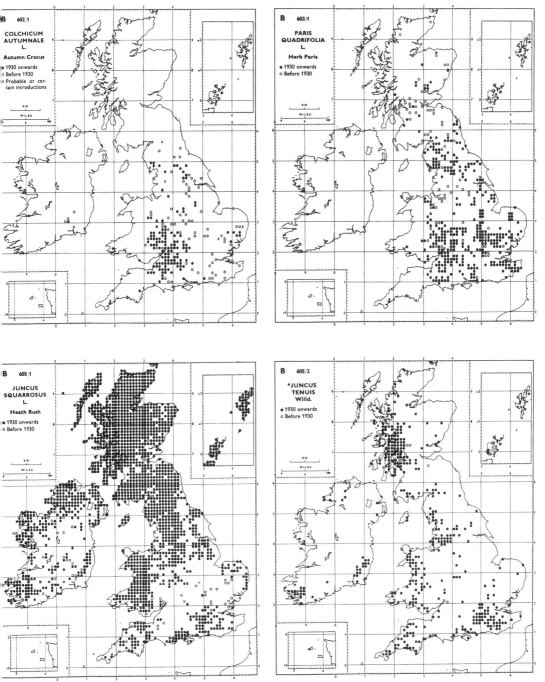

B 602/1

COLCHICUM
AUTUMNALE
L.

Autumn Crocus

• 1930 onwards
○ Before 1930
× Probable or certain introductions

B 603/1

PARIS
QUADRIFOLIA
L.

Herb Paris

• 1930 onwards
○ Before 1930

B 605/1

JUNCUS
SQUARROSUS
L.

Heath Rush

• 1930 onwards
○ Before 1930

B 605/2

*JUNCUS
TENUIS
Willd.

• 1930 onwards
○ Before 1930

317

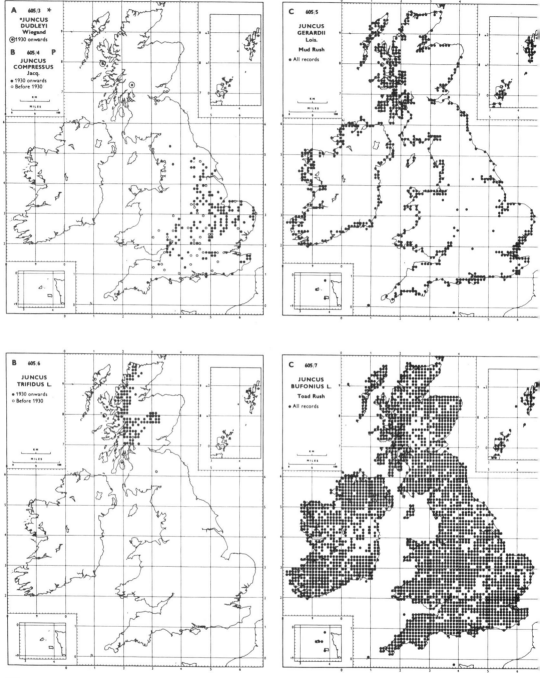

A 605/3 *
*JUNCUS
DUDLEYI
Wiegand
⊙ 1930 onwards

B 605/4 P
JUNCUS
COMPRESSUS
Jacq.
● 1930 onwards
○ Before 1930

C 605/5
JUNCUS
GERARDII
Lois.
Mud Rush
● All records

B 605/6
JUNCUS
TRIFIDUS L.
● 1930 onwards
○ Before 1930

C 605/7
JUNCUS
BUFONIUS L.
Toad Rush
● All records

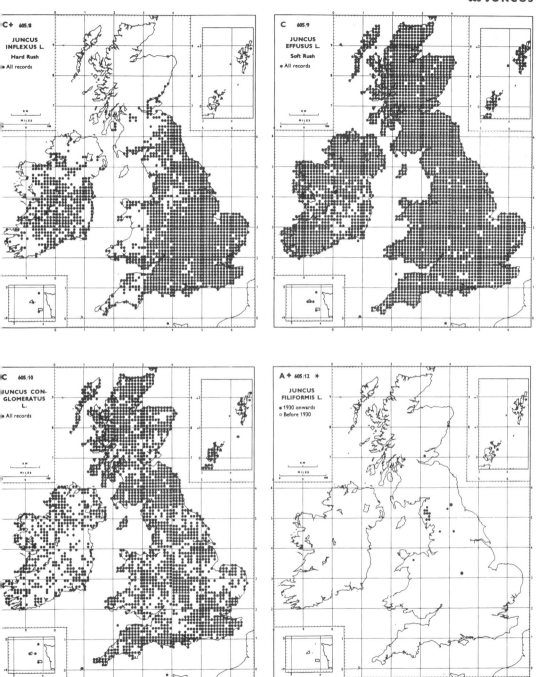

C+ 605/8
JUNCUS
INFLEXUS L.
Hard Rush
• All records

C 605/9
JUNCUS
EFFUSUS L.
Soft Rush
• All records

C 605/10
JUNCUS CON-
GLOMERATUS
L.
• All records

A+ 605/12 ✱
JUNCUS
FILIFORMIS L.
• 1930 onwards
○ Before 1930

319

JUNCACEAE

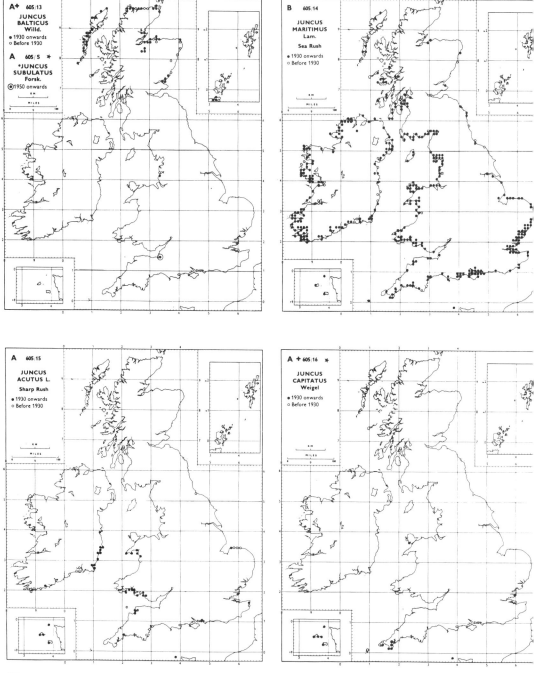

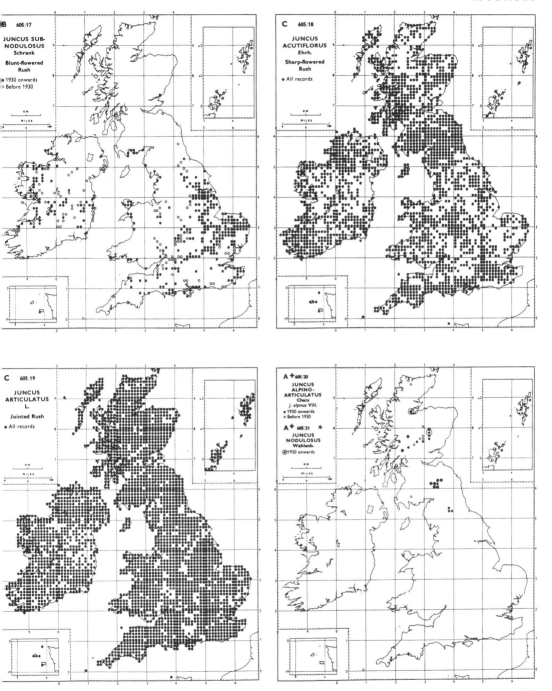

B 605/17

JUNCUS SUB-
NODULOSUS
Schrank

Blunt-flowered
Rush

● 1930 onwards
○ Before 1930

C 605/18

JUNCUS
ACUTIFLORUS
Ehrh.

Sharp-flowered
Rush

● All records

C 605/19

JUNCUS
ARTICULATUS
L.

Jointed Rush

● All records

A ✚ 605/20

JUNCUS
ALPINO-
ARTICULATUS
Chaix
J. alpinus Vill.

● 1930 onwards
○ Before 1930

A ✚ 605/21 ✱

JUNCUS
NODULOSUS
Wahlenb.

⊕1930 onwards

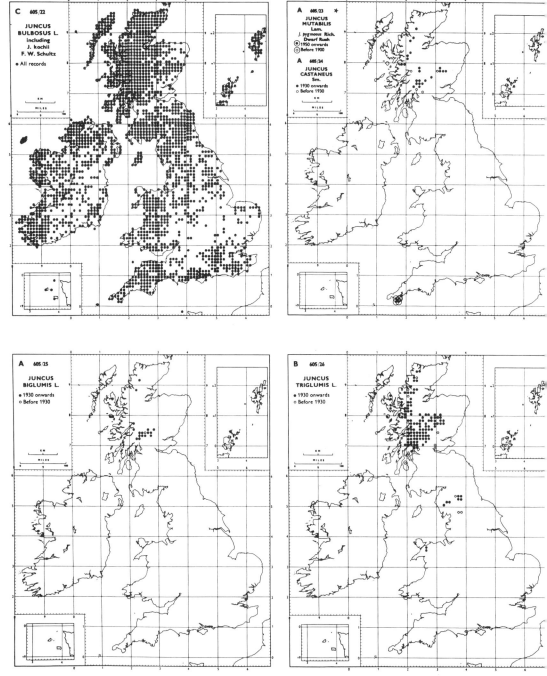

C 605/22

JUNCUS
BULBOSUS L.
including
J. kochii
F. W. Schultz

• All records

A 605/23 ✳

JUNCUS
MUTABILIS
Lam.
J. pygmaeus Rich.
Dwarf Rush
⊕ 1950 onwards
○ Before 1900

A 605/24

JUNCUS
CASTANEUS
Sm.
• 1930 onwards
○ Before 1930

A 605/25

JUNCUS
BIGLUMIS L.
• 1930 onwards
○ Before 1930

B 605/26

JUNCUS
TRIGLUMIS L.
• 1930 onwards
○ Before 1930

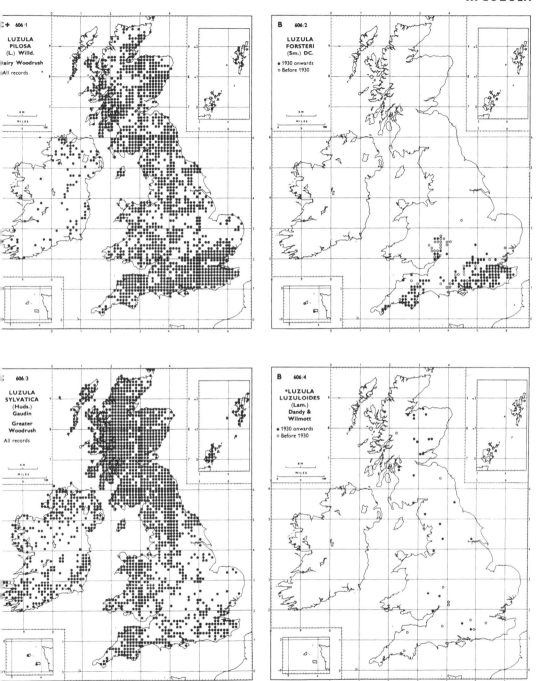

C + 606/1

**LUZULA
PILOSA**
(L.) Willd.
Hairy Woodrush
All records

KM
MILES

B 606/2

**LUZULA
FORSTERI**
(Sm.) DC.

● 1930 onwards
○ Before 1930

KM
MILES

C 606/3

**LUZULA
SYLVATICA**
(Huds.)
Gaudin

**Greater
Woodrush**
All records

KM
MILES

B 606/4

*LUZULA
LUZULOIDES
(Lam.)
Dandy &
Wilmott

● 1930 onwards
○ Before 1930

KM
MILES

323

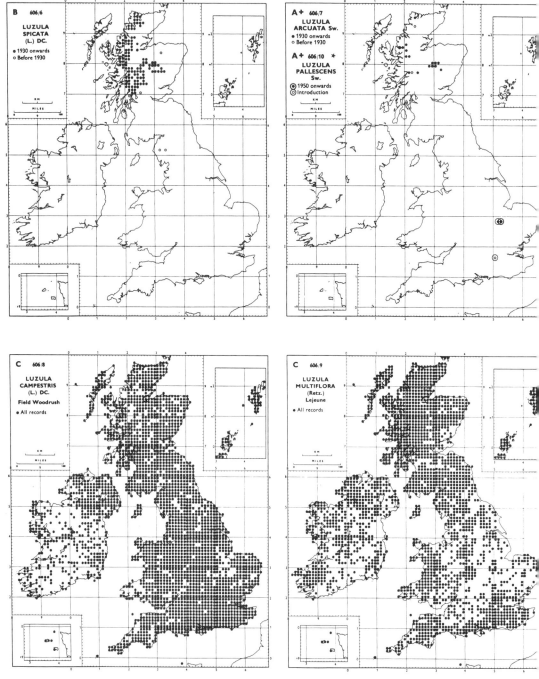

B 606/6
LUZULA
SPICATA
(L.) DC.
• 1930 onwards
o Before 1930

A+ 606/7
LUZULA
ARCUATA Sw.
• 1930 onwards
o Before 1930

A+ 606/10 ✳
LUZULA
PALLESCENS
Sw.
⊙ 1950 onwards
⊙ Introduction

C 606/8
LUZULA
CAMPESTRIS
(L.) DC.
Field Woodrush
• All records

C 606/9
LUZULA
MULTIFLORA
(Retz.)
Lejeune
• All records

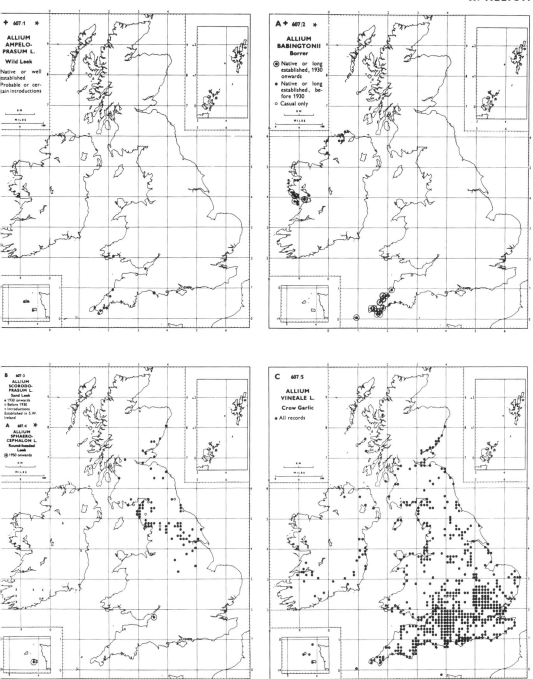

607/1

ALLIUM
AMPELO-
PRASUM L.

Wild Leek

● Native or well
established
◉ Probable or cer-
tain introductions

A 607/2

ALLIUM
BABINGTONII
Borrer

◉ Native or long
established, 1930
onwards
● Native or long
established, be-
fore 1930
○ Casual only

B 607/3

ALLIUM
SCORODO-
PRASUM L.
Sand Leek

● 1930 onwards
□ Before 1930
× Introductions
Established in S.W.
Ireland

A 607/4

ALLIUM
SPHAERO-
CEPHALON L.
Round-headed
Leek

⊕ 1950 onwards

C 607/5

ALLIUM
VINEALE L.

Crow Garlic

● All records

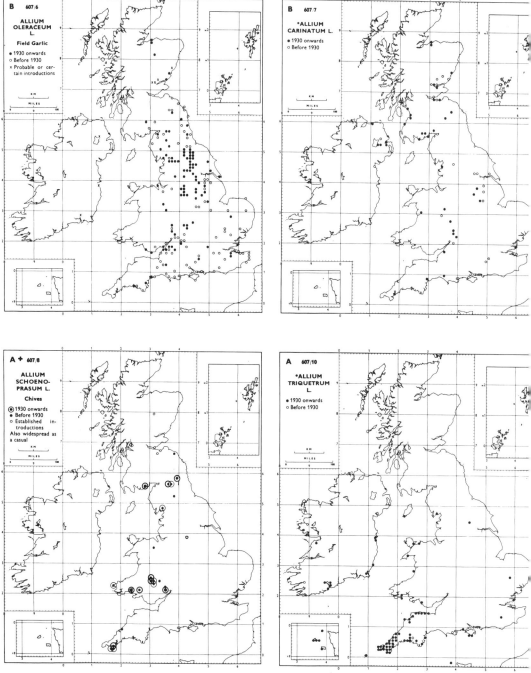

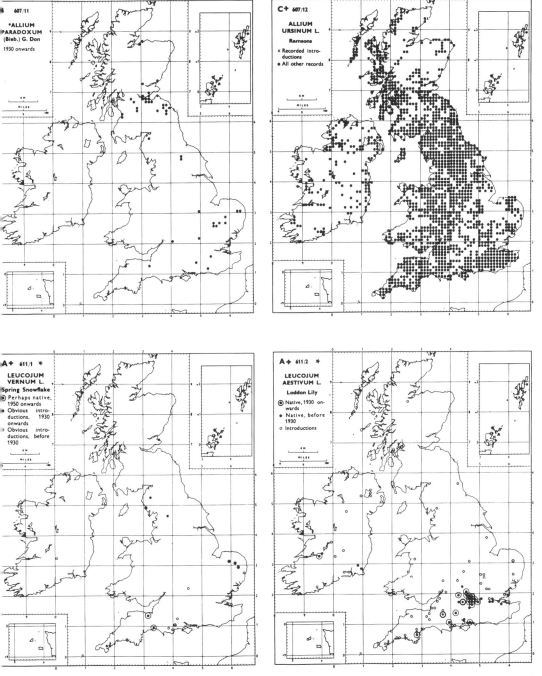

607/11
*ALLIUM
PARADOXUM
(Bieb.) G. Don
1930 onwards

C+ 607/12
ALLIUM
URSINUM L.
Ramsons
× Recorded intro-
ductions
● All other records

A+ 611/1 ✳
LEUCOJUM
VERNUM L.
Spring Snowflake
◉ Perhaps native,
1950 onwards
● Obvious intro-
ductions, 1930
onwards
○ Obvious intro-
ductions, before
1930

A+ 611/2 ✳
LEUCOJUM
AESTIVUM L.
Loddon Lily
◉ Native, 1930 on-
wards
● Native, before
1930
○ Introductions

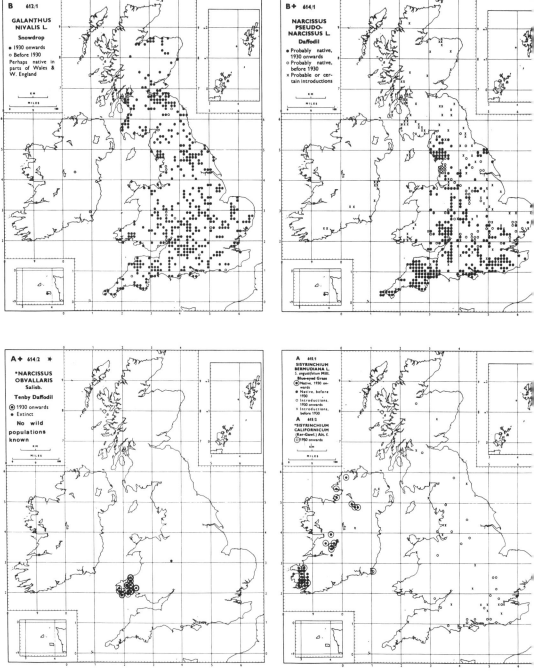

B 612/1

GALANTHUS
NIVALIS L.

Snowdrop

● 1930 onwards
○ Before 1930
Perhaps native in parts of Wales & W. England

KM
MILES

B+ 614/1

NARCISSUS
PSEUDO-
NARCISSUS L.

Daffodil

● Probably native, 1930 onwards
○ Probably native, before 1930
× Probable or certain introductions

KM
MILES

A+ 614/2 ✱

*NARCISSUS
OBVALLARIS
Salisb.

Tenby Daffodil

⊙ 1930 onwards
● Extinct

No wild populations known

KM
MILES

A 615/1

SISYRINCHIUM
BERMUDIANA L.
S. angustifolium Mill.

Blue-eyed Grass

⊙ Native, 1930 onwards
● Native, before 1930
○ Introductions, 1930 onwards
× Introductions, before 1930

A 615/2

*SISYRINCHIUM
CALIFORNICUM
(Ker-Gawl.) Ait. f.

⊙ 1950 onwards

KM
MILES

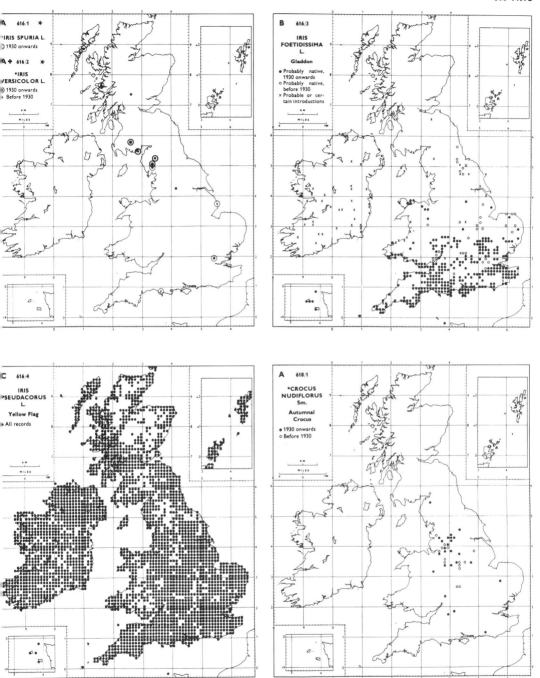

A 616/1 ✳
IRIS SPURIA L.
⦶ 1930 onwards

A 616/2 ✳
*IRIS
VERSICOLOR L.
⦶ 1930 onwards
● Before 1930

B 616/3
IRIS
FOETIDISSIMA
L.
Gladdon
● Probably native,
1930 onwards
○ Probably native,
before 1930
× Probable or cer-
tain introductions

C 616/4
IRIS
PSEUDACORUS
L.
Yellow Flag
● All records

A 618/1
*CROCUS
NUDIFLORUS
Sm.
Autumnal
Crocus
● 1930 onwards
○ Before 1930

329

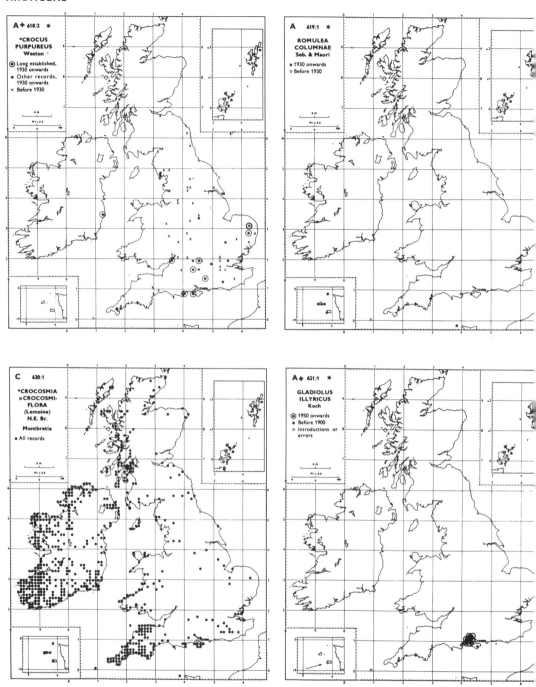

A + 618/2 ✷

*CROCUS
PURPUREUS
Weston ·

⊚ Long established,
1930 onwards
● Other records,
1930 onwards
× Before 1930

A 619/1 ✷

ROMULEA
COLUMNAE
Seb. & Mauri

● 1930 onwards
○ Before 1930

C 620/1

*CROCOSMIA
×CROCOSMI-
FLORA
(Lemoine)
N.E. Br.

Montbretia

● All records

A + 621/1 ✷

GLADIOLUS
ILLYRICUS
Koch

⊚ 1950 onwards
● Before 1900
○ Introductions or
errors

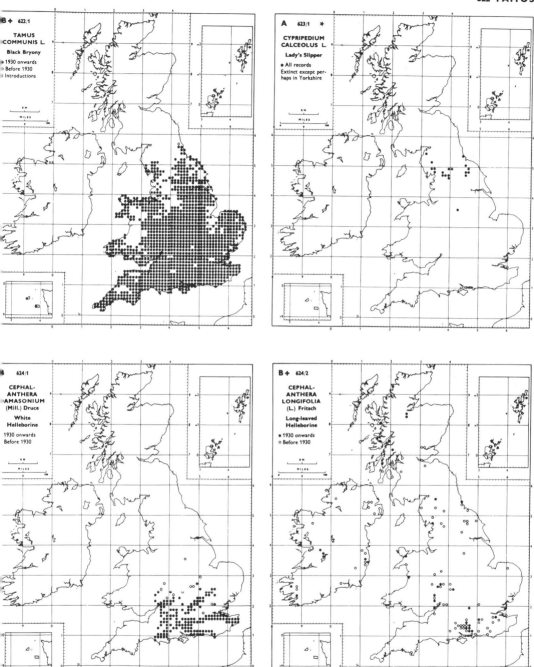

B + 622/1

TAMUS
COMMUNIS L.

Black Bryony

● 1930 onwards
○ Before 1930
⊕ Introductions

KM
MILES

A 623/1 ✳

CYPRIPEDIUM
CALCEOLUS L.

Lady's Slipper

● All records
Extinct except per-
haps in Yorkshire

KM
MILES

624/1

CEPHAL-
ANTHERA
AMASONIUM
(Mill.) Druce

White
Helleborine

1930 onwards
Before 1930

KM
MILES

B + 624/2

CEPHAL-
ANTHERA
LONGIFOLIA
(L.) Fritsch

Long-leaved
Helleborine

● 1930 onwards
○ Before 1930

KM
MILES

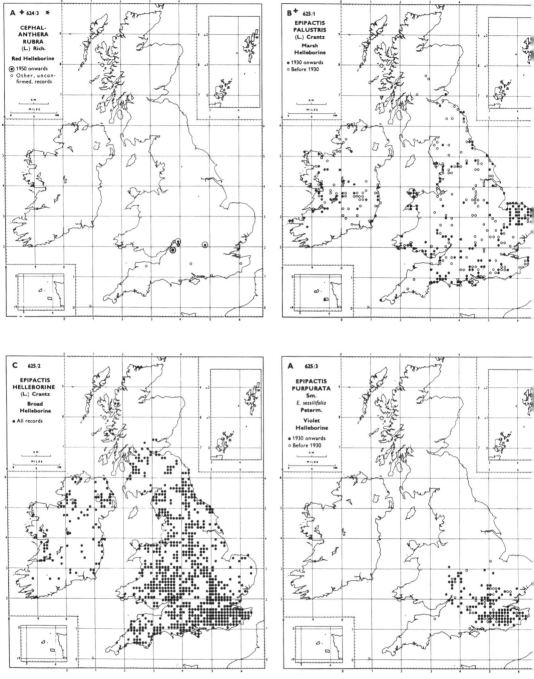

A ✛ 624/3 ✳

CEPHAL-
ANTHERA
RUBRA
(L.) Rich.

Red Helleborine

◉ 1950 onwards
○ Other, uncon-
firmed, records

B✛ 625/1

EPIPACTIS
PALUSTRIS
(L.) Crantz

Marsh
Helleborine

● 1930 onwards
○ Before 1930

C 625/2

EPIPACTIS
HELLEBORINE
(L.) Crantz

Broad
Helleborine

● All records

A 625/3

EPIPACTIS
PURPURATA
Sm.
E. sessilifolia
Peterm.

Violet
Helleborine

● 1930 onwards
○ Before 1930

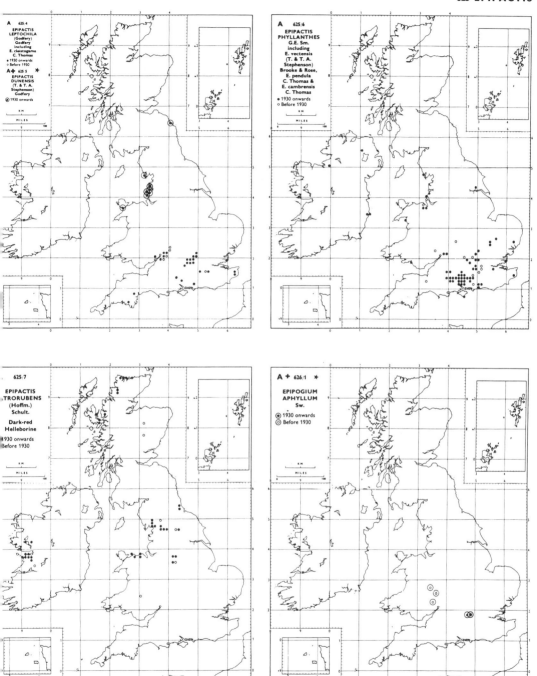

A 625·4
EPIPACTIS
LEPTOCHILA
(Godfery)
Godfery
including
E. cleistogama
C. Thomas
● 1930 onwards
○ Before 1930

A+ 625·5
EPIPACTIS
DUNENSIS
(T. & T. A.
Stephenson)
Godfery
◉ 1930 onwards

A 625·6
EPIPACTIS
PHYLLANTHES
G.E. Sm.
including
E. vectensis
(T. & T. A.
Stephenson)
Brooke & Rose,
E. pendula
C. Thomas &
E. cambrensis
C. Thomas
● 1930 onwards
○ Before 1930

625·7
EPIPACTIS
ATRORUBENS
(Hoffm.)
Schult.

Dark-red
Helleborine

1930 onwards
Before 1930

A+ 626/1 ✴
EPIPOGIUM
APHYLLUM
Sw.
◉ 1930 onwards
◎ Before 1930

333

ORCHIDACEAE

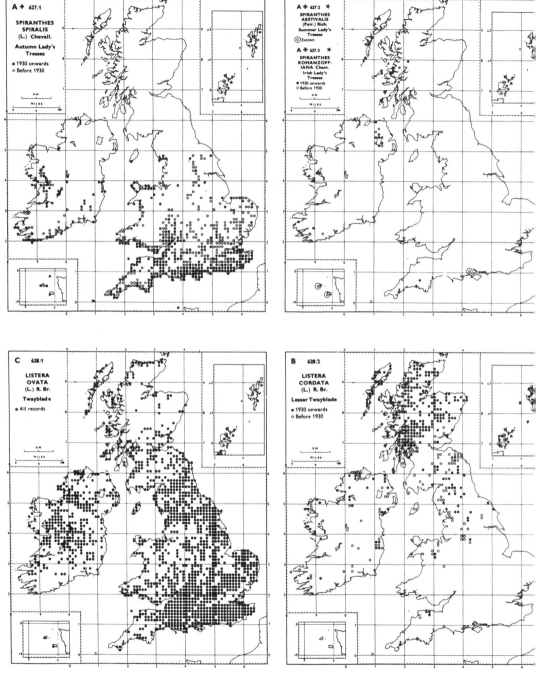

A ✛ 627/1

SPIRANTHES
SPIRALIS
(L.) Chevall.

Autumn Lady's
Tresses

● 1930 onwards
○ Before 1930

A ✛ 627/2 ✶

SPIRANTHES
AESTIVALIS
(Poir.) Rich.
Summer Lady's
Tresses
◎ Extinct

A ✛ 627/3 ✶

SPIRANTHES
ROMANZOFF-
IANA Cham.
Irish Lady's
Tresses

● 1930 onwards
○ Before 1930

C 628/1

LISTERA
OVATA
(L.) R. Br.

Twayblade

● All records

B 628/2

LISTERA
CORDATA
(L.) R. Br.

Lesser Twayblade

● 1930 onwards
○ Before 1930

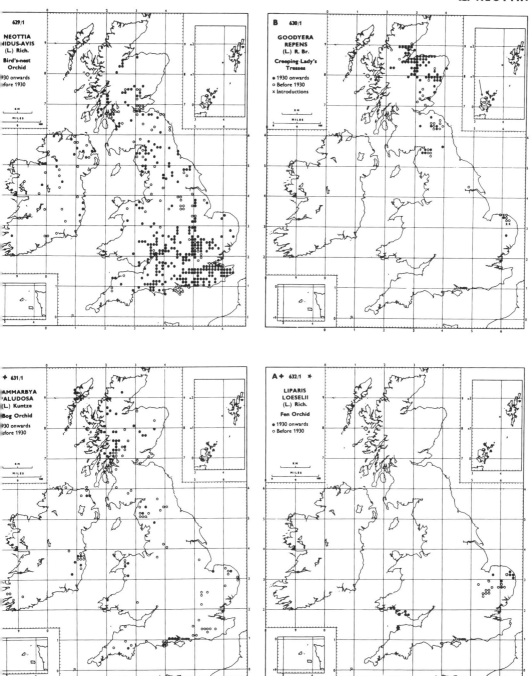

629/1

NEOTTIA
NIDUS-AVIS
(L.) Rich.

Bird's-nest
Orchid

● 1930 onwards
○ Before 1930

B   630/1

GOODYERA
REPENS
(L.) R. Br.

Creeping Lady's
Tresses

● 1930 onwards
○ Before 1930
× Introductions

631/1

HAMMARBYA
PALUDOSA
(L.) Kuntze

Bog Orchid

● 1930 onwards
○ Before 1930

A+   632/1   ✳

LIPARIS
LOESELII
(L.) Rich.

Fen Orchid

● 1930 onwards
○ Before 1930

335

ORCHIDACEAE

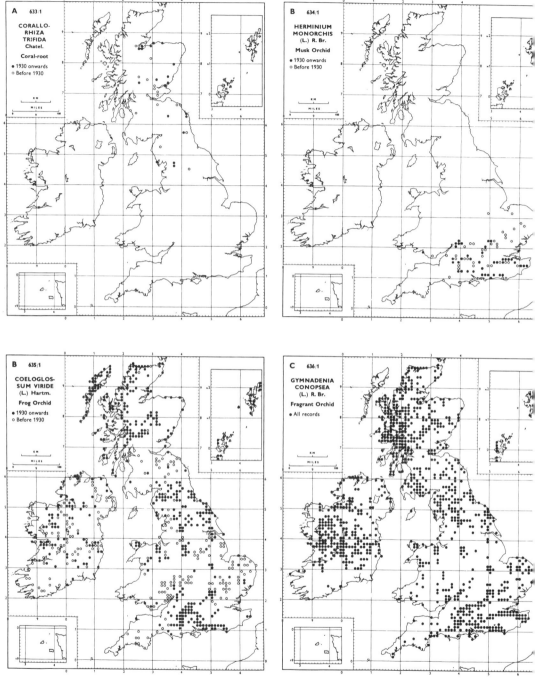

A 633/1

CORALLO-
RHIZA
TRIFIDA Chatel.

Coral-root

● 1930 onwards
○ Before 1930

B 634/1

HERMINIUM
MONORCHIS
(L.) R. Br.

Musk Orchid

● 1930 onwards
○ Before 1930

B 635/1

COELOGLOS-
SUM VIRIDE
(L.) Hartm.

Frog Orchid

● 1930 onwards
○ Before 1930

C 636/1

GYMNADENIA
CONOPSEA
(L.) R. Br.

Fragrant Orchid

● All records

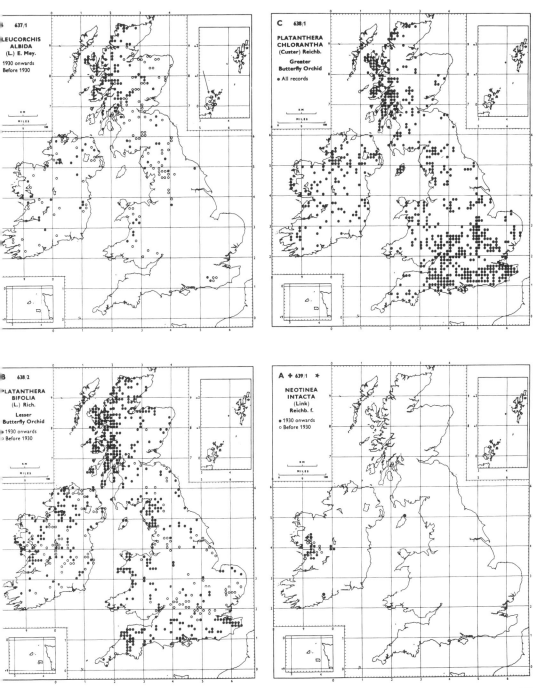

637/1
LEUCORCHIS
ALBIDA
(L.) E. Mey.
• 1930 onwards
○ Before 1930

C 638/1
PLATANTHERA
CHLORANTHA
(Custer) Reichb.
Greater
Butterfly Orchid
• All records

B 638/2
PLATANTHERA
BIFOLIA
(L.) Rich.
Lesser
Butterfly Orchid
• 1930 onwards
○ Before 1930

A ✢ 639/1 ✱
NEOTINEA
INTACTA
(Link)
Reichb. f.
• 1930 onwards
○ Before 1930

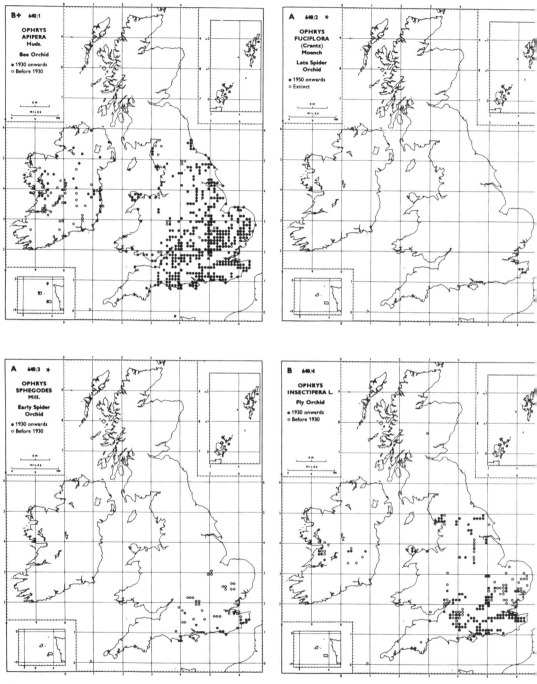

B+ 640/1
OPHRYS APIFERA Huds.
Bee Orchid
● 1930 onwards
○ Before 1930

A 640/2 ✳
OPHRYS FUCIFLORA (Crantz) Moench
Late Spider Orchid
● 1950 onwards
○ Extinct

A 640/3 ✳
OPHRYS SPHEGODES Mill.
Early Spider Orchid
● 1930 onwards
○ Before 1930

B 640/4
OPHRYS INSECTIFERA L.
Fly Orchid
● 1930 onwards
○ Before 1930

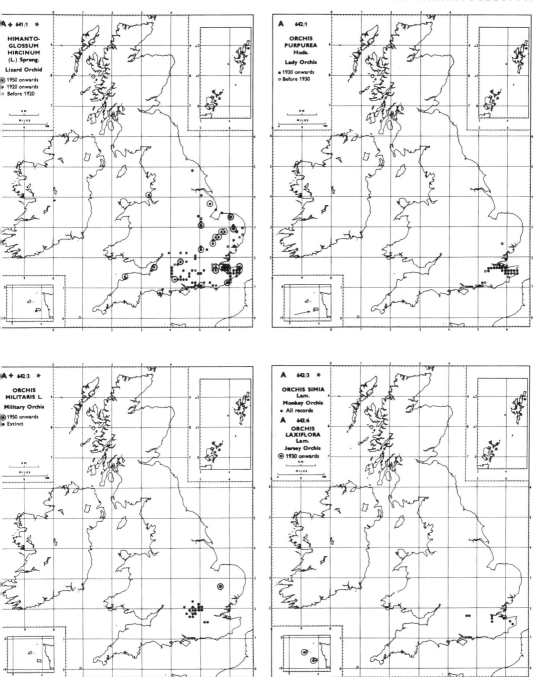

ORCHIDACEAE

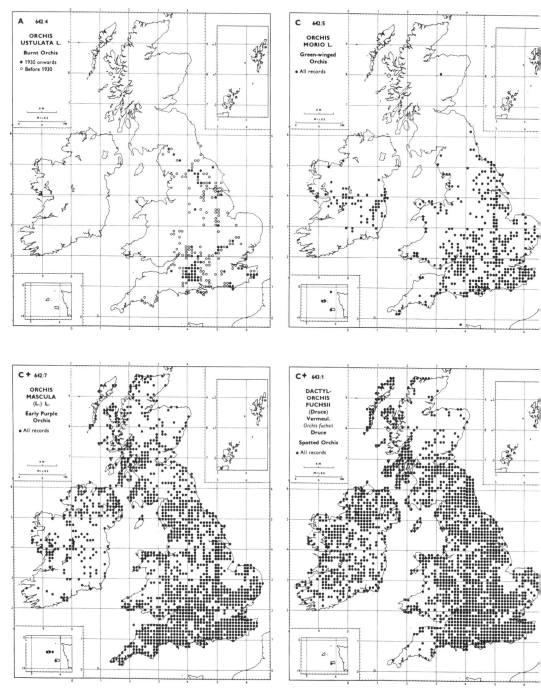

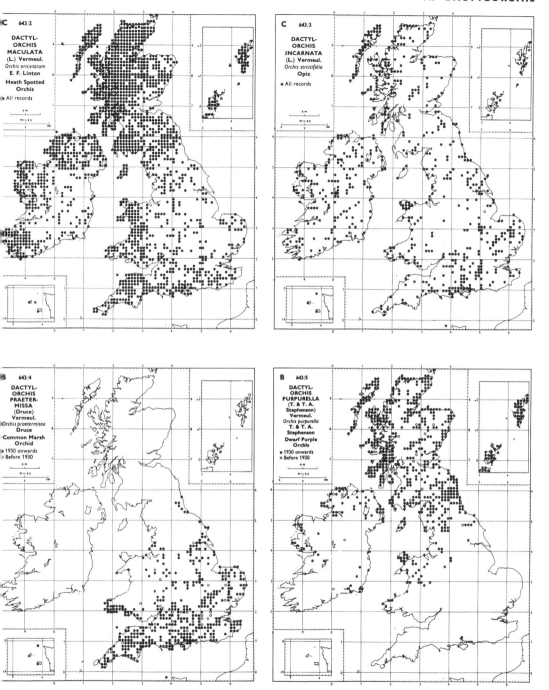

C 643/2

DACTYL-
ORCHIS
MACULATA
(L.) Vermeul.
Orchis ericetorum
E. F. Linton
Heath Spotted
Orchis
● All records

KM
MILES

C 643/3

DACTYL-
ORCHIS
INCARNATA
(L.) Vermeul.
Orchis strictifolia
Opiz
● All records

KM
MILES

B 643/4

DACTYL-
ORCHIS
PRAETER-
MISSA
(Druce)
Vermeul.
Orchis praetermissa
Druce
Common Marsh
Orchid
● 1930 onwards
○ Before 1930

KM
MILES

B 643/5

DACTYL-
ORCHIS
PURPURELLA
(T. & T. A.
Stephenson)
Vermeul.
Orchis purpurella
T. & T. A.
Stephenson
Dwarf Purple
Orchis
● 1930 onwards
○ Before 1930

KM
MILES

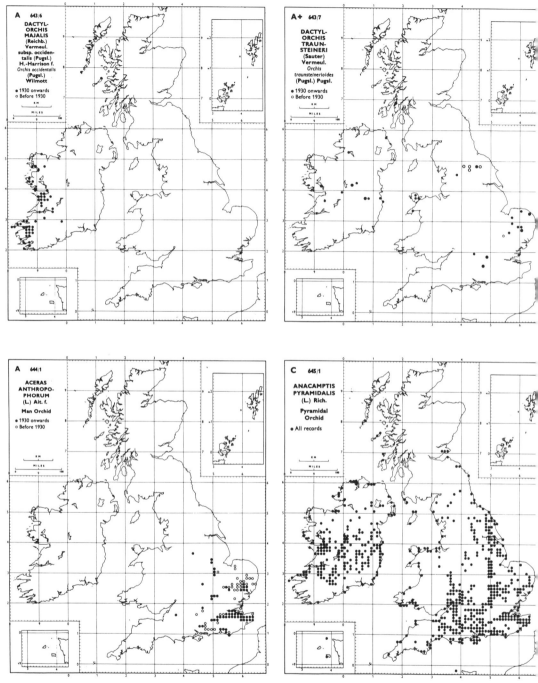

A 643/6
DACTYL-
ORCHIS
MAJALIS
(Reichb.)
Vermeul.
subsp. occiden-
talis (Pugsl.)
H.-Harrison f.
Orchis occidentalis
(Pugsl.)
Wilmott

● 1930 onwards
○ Before 1930

A+ 643/7
DACTYL-
ORCHIS
TRAUN-
STEINERI
(Sauter)
Vermeul.
Orchis
traunsteinerioides
(Pugsl.) Pugsl.

● 1930 onwards
○ Before 1930

A 644/1
ACERAS
ANTHROPO-
PHORUM
(L.) Ait. f.
Man Orchid

● 1930 onwards
○ Before 1930

C 645/1
ANACAMPTIS
PYRAMIDALIS
(L.) Rich.
Pyramidal
Orchid

● All records

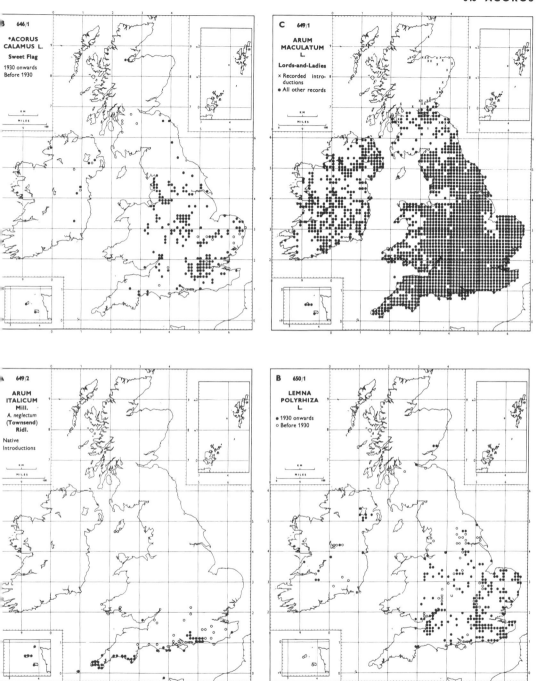

646/1

*ACORUS CALAMUS L.

Sweet Flag

• 1930 onwards
○ Before 1930

KM
MILES

C 649/1

ARUM MACULATUM L.

Lords-and-Ladies

× Recorded intro-
ductions
• All other records

KM
MILES

649/2

ARUM ITALICUM Mill.
A. neglectum
(Townsend) Ridl.

• Native
○ Introductions

KM
MILES

B 650/1

LEMNA POLYRHIZA L.

• 1930 onwards
○ Before 1930

KM
MILES

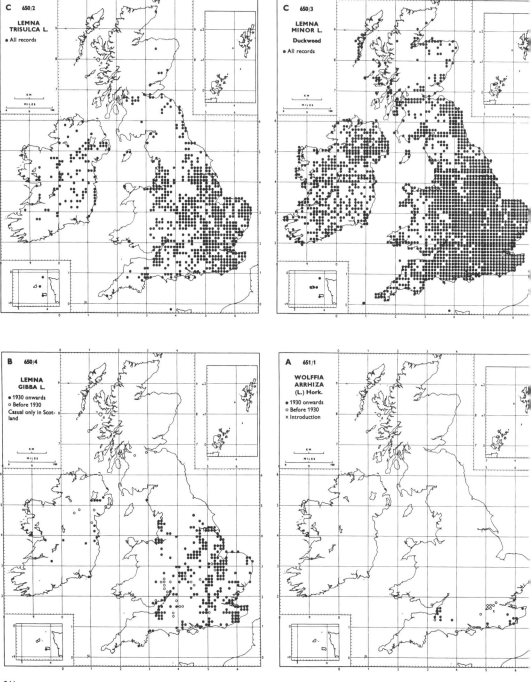

C 650/2
LEMNA
TRISULCA L.
• All records

C 650/3
LEMNA
MINOR L.
Duckweed
• All records

B 650/4
LEMNA
GIBBA L.
• 1930 onwards
○ Before 1930
Casual only in Scotland

A 651/1
WOLFFIA
ARRHIZA
(L.) Hork.
• 1930 onwards
○ Before 1930
× Introduction

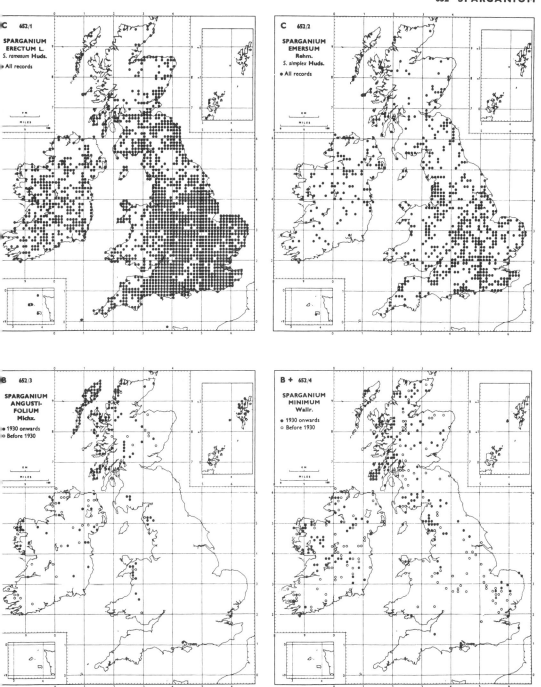

C 652/1

SPARGANIUM
ERECTUM L.
S. ramosum Huds.

● All records

C 652/2

SPARGANIUM
EMERSUM
Rehm.
S. simplex Huds.

● All records

B 652/3

SPARGANIUM
ANGUSTI-
FOLIUM
Michx.

● 1930 onwards
○ Before 1930

B + 652/4

SPARGANIUM
MINIMUM
Wallr.

● 1930 onwards
○ Before 1930

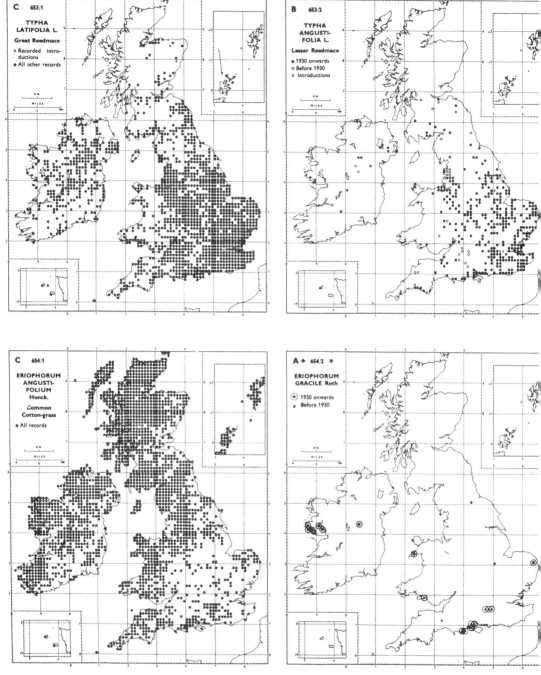

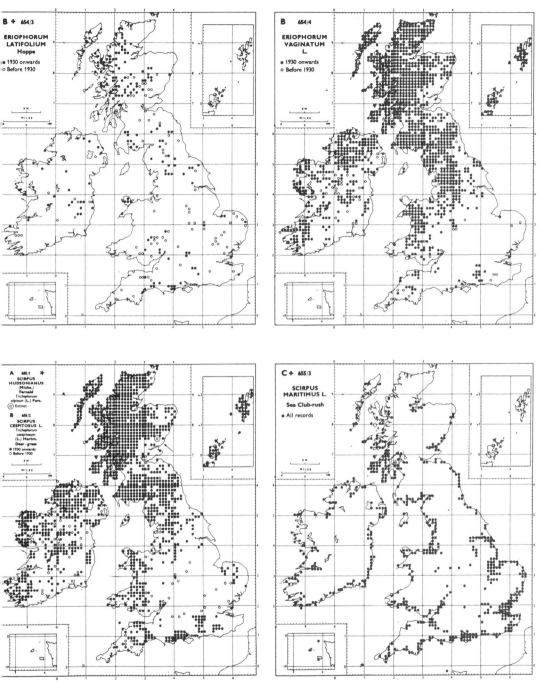

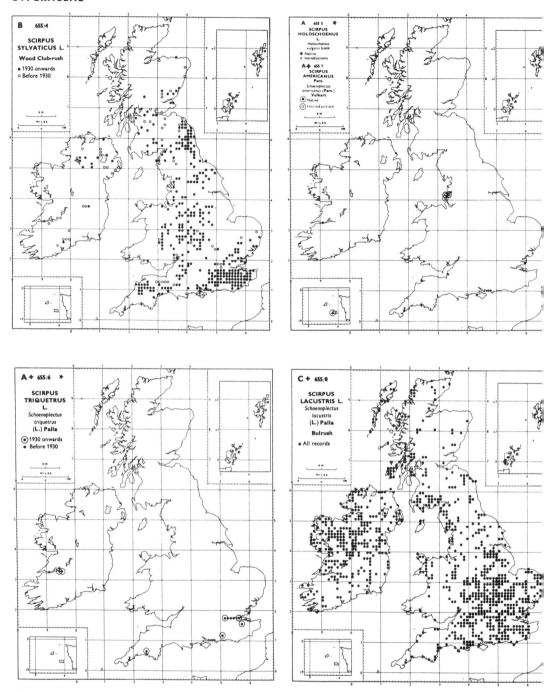

B 655/4

SCIRPUS
SYLVATICUS L.

Wood Club-rush

● 1930 onwards
○ Before 1930

A 655 5 ✳

SCIRPUS
HOLOSCHOENUS
L.
Holoschoenus
vulgaris Link
● Native
✕ Introductions

A 655 7
SCIRPUS
AMERICANUS
Pers.
Schoenoplectus
americanus (Pers.)
Volkart
⊙ Native
⊚ Introduction

A 655/6 ✳

SCIRPUS
TRIQUETRUS
L.
Schoenoplectus
triquetrus
(L.) Palla

⊙ 1930 onwards
● Before 1930

C 655/8

SCIRPUS
LACUSTRIS L.
Schoenoplectus
lacustris
(L.) Palla

Bulrush

● All records

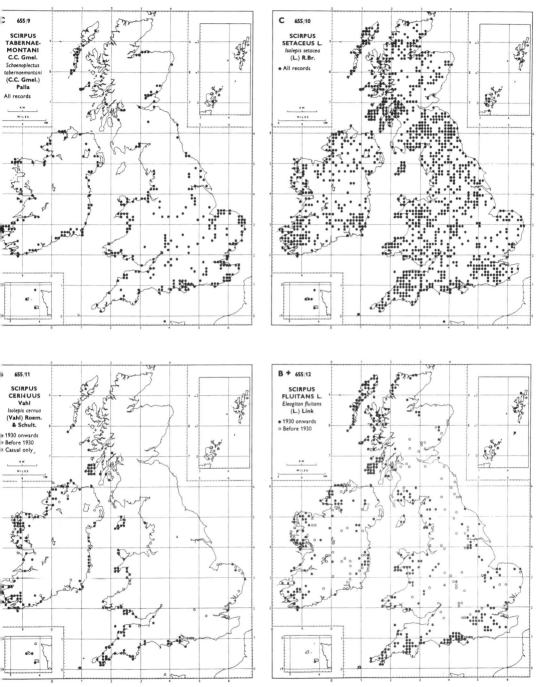

655/9

SCIRPUS
TABERNAE-
MONTANI
C.C. Gmel.
*Schoenoplectus
tabernaemontani
(C.C. Gmel.)
Palla*
All records

C 655/10

SCIRPUS
SETACEUS L.
*Isolepis setacea
(L.) R.Br.*
● All records

655/11

SCIRPUS
CERNUUS
Vahl
Isolepis cernua
(Vahl) Roem.
& Schult.
● 1930 onwards
○ Before 1930
x Casual only

B + 655/12

SCIRPUS
FLUITANS L.
Eleogiton fluitans
(L.) Link
● 1930 onwards
○ Before 1930

CYPERACEAE

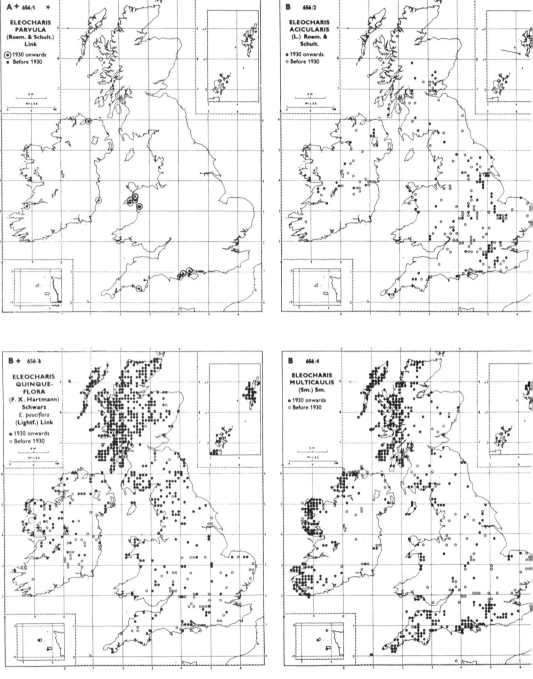

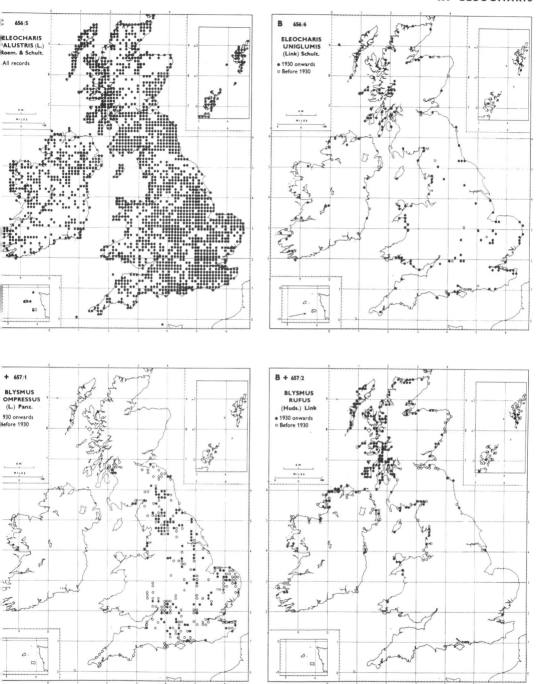

C 656/5

ELEOCHARIS
PALUSTRIS (L.)
Roem. & Schult.

All records

KM
MILES

B 656/6

ELEOCHARIS
UNIGLUMIS
(Link) Schult.

● 1930 onwards
○ Before 1930

KM
MILES

+ 657/1

BLYSMUS
COMPRESSUS
(L.) Panz.

1930 onwards
Before 1930

KM
MILES

B + 657/2

BLYSMUS
RUFUS
(Huds.) Link

● 1930 onwards
○ Before 1930

KM
MILES

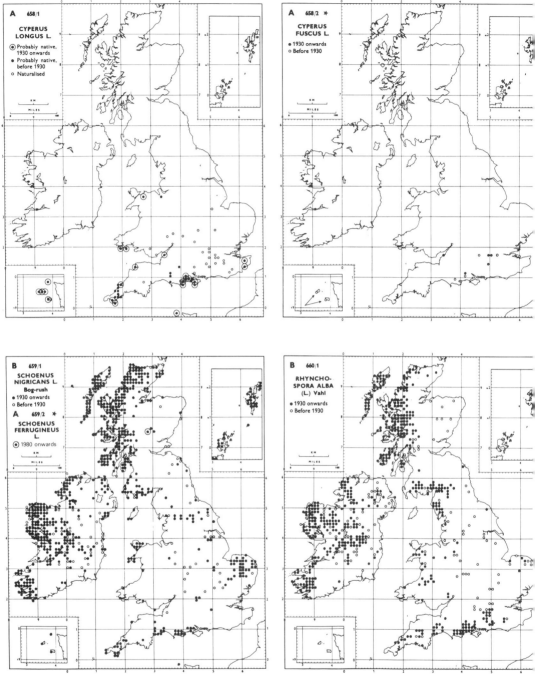

A 658/1

CYPERUS
LONGUS L.
◉ Probably native,
   1930 onwards
● Probably native,
   before 1930
○ Naturalised

A 658/2 ✳

CYPERUS
FUSCUS L.
● 1930 onwards
○ Before 1930

B 659/1

SCHOENUS
NIGRICANS L.
Bog-rush
● 1930 onwards
○ Before 1930

A 659/2 ✳

SCHOENUS
FERRUGINEUS
L.
◉ 1980 onwards

B 660/1

RHYNCHO-
SPORA ALBA
(L.) Vahl
● 1930 onwards
○ Before 1930

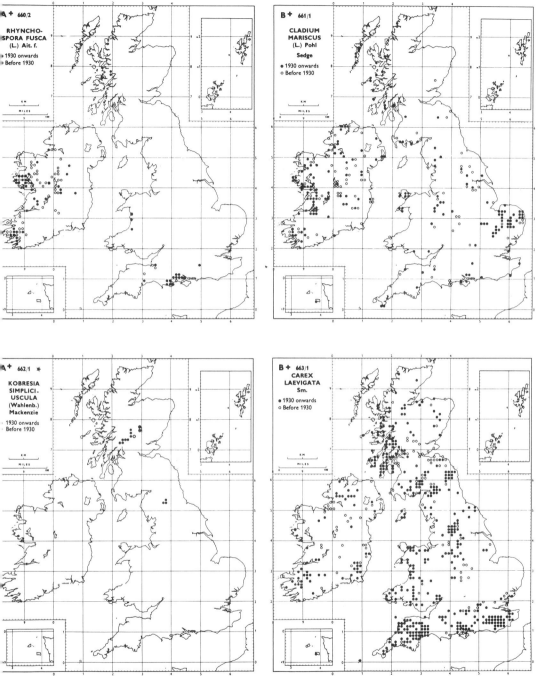

A + 660/2

**RHYNCHO-
SPORA FUSCA
(L.) Ait. f.**

● 1930 onwards
◐ Before 1930

B + 661/1

**CLADIUM
MARISCUS
(L.) Pohl**

Sedge

● 1930 onwards
○ Before 1930

A + 662/1 ✱

**KOBRESIA
SIMPLICI-
USCULA
(Wahlenb.)
Mackenzie**

1930 onwards
Before 1930

B + 663/1

**CAREX
LAEVIGATA
Sm.**

● 1930 onwards
○ Before 1930

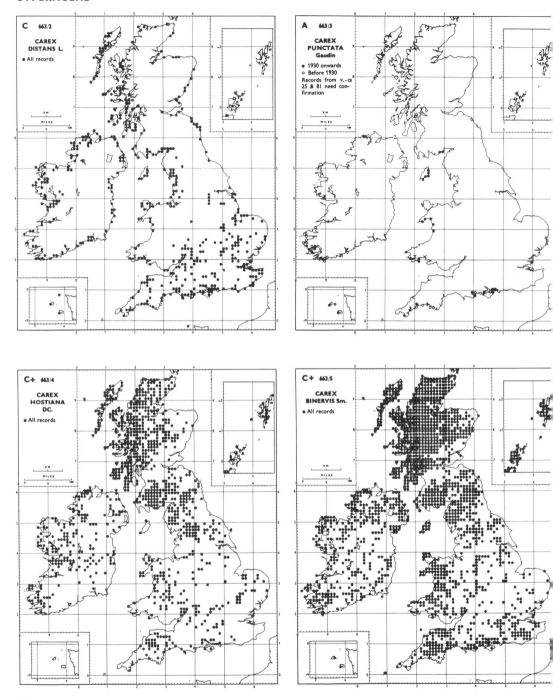

C 663/2
CAREX
DISTANS L.
• All records

A 663/3
CAREX
PUNCTATA
Gaudin
• 1930 onwards
○ Before 1930
Records from v.-cs
25 & 81 need con-
firmation

C+ 663/4
CAREX
HOSTIANA
DC.
• All records

C+ 663/5
CAREX
BINERVIS Sm.
• All records

354

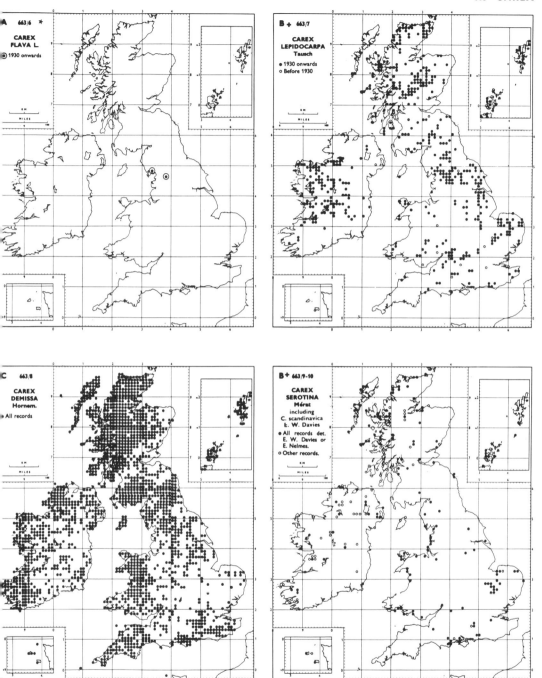

A 663/6 ✳

CAREX
FLAVA L.

⊙ 1930 onwards

B + 663/7

CAREX
LEPIDOCARPA
Tausch

● 1930 onwards
○ Before 1930

C 663/8

CAREX
DEMISSA
Hornem.

● All records

B + 663/9-10

CAREX
SEROTINA
Mérat
including
C. scandinavica
E. W. Davies

● All records det.
  E. W. Davies or
  E. Nelmes.
○ Other records.

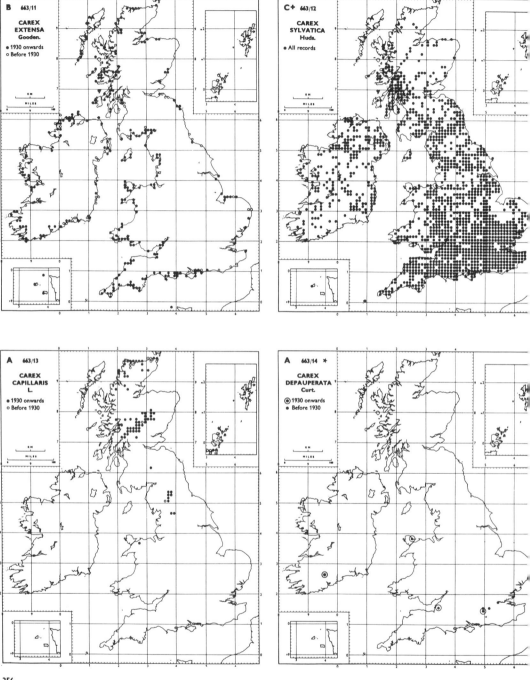

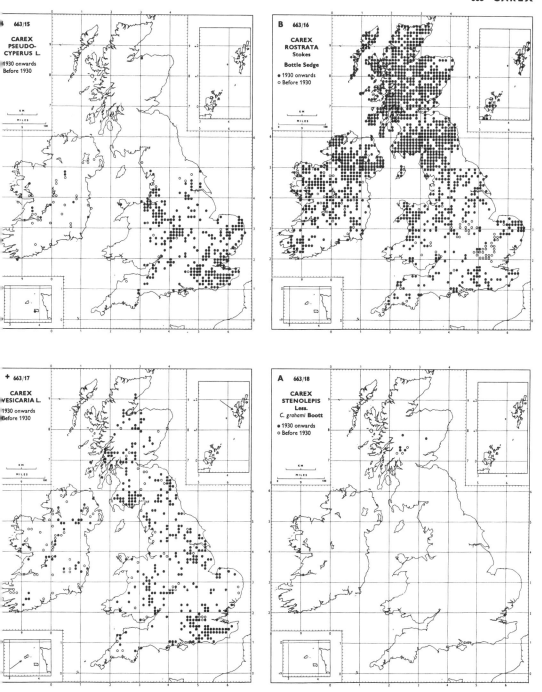

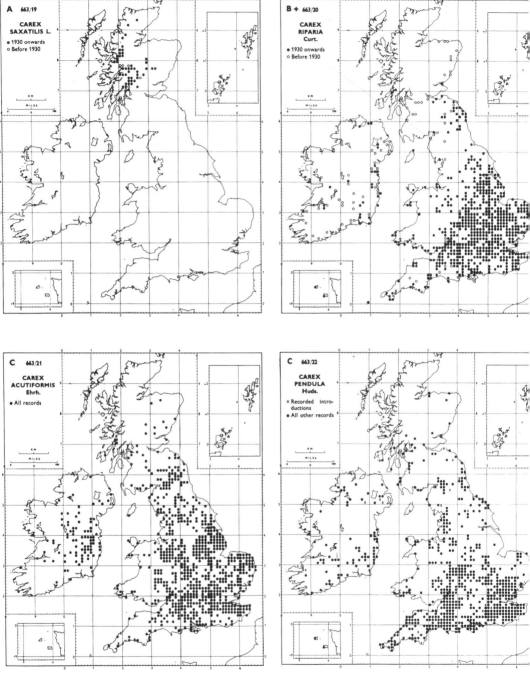

A 663/19
CAREX
SAXATILIS L.
● 1930 onwards
○ Before 1930

B + 663/20
CAREX
RIPARIA
Curt.
● 1930 onwards
○ Before 1930

C 663/21
CAREX
ACUTIFORMIS
Ehrh.
● All records

C 663/22
CAREX
PENDULA
Huds.
× Recorded intro-
ductions
● All other records

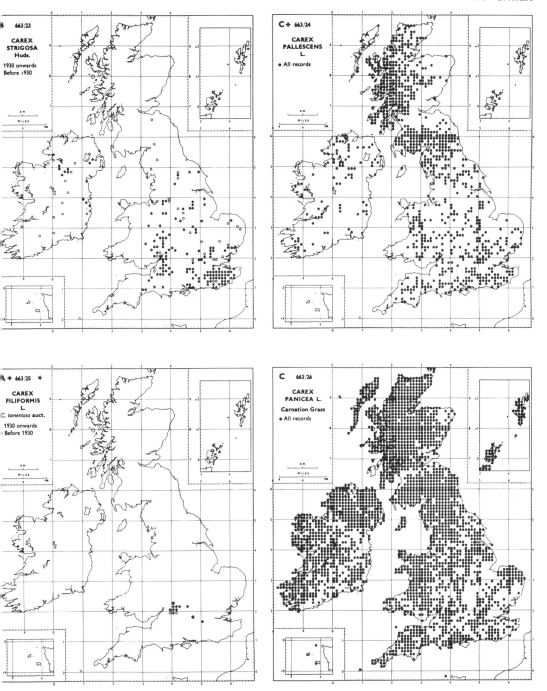

663/23
CAREX
STRIGOSA
Huds.
1930 onwards
Before 1930

C+ 663/24
CAREX
PALLESCENS
L.
• All records

▲+ 663/25 ✱
CAREX
FILIFORMIS
L.
C. tomentosa auct.
1930 onwards
Before 1930

C 663/26
CAREX
PANICEA L.
Carnation Grass
• All records

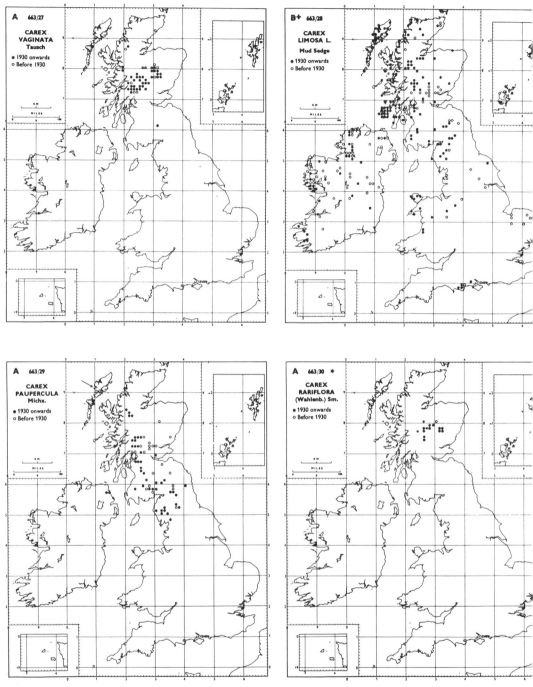

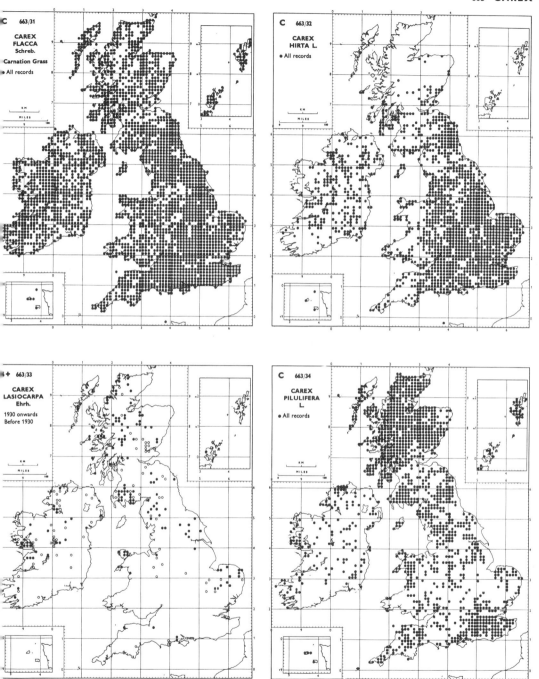

C 663/31
CAREX
FLACCA Schreb.
Carnation Grass
● All records

C 663/32
CAREX
HIRTA L.
● All records

+ 663/33
CAREX
LASIOCARPA Ehrh.
1930 onwards
Before 1930

C 663/34
CAREX
PILULIFERA L.
● All records

CYPERACEAE

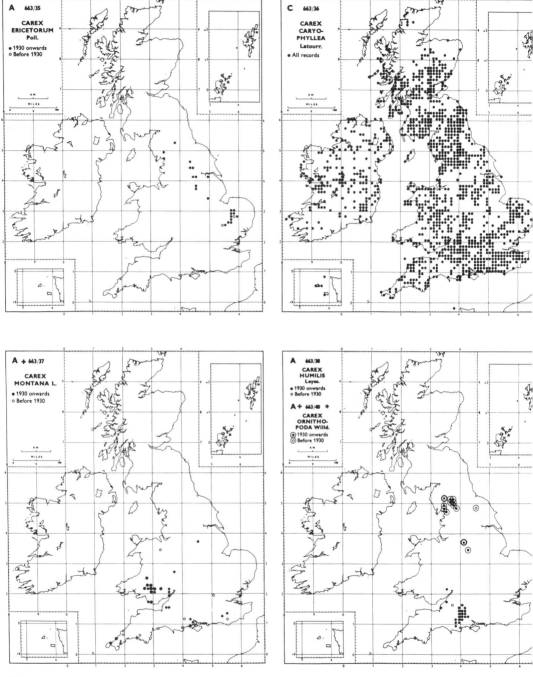

A 663/35
CAREX
ERICETORUM
Poll.
● 1930 onwards
○ Before 1930

C 663/36
CAREX
CARYO-
PHYLLEA
Latourr.
● All records

A + 663/37
CAREX
MONTANA L.
● 1930 onwards
○ Before 1930

A 663/38
CAREX
HUMILIS
Leyss.
● 1930 onwards
○ Before 1930

A + 663/40 ✳
CAREX
ORNITHO-
PODA Willd.
⊙ 1930 onwards
◎ Before 1930

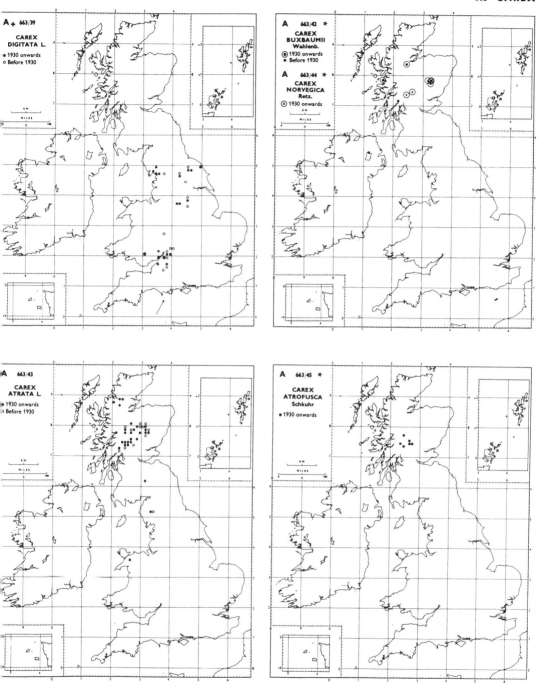

A + 663/39
CAREX
DIGITATA L.
● 1930 onwards
○ Before 1930

A 663/42 ✱
CAREX
BUXBAUMII
Wahlenb.
⊙ 1930 onwards
● Before 1930

A 663/44 ✱
CAREX
NORVEGICA
Retz.
⊗ 1930 onwards

A 663/43
CAREX
ATRATA L.
● 1930 onwards
○ Before 1930

A 663/45 ✱
CAREX
ATROFUSCA
Schkuhr
● 1930 onwards

CYPERACEAE

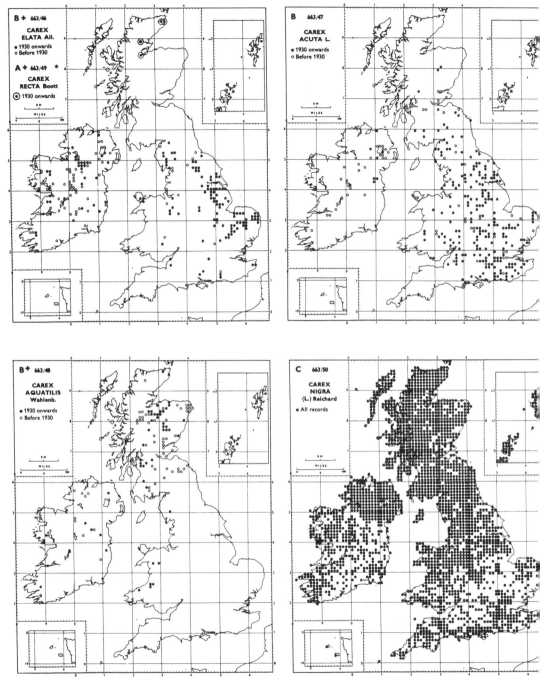

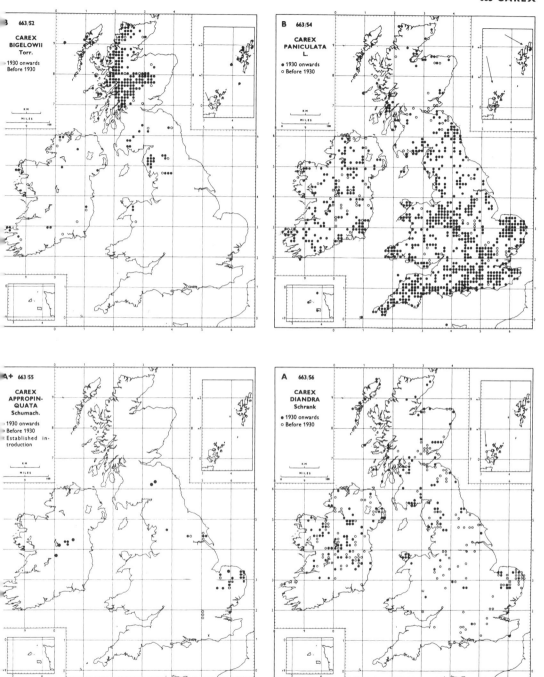

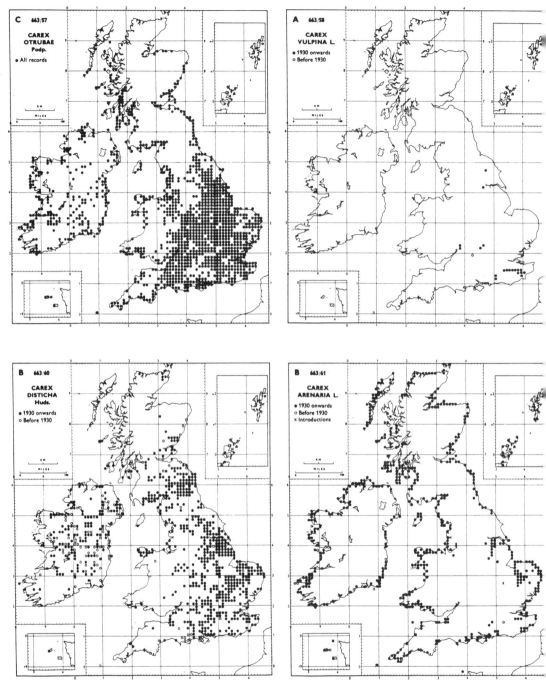

C 663/57
CAREX
OTRUBAE
Podp.
• All records

A 663/58
CAREX
VULPINA L.
• 1930 onwards
○ Before 1930

B 663/60
CAREX
DISTICHA
Huds.
• 1930 onwards
○ Before 1930

B 663/61
CAREX
ARENARIA L.
• 1930 onwards
○ Before 1930
× Introductions

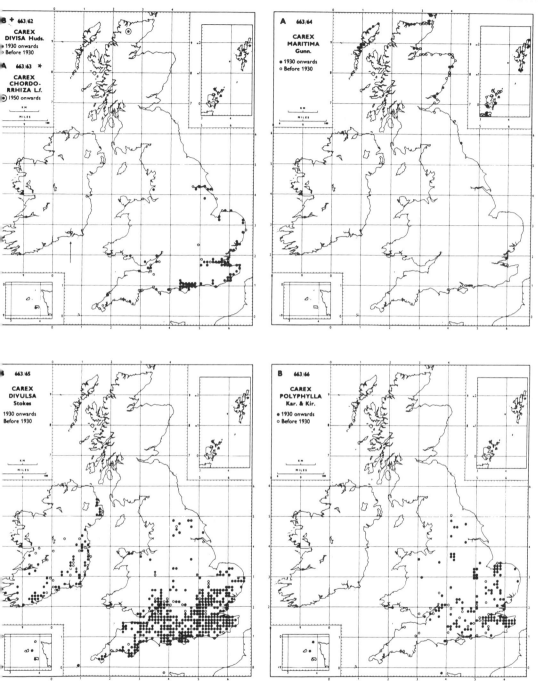

CYPERACEAE

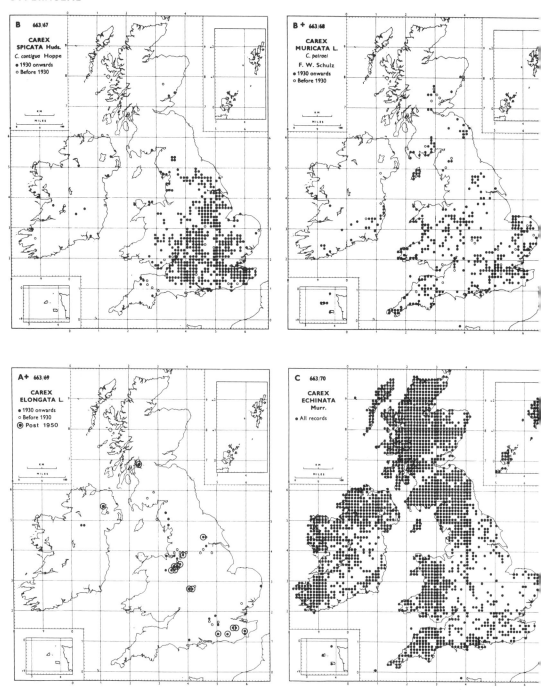

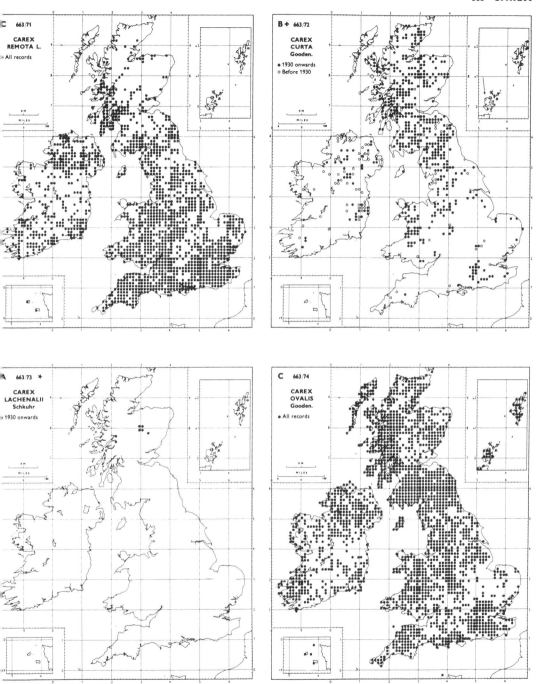

C 663/71
CAREX
REMOTA L.
• All records

B+ 663/72
CAREX
CURTA
Gooden.
• 1930 onwards
○ Before 1930

A 663/73 ＊
CAREX
LACHENALII
Schkuhr
• 1930 onwards

C 663/74
CAREX
OVALIS
Gooden.
• All records

CYPERACEAE

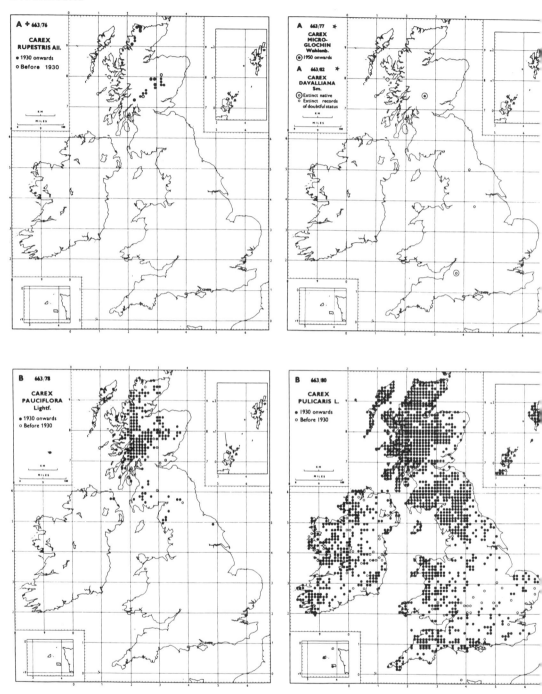

A + 663/76

CAREX RUPESTRIS All.

● 1930 onwards
○ Before 1930

A 663/77 ✳

CAREX MICRO-GLOCHIN Wahlenb.

◉ 1950 onwards

A 663/82 ✳

CAREX DAVALLIANA Sm.

◉ Extinct native
○ Extinct records of doubtful status

B 663/78

CAREX PAUCIFLORA Lightf.

● 1930 onwards
○ Before 1930

B 663/80

CAREX PULICARIS L.

● 1930 onwards
○ Before 1930

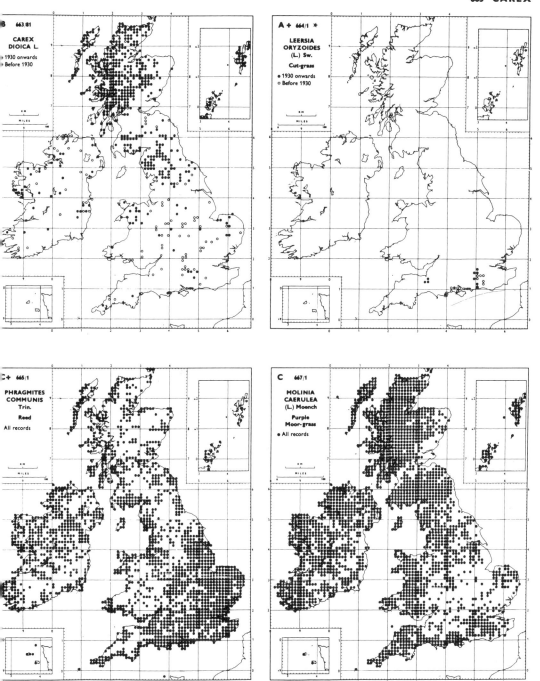

B 663/81

**CAREX
DIOICA L.**

• 1930 onwards
○ Before 1930

A + 664/1 ✱

**LEERSIA
ORYZOIDES**
(L.) Sw.

Cut-grass

• 1930 onwards
○ Before 1930

C + 665/1

**PHRAGMITES
COMMUNIS**
Trin.

Reed

All records

C 667/1

**MOLINIA
CAERULEA**
(L.) Moench

**Purple
Moor-grass**

• All records

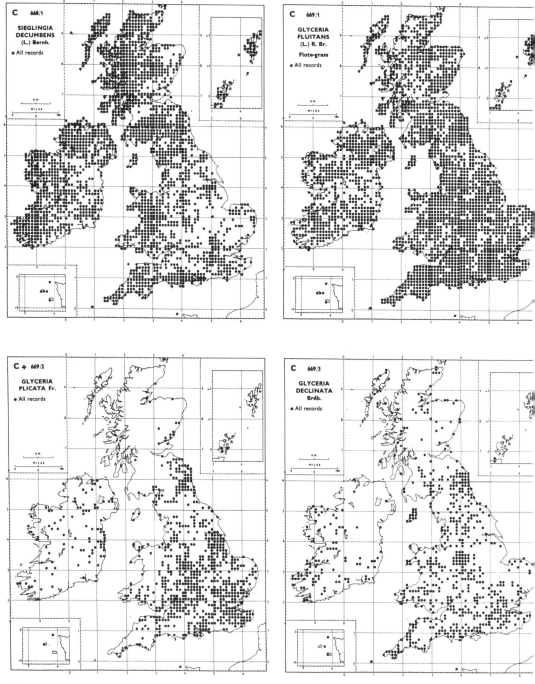

C   668/1
SIEGLINGIA
DECUMBENS
(L.) Bernh.
● All records

C   669/1
GLYCERIA
FLUITANS
(L.) R. Br.
Flote-grass
● All records

C ✛ 669/2
GLYCERIA
PLICATA Fr.
● All records

C   669/3
GLYCERIA
DECLINATA
Bréb.
● All records

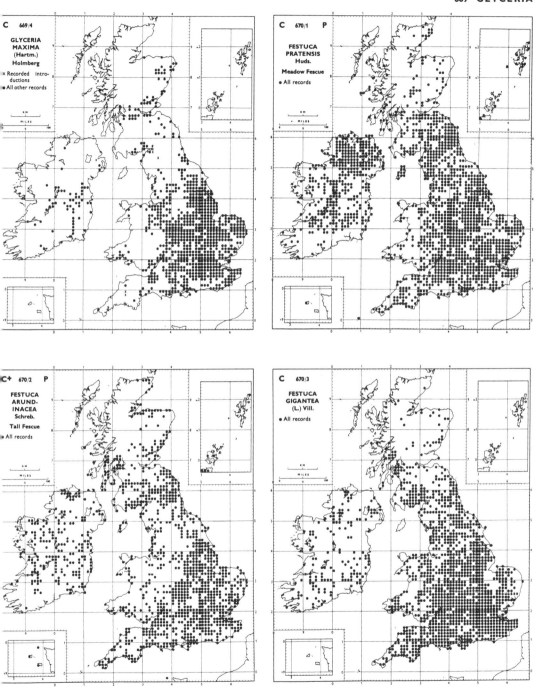

C 669/4

GLYCERIA
MAXIMA
(Hartm.)
Holmberg

x Recorded introductions
• All other records

C 670/1 P

FESTUCA
PRATENSIS
Huds.

Meadow Fescue

• All records

C+ 670/2 P

FESTUCA
ARUNDINACEA
Schreb.

Tall Fescue

• All records

C 670/3

FESTUCA
GIGANTEA
(L.) Vill.

• All records

GRAMINEAE

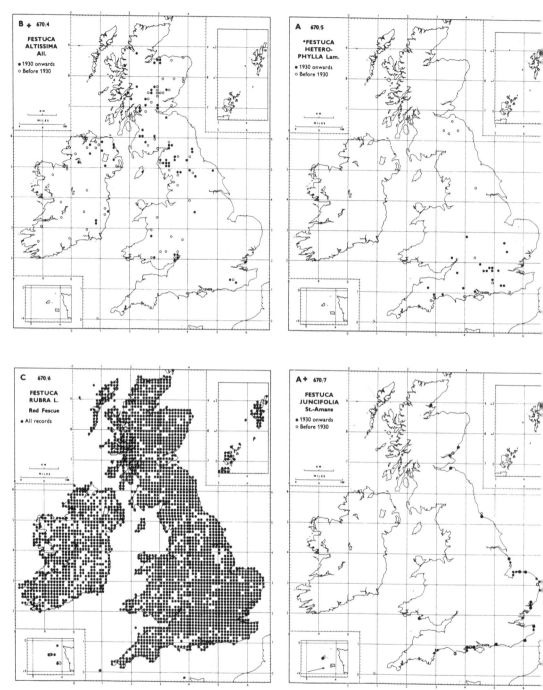

B ✚ 670/4

FESTUCA
ALTISSIMA
All.

● 1930 onwards
○ Before 1930

A 670/5

*FESTUCA
HETERO-
PHYLLA Lam.

● 1930 onwards
○ Before 1930

C 670/6

FESTUCA
RUBRA L.

Red Fescue

● All records

A ✚ 670/7

FESTUCA
JUNCIFOLIA
St.-Amans

● 1930 onwards
○ Before 1930

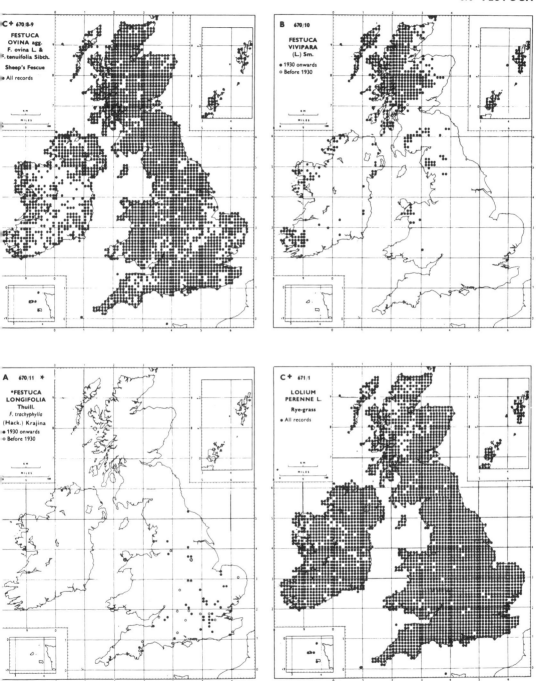

C+ 670/8-9
FESTUCA
OVINA agg.
F. ovina L. &
F. tenuifolia Sibth.
Sheep's Fescue
● All records

B 670/10
FESTUCA
VIVIPARA
(L.) Sm.
● 1930 onwards
○ Before 1930

A 670/11 ✶
*FESTUCA
LONGIFOLIA
Thuil.
F. trachyphylla
(Hack.) Krajina
● 1930 onwards
○ Before 1930

C+ 671/1
LOLIUM
PERENNE L.
Rye-grass
● All records

375

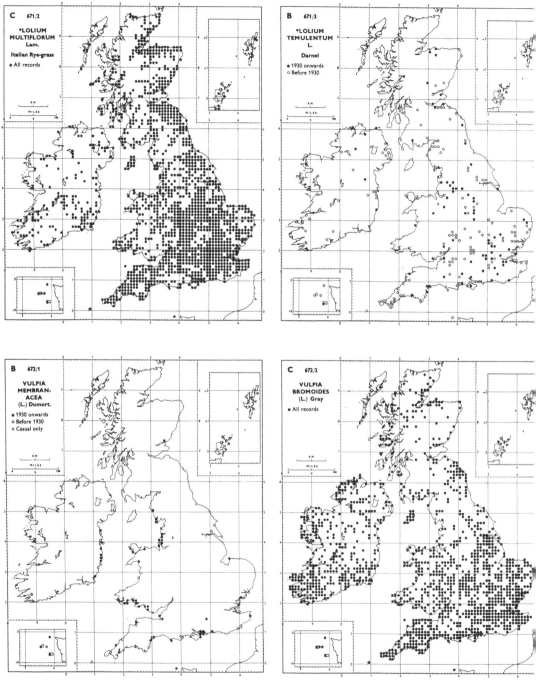

C 671/2
*LOLIUM
MULTIFLORUM
Lam.
Italian Rye-grass
● All records

B 671/3
*LOLIUM
TEMULENTUM
L.
Darnel
● 1930 onwards
○ Before 1930

B 672/1
VULPIA
MEMBRAN-
ACEA
(L.) Dumort.
● 1930 onwards
○ Before 1930
× Casual only

C 672/2
VULPIA
BROMOIDES
(L.) Gray
● All records

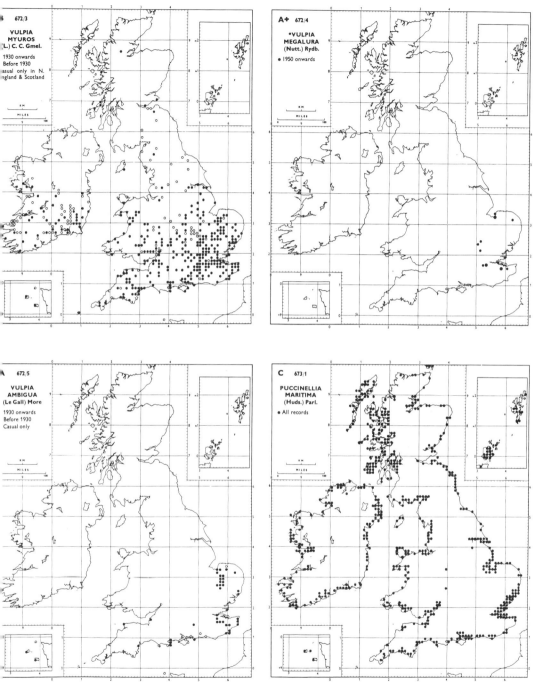

672/3

VULPIA
MYUROS
(L.) C. C. Gmel.

1930 onwards
Before 1930
casual only in N.
England & Scotland

KM
MILES

A+ 672/4

*VULPIA
MEGALURA
(Nutt.) Rydb.

● 1950 onwards

KM
MILES

672/5

VULPIA
AMBIGUA
(Le Gall) More

1930 onwards
Before 1930
Casual only

KM
MILES

C 673/1

PUCCINELLIA
MARITIMA
(Huds.) Parl.

● All records

KM
MILES

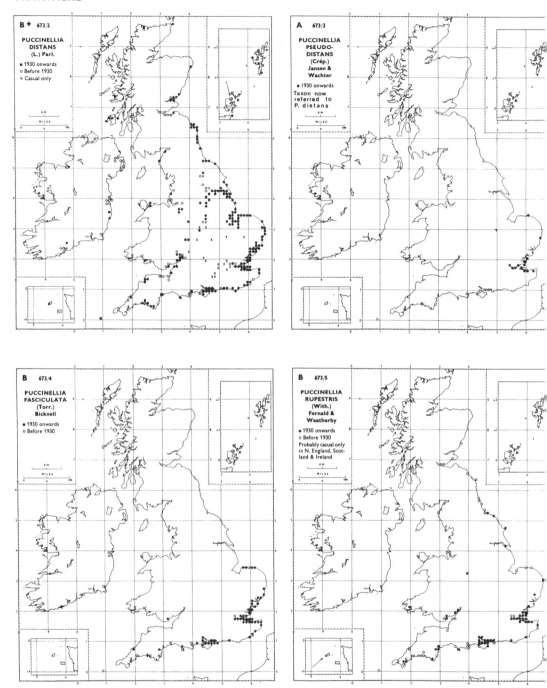

B + 673/2

PUCCINELLIA
DISTANS
(L.) Parl.

● 1930 onwards
○ Before 1930
× Casual only

A 673/3

PUCCINELLIA
PSEUDO-
DISTANS
(Crép.)
Jansen &
Wachter

● 1930 onwards
Taxon now
referred to
P. distans

B 673/4

PUCCINELLIA
FASCICULATA
(Torr.)
Bicknell

● 1930 onwards
○ Before 1930

B 673/5

PUCCINELLIA
RUPESTRIS
(With.)
Fernald &
Weatherby

● 1930 onwards
○ Before 1930
Probably casual only
in N. England, Scot-
land & Ireland

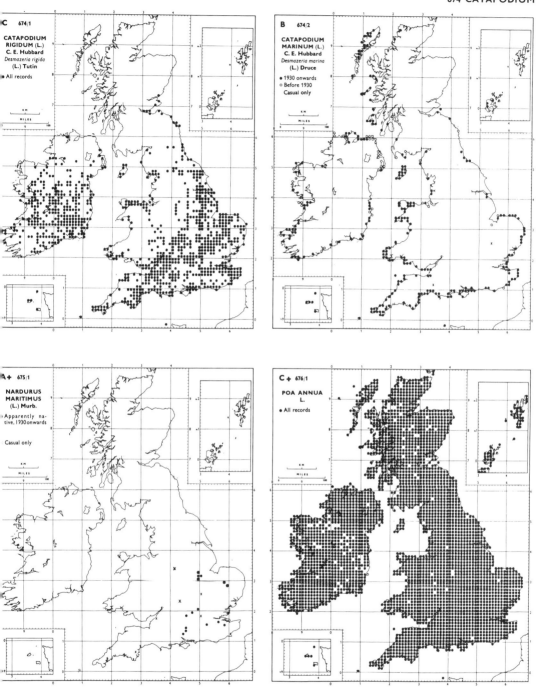

C 674/1
CATAPODIUM
RIGIDUM (L.)
C. E. Hubbard
Desmazeria rigida
(L.) Tutin
● All records

B 674/2
CATAPODIUM
MARINUM (L.)
C. E. Hubbard
Desmazeria marina
(L.) Druce
● 1930 onwards
○ Before 1930
Casual only

A + 675/1
NARDURUS
MARITIMUS
(L.) Murb.
Apparently na-
tive, 1930 onwards
Casual only

C + 676/1
POA ANNUA
L.
● All records

379

GRAMINEAE

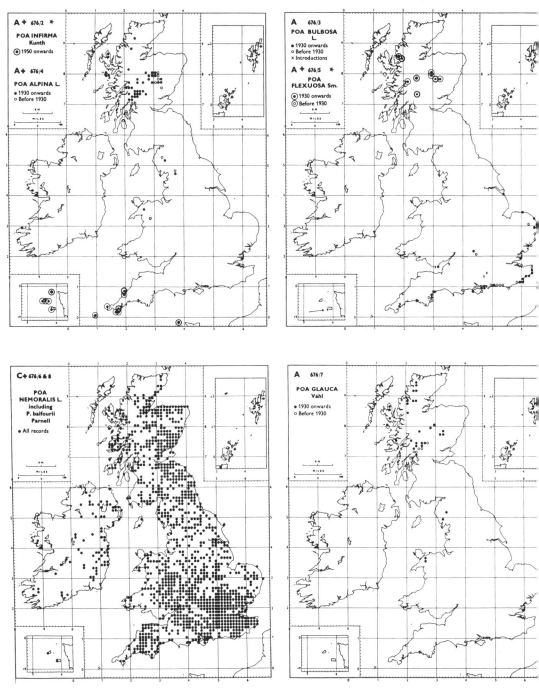

A + 676/2 *

POA INFIRMA
Kunth
⊙ 1950 onwards

A + 676/4

POA ALPINA L.
● 1930 onwards
○ Before 1930

A 676/3

POA BULBOSA
L.
● 1930 onwards
○ Before 1930
✕ Introductions

A + 676/5 *

POA
FLEXUOSA Sm.
⊙ 1930 onwards
⊙ Before 1930

C + 676/6 & 8

POA
NEMORALIS L.
including
P. balfourii
Parnell
● All records

A 676/7

POA GLAUCA
Vahl
● 1930 onwards
○ Before 1930

380

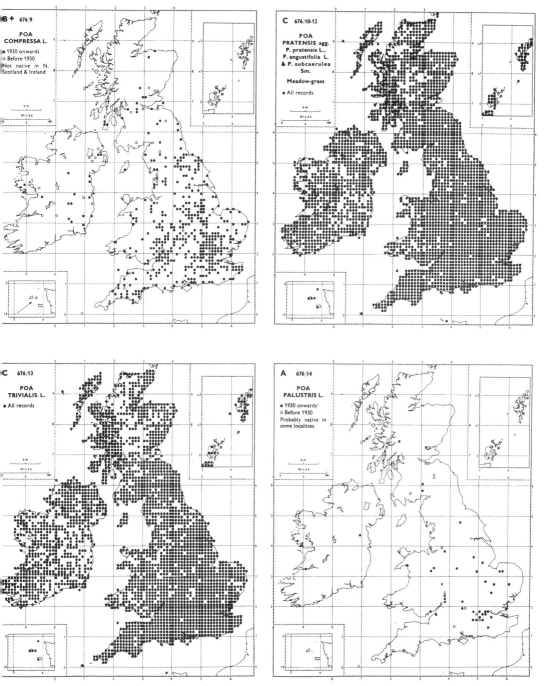

B+ 676/9

POA
COMPRESSA L.
• 1930 onwards
○ Before 1930
Not native in N.
Scotland & Ireland

C 676/10-12

POA
PRATENSIS agg.
P. pratensis L.,
P. angustifolia L.
& P. subcaerulea
Sm.

Meadow-grass

• All records

C 676/13

POA
TRIVIALIS L.
• All records

A 676/14

POA PALUSTRIS L.
• 1930 onwards·
○ Before 1930
Probably native in
some localities

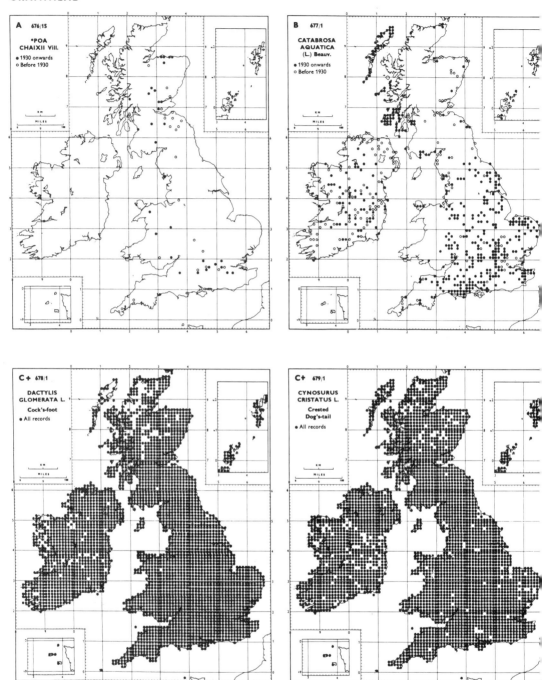

A 676/15
*POA
CHAIXII Vill.
● 1930 onwards
○ Before 1930

B 677/1
CATABROSA
AQUATICA
(L.) Beauv.
● 1930 onwards
○ Before 1930

C+ 678/1
DACTYLIS
GLOMERATA L.
Cock's-foot
● All records

C+ 679/1
CYNOSURUS
CRISTATUS L.
Crested
Dog's-tail
● All records

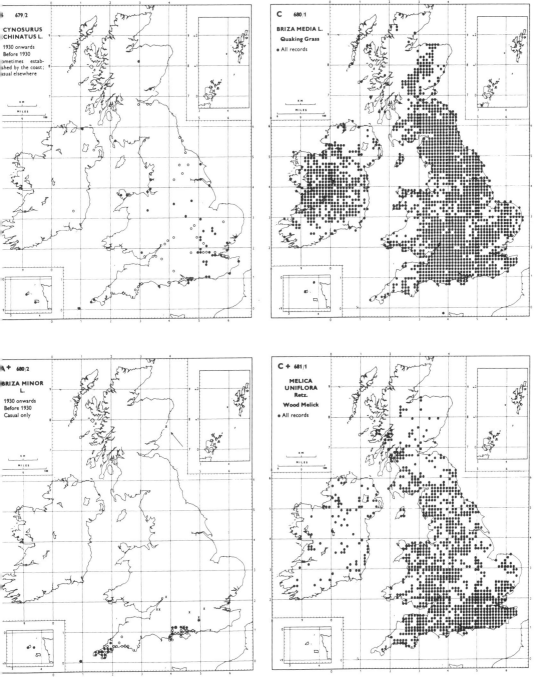

679/2

CYNOSURUS
ECHINATUS L.

1930 onwards
Before 1930
Sometimes estab-
lished by the coast;
casual elsewhere

C   680/1

BRIZA MEDIA L.
Quaking Grass

● All records

680/2

BRIZA MINOR
L.

1930 onwards
Before 1930
Casual only

C +   681/1

MELICA
UNIFLORA
Retz.

Wood Melick

● All records

383

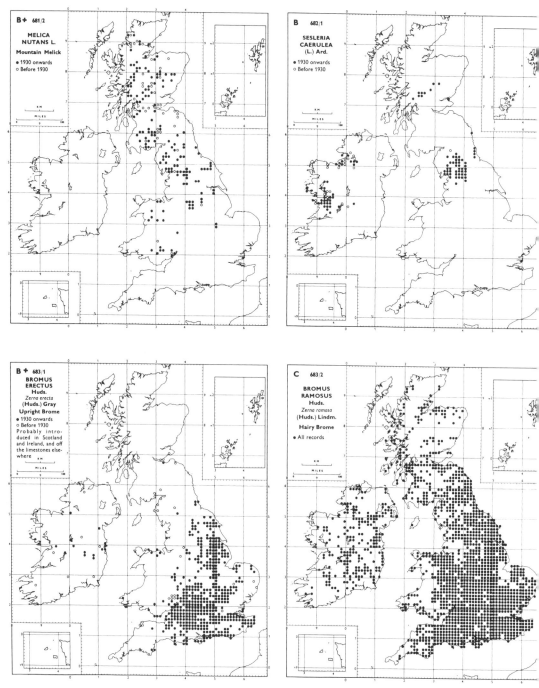

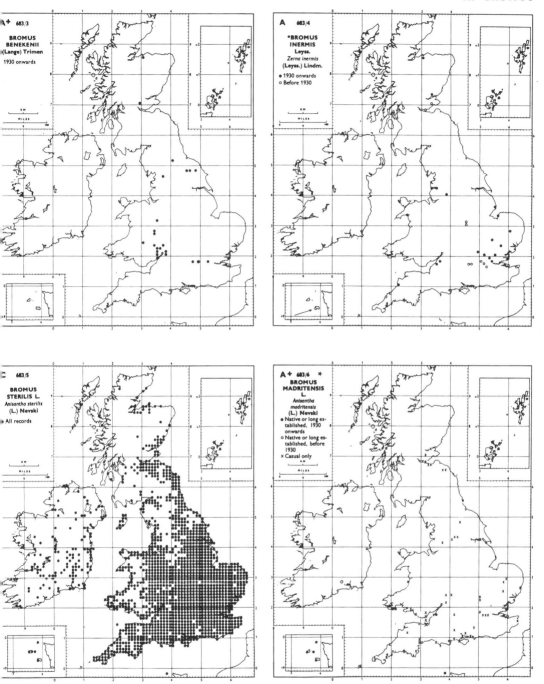

A+ 683/3
**BROMUS
BENEKENII
(Lange) Trimen**
1930 onwards

A 683/4
**°BROMUS
INERMIS
Leyss.
Zerna inermis
(Leyss.) Lindm.**
● 1930 onwards
○ Before 1930

C 683/5
**BROMUS
STERILIS L.**
Anisantha sterilis
(L.) Nevski
● All records

A+ 683/6 ✳
**BROMUS
MADRITENSIS
L.**
*Anisantha
madritensis*
(L.) Nevski
● Native or long es-
tablished, 1930
onwards
○ Native or long es-
tablished, before
1930
✕ Casual only

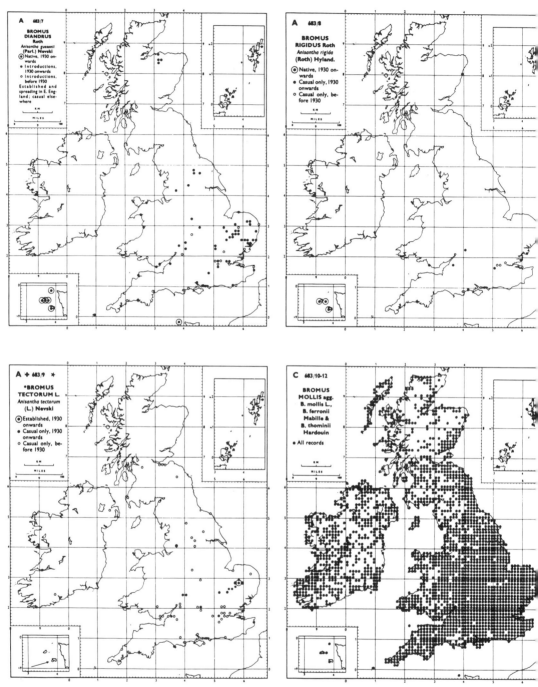

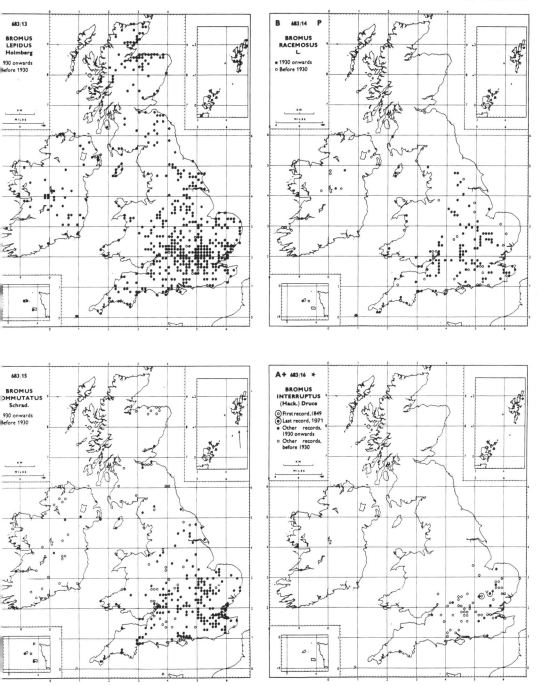

683/13

**BROMUS
LEPIDUS
Holmberg**

930 onwards
efore 1930

B 683/14 P

**BROMUS
RACEMOSUS
L.**

● 1930 onwards
○ Before 1930

683/15

**BROMUS
OMMUTATUS
Schrad.**

930 onwards
efore 1930

A+ 683/16 ✷

**BROMUS
INTERRUPTUS
(Hack.) Druce**

⊚ First record, 1849
◉ Last record, 1971
● Other records,
1930 onwards
○ Other records,
before 1930

387

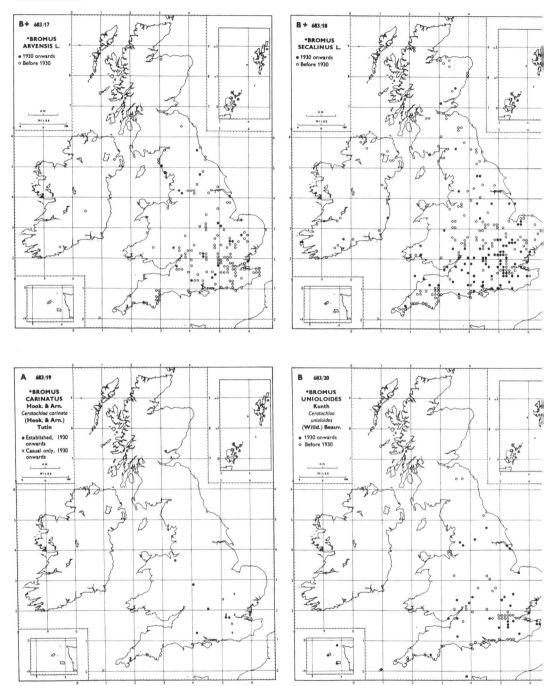

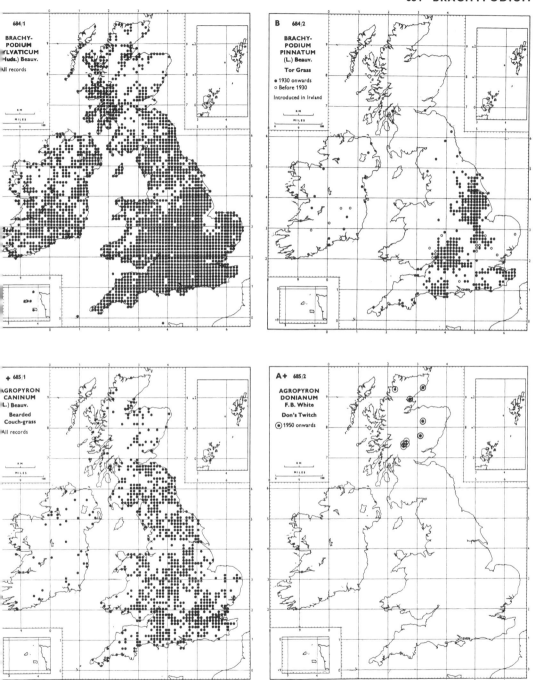

684/1

BRACHY-
PODIUM
SYLVATICUM
(Huds.) Beauv.

All records

KM
MILES

B   684/2

BRACHY-
PODIUM
PINNATUM
(L.) Beauv.

Tor Grass

● 1930 onwards
○ Before 1930

Introduced in Ireland

KM
MILES

684/1

AGROPYRON
CANINUM
(L.) Beauv.

Bearded
Couch-grass

All records

KM
MILES

A+   685/2

AGROPYRON
DONIANUM
F.B. White

Don's Twitch

◉ 1950 onwards

KM
MILES

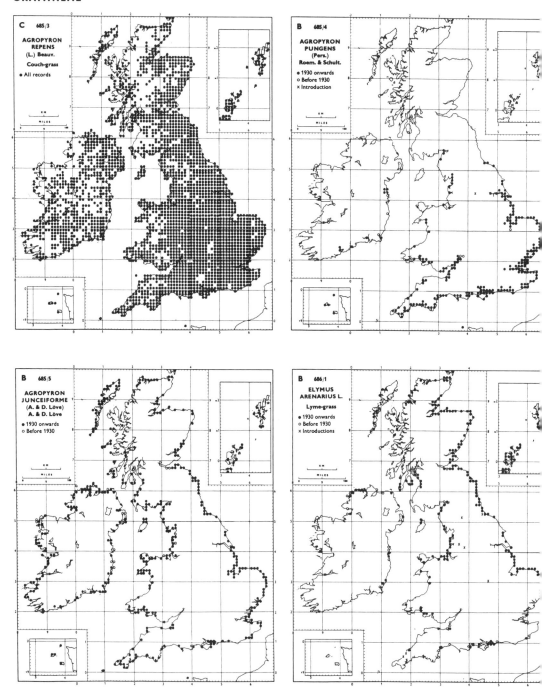

C 685/3

AGROPYRON
REPENS
(L.) Beauv.

Couch-grass

● All records

B 685/4

AGROPYRON
PUNGENS
(Pers.)
Roem. & Schult.

● 1930 onwards
○ Before 1930
× Introduction

B 685/5

AGROPYRON
JUNCEIFORME
(A. & D. Löve)
A. & D. Löve

● 1930 onwards
○ Before 1930

B 686/1

ELYMUS
ARENARIUS L.

Lyme-grass

● 1930 onwards
○ Before 1930
× Introductions

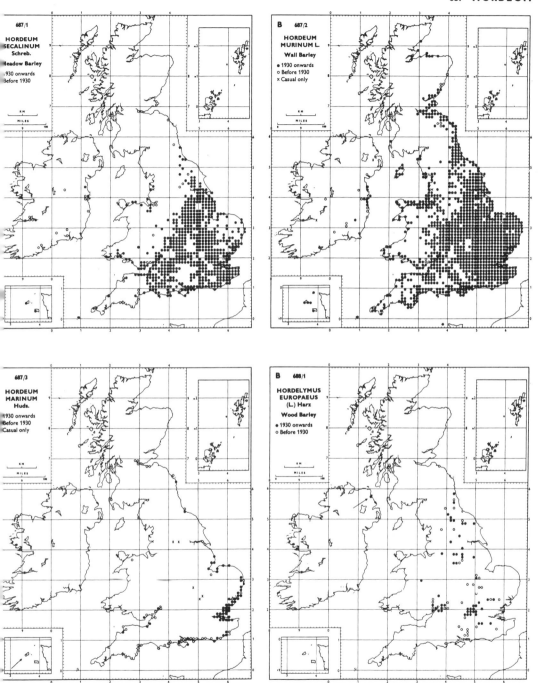

687/1

HORDEUM
SECALINUM
Schreb.

Meadow Barley

• 1930 onwards
○ Before 1930

B   687/2

HORDEUM
MURINUM L.

Wall Barley

• 1930 onwards
○ Before 1930
× Casual only

687/3

HORDEUM
MARINUM
Huds.

• 1930 onwards
○ Before 1930
× Casual only

B   688/1

HORDELYMUS
EUROPAEUS
(L.) Harz

Wood Barley

• 1930 onwards
○ Before 1930

391

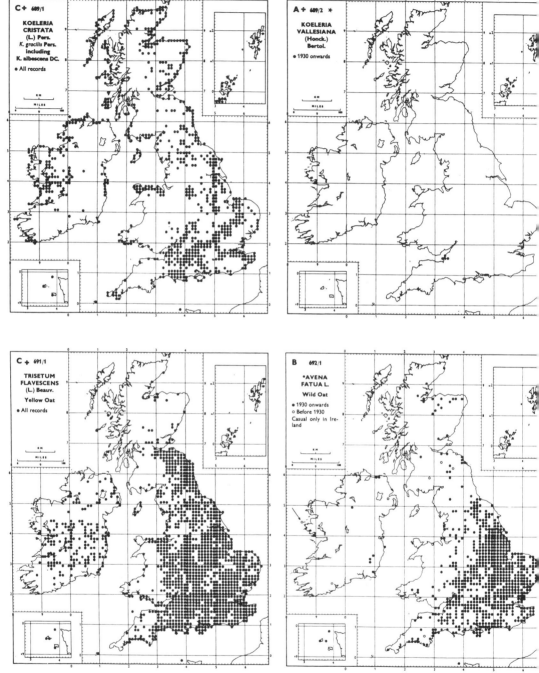

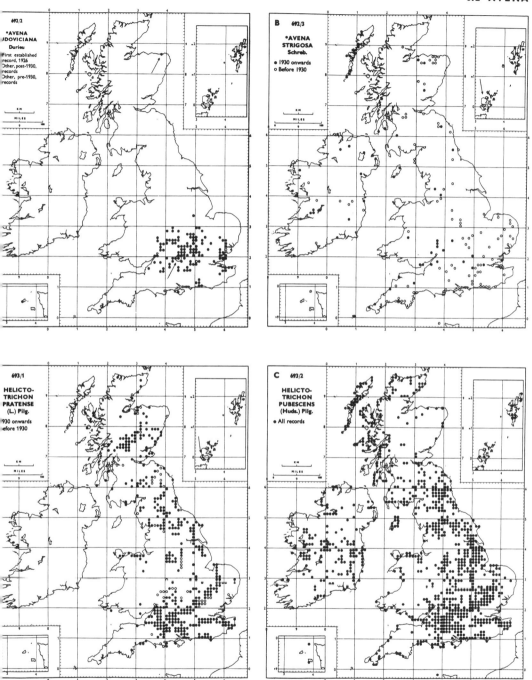

692/2
*AVENA
UDOVICIANA
Durieu
First established
record, 1926
Other, post-1930,
records
Other, pre-1930,
records

B   692/3
*AVENA
STRIGOSA
Schreb.
● 1930 onwards
○ Before 1930

693/1
HELICTO-
TRICHON
PRATENSE
(L.) Pilg.
930 onwards
efore 1930

C   693/2
HELICTO-
TRICHON
PUBESCENS
(Huds.) Pilg.
● All records

393

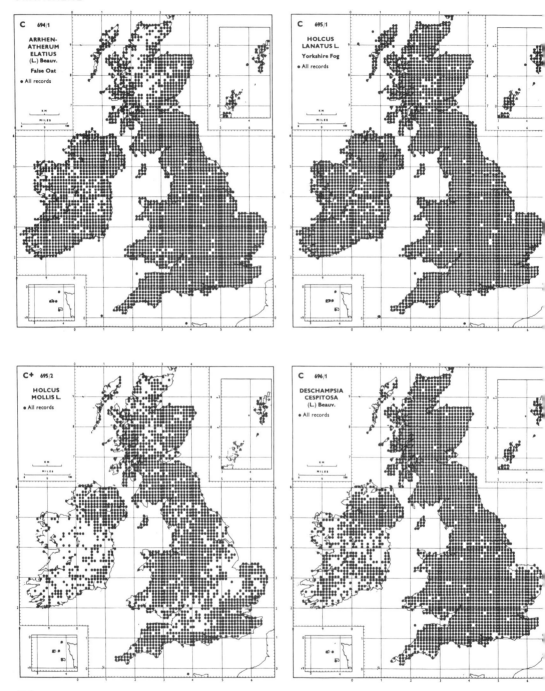

C 694/1

ARRHEN-
ATHERUM
ELATIUS
(L.) Beauv.

False Oat

● All records

C 695/1

HOLCUS
LANATUS L.

Yorkshire Fog

● All records

C+ 695/2

HOLCUS
MOLLIS L.

● All records

C 696/1

DESCHAMPSIA
CESPITOSA
(L.) Beauv.

● All records

394

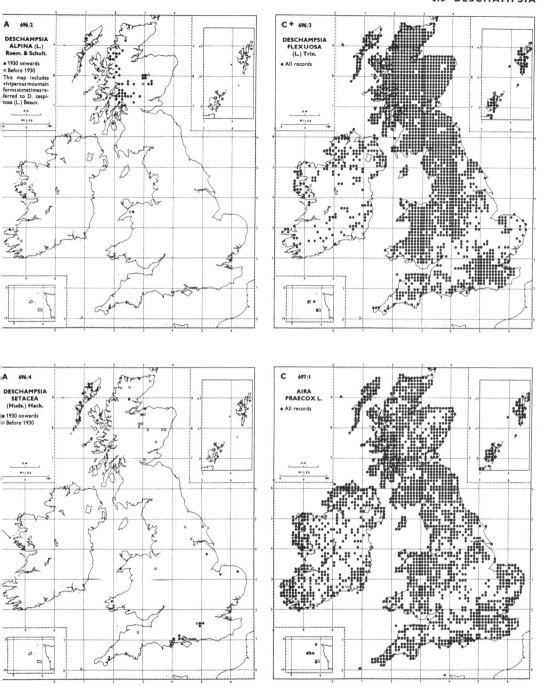

A 696/2

DESCHAMPSIA
ALPINA (L.)
Roem. & Schult.

● 1930 onwards
○ Before 1930
This map includes
viviparous mountain
forms sometimes re-
ferred to D. cespi-
tosa (L.) Beauv.

C + 696/3

DESCHAMPSIA
FLEXUOSA
(L.) Trin.

● All records

A 696/4

DESCHAMPSIA
SETACEA
(Huds.) Hack.

● 1930 onwards
○ Before 1930

C 697/1

AIRA
PRAECOX L.

● All records

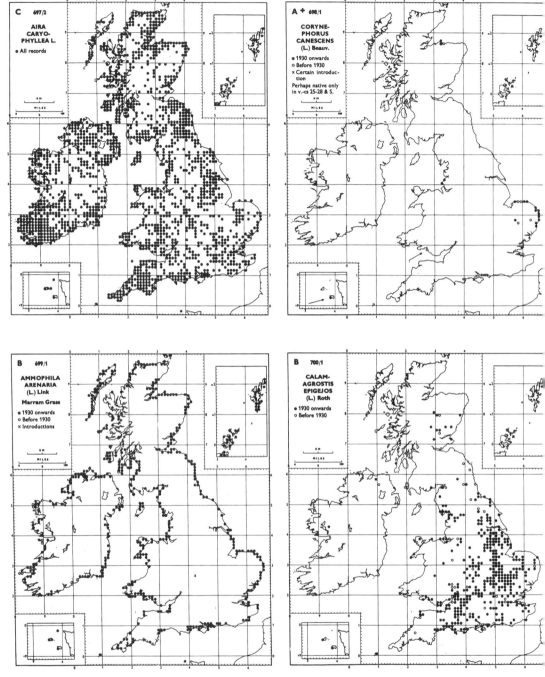

C 697/2
AIRA CARYO-PHYLLEA L.
• All records

A + 698/1
CORYNE-PHORUS CANESCENS (L.) Beauv.
• 1930 onwards
○ Before 1930
× Certain Introduc-tion
Perhaps native only in v.-cs 25-28 & S.

B 699/1
AMMOPHILA ARENARIA (L.) Link
Marram Grass
• 1930 onwards
○ Before 1930
× Introductions

B 700/1
CALAM-AGROSTIS EPIGEJOS (L.) Roth
• 1930 onwards
○ Before 1930

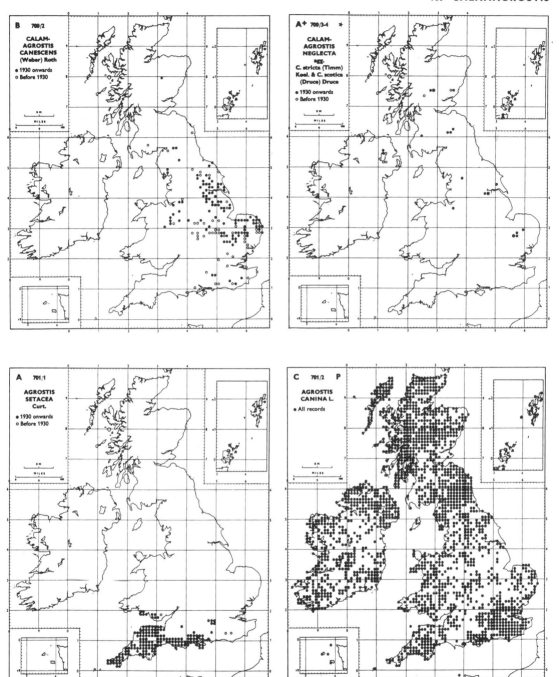

B 700/2
CALAM-
AGROSTIS
CANESCENS
(Weber) Roth

● 1930 onwards
○ Before 1930

A+ 700/3-4
CALAM-
AGROSTIS
NEGLECTA
agg.
C. stricta (Timm)
Koel. & C. scotica
(Druce) Druce

● 1930 onwards
○ Before 1930

A 701/1
AGROSTIS
SETACEA
Curt.

● 1930 onwards
○ Before 1930

C 701/2 **P**
AGROSTIS
CANINA L.

● All records

397

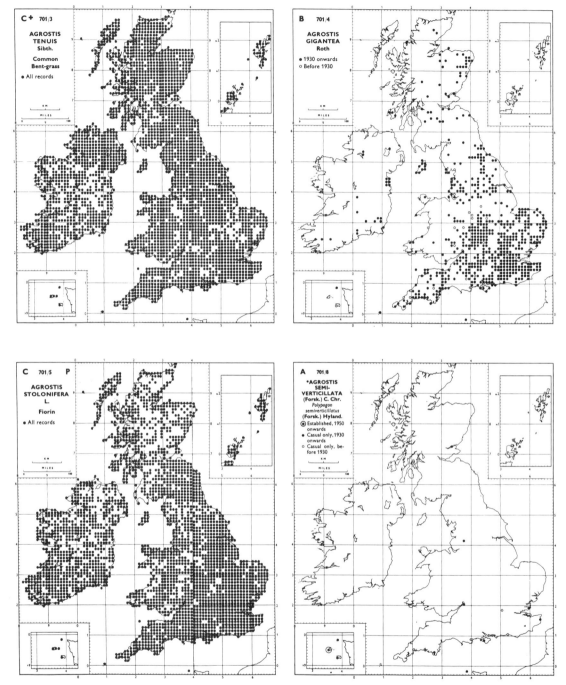

C+ 701/3

AGROSTIS TENUIS
Sibth.

Common
Bent-grass

• All records

B 701/4

AGROSTIS GIGANTEA
Roth

• 1930 onwards
○ Before 1930

C 701/5 P

AGROSTIS STOLONIFERA
L.

Fiorin

• All records

A 701/8

***AGROSTIS SEMI-VERTICILLATA**
(Forsk.) C. Chr.
Polypogon semiverticillatus
(Forsk.) Hyland.

⊙ Established, 1950 onwards
• Casual only, 1930 onwards
○ Casual only, before 1930

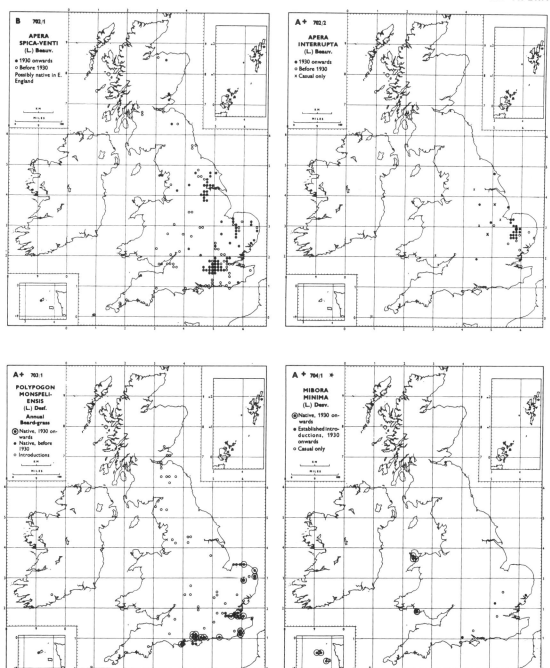

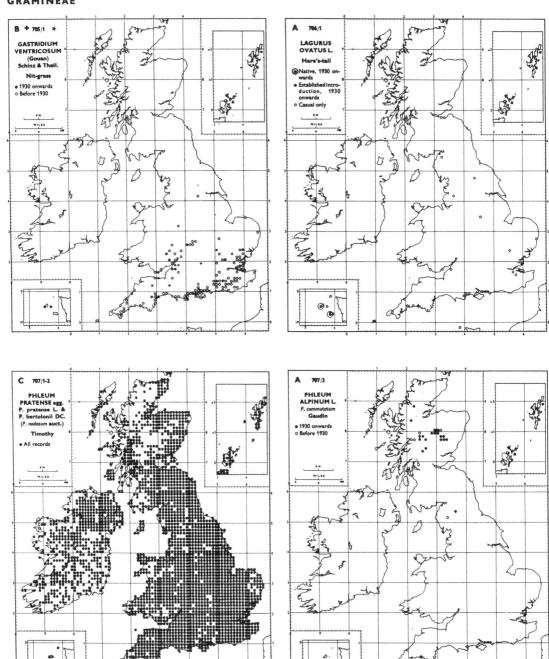

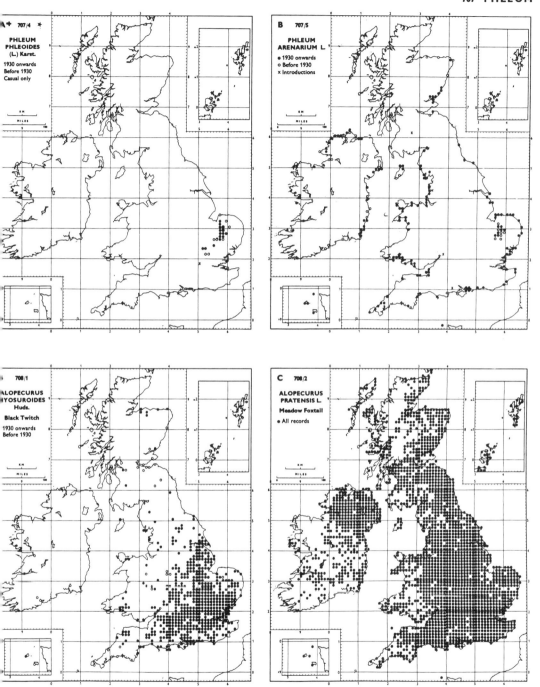

A+ 707/4 *

PHLEUM PHLEOIDES (L.) Karst.

1930 onwards
Before 1930
Casual only

B 707/5

PHLEUM ARENARIUM L.

● 1930 onwards
○ Before 1930
✕ Introductions

708/1

ALOPECURUS MYOSUROIDES Huds.

Black Twitch

1930 onwards
Before 1930

C 708/2

ALOPECURUS PRATENSIS L.

Meadow Foxtail

● All records

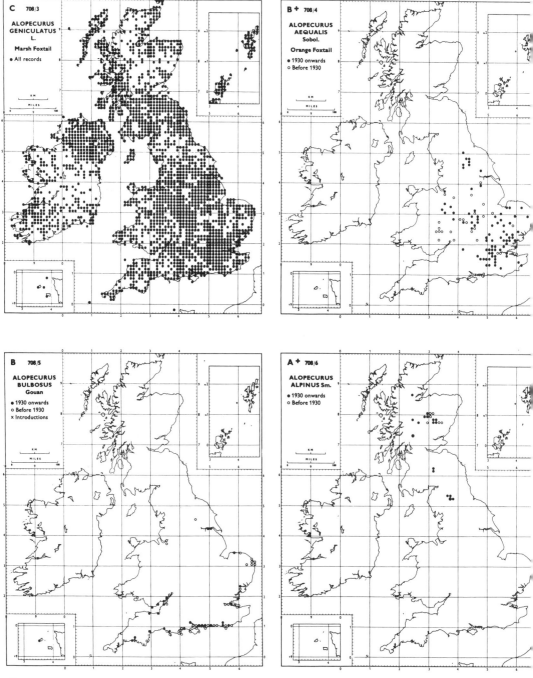

C 708/3

ALOPECURUS
GENICULATUS
L.

Marsh Foxtail

● All records

B + 708/4

ALOPECURUS
AEQUALIS
Sobol.

Orange Foxtail

● 1930 onwards
○ Before 1930

B 708/5

ALOPECURUS
BULBOSUS
Gouan

● 1930 onwards
○ Before 1930
× Introductions

A + 708/6

ALOPECURUS
ALPINUS Sm.

● 1930 onwards
○ Before 1930

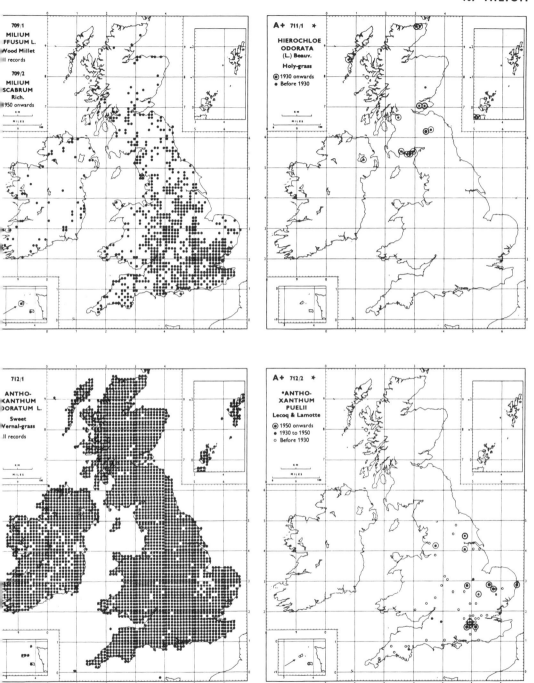

709/1
MILIUM
EFFUSUM L.
Wood Millet
all records

709/2
MILIUM
SCABRUM
Rich.
1950 onwards

A+ 711/1 *
HIEROCHLOE
ODORATA
(L.) Beauv.
Holy-grass
⊙ 1930 onwards
● Before 1930

712/1
ANTHO-
XANTHUM
ODORATUM L.
Sweet
Vernal-grass
all records

A+ 712/2 *
*ANTHO-
XANTHUM
PUELII
Lecoq & Lamotte
⊙ 1950 onwards
● 1930 to 1950
○ Before 1930

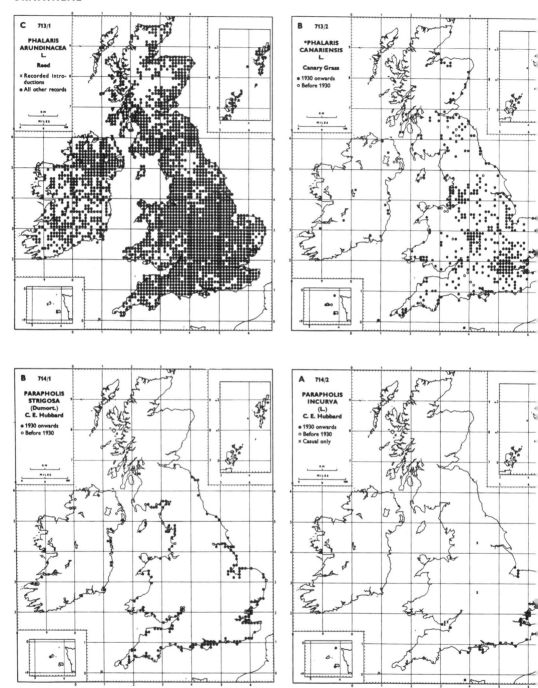

C 713/1

PHALARIS
ARUNDINACEA
L.

Reed

× Recorded intro-
   ductions
● All other records

B 713/2

*PHALARIS
CANARIENSIS
(L.)

Canary Grass

● 1930 onwards
○ Before 1930

B 714/1

PARAPHOLIS
STRIGOSA
(Dumort.)
C. E. Hubbard

● 1930 onwards
○ Before 1930

A 714/2

PARAPHOLIS
INCURVA
(L.)
C. E. Hubbard

● 1930 onwards
○ Before 1930
× Casual only

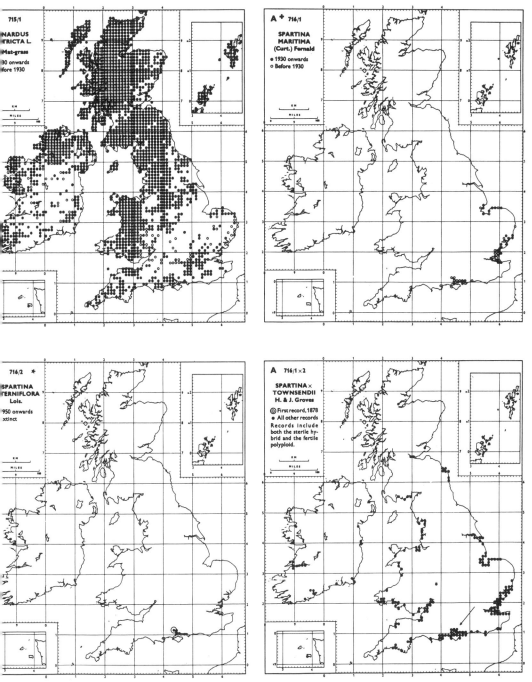

715/1
NARDUS
STRICTA L.
Mat-grass
30 onwards
fore 1930

A + 716/1
SPARTINA
MARITIMA
(Curt.) Fernald
● 1930 onwards
○ Before 1930

716/2 ✳
SPARTINA
ALTERNIFLORA
Lois.
950 onwards
xtinct

A 716/1 × 2
SPARTINA ×
TOWNSENDII
H. & J. Groves
⊚ First record, 1878
● All other records
Records include
both the sterile hy-
brid and the fertile
polyploid.

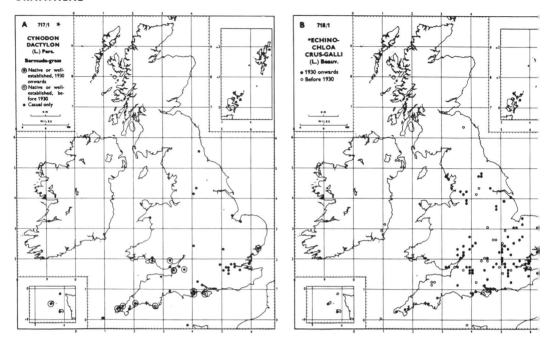

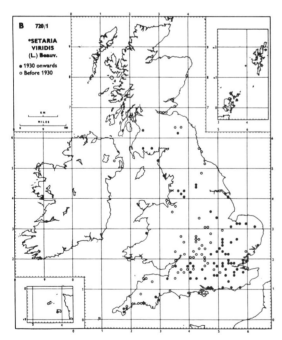

Appendix I

LIST OF AGGREGATE AND PROVISIONAL MAPS

AGGREGATE MAPS

21/1–2 Dryopteris filix-mas (+D. borreri). *D. borreri* Newm. is also mapped separately.
58/2–3 Papaver dubium (+P. lecoqii).
88/1–3 Cochlearia officinalis (+C. alpina & C. micacea).
95/1–3 Erophila verna (+E. spathulata & E. praecox).
102/1–2 Rorippa nasturtium-aquaticum (+R. microphylla).
120/1–2 Tamarix gallica (+T. anglica).
141/1–2 Arenaria serpyllifolia (+A. leptoclados).
160/2–4 Salicornia (S. dolichostachya, S. europaea & S. ramosissima).
222/1–2 Aphanes arvensis (+A. microcarpa).
225/8–10 Rosa canina (+R. dumalis & R. obtusifolia).
225/11–13 Rosa villosa (+R. tomentosa & R. sherardii).
225/14–17 Rosa rubiginosa (+R. micrantha, R. elliptica & R. agrestis).
258/2–3 Circaea alpina (+C. intermedia).
262/1–2 Callitriche stagnalis (+C. platycarpa).
319/14–15 Euphorbia esula (+E. uralensis).
330/3–6 Ulmus carpinifolia (+U. angustifolia, U. coritana & U. plotii).
342/4–5 Populus nigra (+P.×canadensis).
344/1–2 Ledum palustre (+L. groenlandicum).
409/1–2 Lycium chinense (+L. halimifolium).
442/1–2 Utricularia vulgaris (+U. neglecta).
538/2–4 Arctium minus (+A. nemorosum & A. pubens).
544/6–7 Centaurea nigra (+C. nemoralis).
663/9–10 Carex serotina (+C. scandinavica).
670/8–9 Festuca ovina (+F. tenuifolia).
676/10–12 Poa pratensis (+P. angustifolia & P. subcaerulea).
683/10–12 Bromus mollis (+B. ferronii & B. thominii).
700/3–4 Calamagrostis neglecta (C. stricta & C. scotica).
707/1–2 Phleum pratense (+P. bertolonii).

PROVISIONAL MAPS

21/6 Dryopteris lanceolatocristata.
Some forms of *D. dilatata* have more or less concolorous scales on the rachis, and such plants are often recorded as *D. lanceolatocristata*.
43/1 Anemone nemorosa &
46/24 Ranunculus ficaria.
These are examples of a group of species probably under-recorded in south-west Ireland (see Introduction, page xv).
67/2–3 Brassica napus & B. rapa.
Many recorders have ignored plants referable to one or other of these two species because of their frequent occurrence as relics of cultivated crops. In addition these species are not always readily distinguishable.
88/4 Cochlearia scotica.
Some plants referred to this species differ considerably from the type, and the limits of the species are very uncertain.
136/2 Sagina ciliata.
The taxonomy of this species and *S. apetala* is not properly worked out. Most competent recorders have distinguished two species but plants occur with various combinations of the diagnostic characters.
156/3–4 Atriplex hastata & A. glabriuscula.
The distinction between these two species in maritime habitats is very unsatisfactory.
175/1 Aesculus hippocastanum &

184/1 Laburnum anagyroides.
These are examples of alien trees which have been omitted from lists sent in by some recorders (see Introduction, page xix).
211/9 Rubus caesius.
It is probable that this species has been over-recorded in error for taxa of *Rubus fruticosus* agg. particularly in north and west Britain.
254/5 Epilobium roseum.
This species is perhaps over-recorded in error for *E. adenocaulon* and other species and hybrids.
254/7–8 Epilobium adnatum & E. lamyi.
These two species differ in leaf shape and in size of floral parts but lack the precise qualitative distinction present in most other species of *Epilobium*. Continental authorities often treat them as sub-species. Our records reflect this taxonomic difficulty.
254/9 Epilobium obscurum.
One of the commonest species of *Epilobium* in the country except in parts of eastern England, but apparently overlooked and under-recorded.
256/1–2 Oenothera biennis & O. erythrosepala.
There is acknowledged taxonomic confusion here and the records should be treated with caution.
262/4 Callitriche intermedia.
This species is often recorded in error for submerged forms of *C. stagnalis* agg.
319/14–16 Euphorbia esula agg. & E. cyparissias.
The treatment of this group in Clapham, Tutin & Warburg (1952) is acknowledged to be unsatisfactory, and the boundaries of the taxa are not clear. Even *E. cyparissias* is not always clearly distinguishable from plants of the *E. esula* agg.
320/11 Polygonum nodosum.
The group of taxa to which *P. nodosum* and *P. lapathifolium* belong is difficult in Europe and the British material is not yet clearly understood.
343/14 Salix nigricans.
There is no unanimity amongst the experts as to the limits of this species.
378/1 Ligustrum vulgare.
This is a widely planted species, therefore the status of the records is often uncertain. In addition there is confusion between the native *L. vulgare* and garden species.
392/1 Symphytum officinale.
The limits of this species are disputed. The cream-flowered plant of fens and marshes in eastern England is generally regarded as native, but many, if not all, purple-flowered variants referred to this species are doubtfully native and possibly of hybrid origin. It has therefore not been possible to separate records of *S. officinale* from the hybrid (*S.×uplandicum* Nyman).
400/1 & 4 Myosotis scorpioides (M. palustris) & M. caespitosa.
It is probable that *M. scorpioides* has been over-recorded in error for *M. caespitosa*, particularly in north and west Britain.
406/1 & 3 Calystegia sepium & C. silvatica.
Although there is little difficulty in distinguishing good material of these two species, three factors may have confused the records.
(i) The occasional occurrence of intermediate populations.
(ii) The existence of a third and, until recently, overlooked taxon, *C. pulchra*.
(iii) The continued use of *C. sepium* as an aggregate name.
440/9 Orobanche picridis.
The taxonomic distinction between this species and varieties of *O. minor* is not understood, and old herbarium material cannot be used satisfactorily.

538/1 Arctium lappa.
    In spite of careful checking of the records the species is perhaps still over-recorded because of the continued use of the name in the aggregate sense. However, the map probably indicates correctly the limits of distribution.

550/2–3 Leontodon hispidus.& L. taraxacoides (*L. leysseri*).
    Although these two species are not in any sense critical the number of erroneous records received suggests that both may have been over-recorded in error for other yellow composites.

559/2 Crepis vesicaria (*C. taraxacifolia*).
    Large-flowered plants of *C. capillaris*, which occur in north and west Britain, have been recorded as this species in error.

559/5 Crepis biennis.
    A distinct, but rather local, species which has probably been over-recorded in error for *C. vesicaria*.

605/4 Juncus compressus.
    There are two factors which may have confused the records for this species.
    (i) Misidentification of specimens of *J. gerardii* near the coast. Northern records are particularly suspect.

(ii) "Juncu com" on the Regional Record Card has been interpreted as *Juncus communis* by some recorders, an aggregate name for *J. effusus* and *J. conglomeratus*. Most errors of this kind have been eliminated.

670/1–2 Festuca pratensis & F. arundinacea.
    Although typical specimens of these two species are easily identified some material is difficult: the determination of such material has probably been inconsistent. However, the general features of the two maps seem reasonable.

683/14 Bromus racemosus.
    A rare and somewhat critical species which has probably been over-recorded in error for *B. commutatus*.

701/2 Agrostis canina &
701/5 Agrostis stolonifera.
    Probably rather seriously under-recorded because of extra taxonomic difficulty, but a special case of the general tendency for the common sedges and grasses to be overlooked (see Introduction, page xvii).

UNDER-RECORDED SQUARES

Great Britain

08/77	33/49	42/68
08/96	33/58	42/72
08/97	34/40	42/73
16/96	34/41	42/74
18/78	34/79	42/75
18/86	34/80	42/77
19/01	34/81	42/78
19/02	34/91	43/47
20/05	35/51	43/49
21/89	35/81	44/07
21/90	35/93	44/32
21/99	35/94	44/59
26/35	37/50	44/69
26/58	38/22	44/96
26/68	38/31	45/03
26/78	38/62	45/14
26/79	38/63	45/61
26/88	39/36	54/07
26/96	39/39	57/53
27/56	39/48	57/62
27/88	41/81	57/63
28/16	41/86	57/64
28/26	41/89	57/74
28/40	42/00	57/75
29/27	42/10	62/02
29/95	42/11	62/11
31/19	42/30	62/12
31/31	42/62	68/48
31/41	42/63	68/49
31/52	42/64	68/58
31/63	42/65	68/59
31/72	42/66	69/50
32/95	42/67	

Ireland

84/83	03/00	13/06
84/94	03/01	13/16
84/98	03/34	13/19
85/81	03/39	14/00
92/10	03/49	14/01
92/95	03/56	14/02
93/25	03/58	14/10
93/45	03/67	14/11
93/66	03/74	
93/82	03/76	
94/01	03/79	
94/08	03/86	
94/10	03/87	
94/15	03/94	
94/16	03/96	
94/17	04/07	
94/40	04/08	
94/42	04/18	
94/43	04/23	
94/46	04/27	
94/52	04/28	
94/61	04/31	
94/69	04/33	
94/82	04/45	
94/83	04/53	
94/88	04/55	
94/93	04/74	
94/99	04/75	
95/01	04/81	
95/10	04/84	
95/90	04/90	
02/29	13/04	
02/38	13/05	

Appendix II

VICE-COUNTY RECORDS OMITTED FROM THE MAPS

THE VICE-COUNTY NUMBERS AND CORRESPONDING VICE-COUNTIES

The vice-county numbers follow Watson (1873) and Praeger (1901). The names have been altered to agree with current usage, but the old names have been added in brackets where they differ significantly from modern ones.

ENGLAND AND WALES

1. West Cornwall (with Scilly)
2. East Cornwall
3. South Devon
4. North Devon
5. South Somerset
6. North Somerset
7. North Wiltshire
8. South Wiltshire
9. Dorset
10. Isle of Wight
11. South Hampshire
12. North Hampshire
13. West Sussex
14. East Sussex
15. East Kent
16. West Kent
17. Surrey
18. South Essex
19. North Essex
20. Hertfordshire
21. Middlesex
22. Berkshire
23. Oxfordshire
24. Buckinghamshire
25. East Suffolk

26. West Suffolk
27. East Norfolk
28. West Norfolk
29. Cambridgeshire
30. Bedfordshire
31. Huntingdonshire
32. Northamptonshire
33. East Gloucestershire
34. West Gloucestershire
35. Monmouthshire
36. Herefordshire
37. Worcestershire
38. Warwickshire
39. Staffordshire
40. Shropshire (Salop)
41. Glamorgan
42. Breconshire
43. Radnorshire
44. Carmarthenshire
45. Pembrokeshire
46. Cardiganshire
47. Montgomeryshire
48. Merionethshire
49. Caernarvonshire
50. Denbighshire

51. Flintshire
52. Anglesey
53. South Lincolnshire
54. North Lincolnshire
55. Leicestershire (with Rutland)
56. Nottinghamshire
57. Derbyshire
58. Cheshire
59. South Lancashire
60. West Lancashire
61. South-east Yorkshire
62. North-east Yorkshire
63. South-west Yorkshire
64. Mid-west Yorkshire
65. North-west Yorkshire
66. Durham
67. South Northumberland
68. North Northumberland (Cheviot)
69. Westmorland with North Lancashire
70. Cumberland
71. Isle of Man
S. Channel Isles

SCOTLAND

72. Dumfriesshire
73. Kirkcudbrightshire
74. Wigtownshire
75. Ayrshire
76. Renfrewshire
77. Lanarkshire
78. Peeblesshire
79. Selkirkshire
80. Roxburghshire
81. Berwickshire
82. East Lothian (Haddington)
83. Midlothian (Edinburgh)
84. West Lothian (Linlithgow)
85. Fifeshire (with Kinross)
86. Stirlingshire

87. West Perthshire (with Clackmannan)
88. Mid Perthshire
89. East Perthshire
90. Angus (Forfar)
91. Kincardineshire
92. South Aberdeenshire
93. North Aberdeenshire
94. Banffshire
95. Moray (Elgin)
96. East Inverness-shire (with Nairn)
97. West Inverness-shire
98. Argyll Main

99. Dunbartonshire
100. Clyde Isles
101. Kintyre
102. South Ebudes
103. Mid Ebudes
104. North Ebudes
105. West Ross
106. East Ross
107. East Sutherland
108. West Sutherland
109. Caithness
110. Outer Hebrides
111. Orkney Islands
112. Shetland Islands (Zetland)

IRELAND

H.1	South Kerry	H.15	South-east Galway	H.28	Sligo
H.2	North Kerry	H.16	West Galway	H.29	Leitrim
H.3	West Cork	H.17	North-east Galway	H.30	Cavan
H.4	Mid Cork	H.18	Offaly (King's County)	H.31	Louth
H.5	East Cork	H.19	Kildare	H.32	Monaghan
H.6	Waterford	H.20	Wicklow	H.33	Fermanagh
H.7	South Tipperary	H.21	Dublin	H.34	East Donegal
H.8	Limerick	H.22	Meath	H.35	West Donegal
H.9	Clare	H.23	West Meath	H.36	Tyrone
H.10	North Tipperary	H.24	Longford	H.37	Armagh
H.11	Kilkenny	H.25	Roscommon	H.38	Down
H.12	Wexford	H.26	East Mayo	H.39	Antrim
H.13	Carlow	H.27	West Mayo	H.40	Londonderry
H.14	Leix (Queen's County)				

THE OMITTED RECORDS

The species numbers and nomenclature follow Dandy (1958). Numbers outside brackets represent vice-counties from which species have been reliably recorded in the past but for which no localizable record has been traced. Numbers in round brackets represent vice-counties from which native species have been reliably recorded as an introduction but for which no localized record has been traced. Non-native species are marked with an asterisk before the species number. Numbers in square brackets represent vice-counties from which species are known or suspected to have been recorded in error.

The list of numbers is mainly based upon a comparison between the records received by the Maps Scheme and the *Comital Flora* (Druce, 1932). Certain critical genera have been omitted from the list because experience has shown that vice-county records previously published are so liable to error as to be useless.

1/1 Lycopodium selago 45, 74, 79, 82, [7, 13]
1/2 inundatum 5, 31, 61, 108, [7, 49]
1/3 annotinum 78, 100, 103, 109, [55, 59, 64, 66]
1/4 clavatum 52, 66, 79, 86, [7]
1/5 alpinum 51, 59, 102, 103, 110, [58, H.28]
2/1 Selaginella selaginoides 51, 74
3/1 Isoetes lacustris 42, 50, 52, 85, 94
3/2 echinospora 87, 97, 107, [H.40]
4/1 Equisetum hyemale 15, 40, 50, 59, [4, 7, 8, 21, 51, 87]
4/4 variegatum 100, [20, 28, 42, H 16, 26]
4/7 sylvaticum 51, 78, [1, 25, 26]
4/8 pratense 78, [59]
4/10 telmateia H.3, 26, [90–2, 111]
5/1 Osmunda regalis 51, 84, 93, [7, 23, 29, 33, 94]
6/1 Trichomanes speciosum 70, [101]
7/1 Hymenophyllum tunbrigense 78, [13, 69]
7/2 wilsonii 40, 52, 109, [6, 9]
9/1 Cryptogramma crispa 93, 95, 109, [4, 36, 55]
11/1 Adiantum capillus-veneris [48]
14/1 Phyllitis scolopendrium 96
15/2 Asplenium obovatum [15, 40, 42, 71, H.1]
15/4 marinum 99, [15, 50, 72]
15/6 viride 106, [3, 13, 15, 17, 38, 55, 58, 63, 83, 84, 103]
15/8 septentrionale 78
16/1 Ceterach officinarum 77, [58]
18/2 Athyrium alpestre 86, 89, 107
19/1 Cystopteris fragilis 12, 52, 74, H.32, [21, 26]
19/3 montana 94, 104, [49]
20/1 Woodsia ilvensis 103, [88]
20/2 alpina 104
21/4 Dryopteris villarii (H.31), [59]
21/5 cristata [8, 9, 30, 37, 74, 84, 92]
21/6 lanceolatocristata 76, 79, 85, 94–6, 106 [111]
21/8 aemula 51, 89, 99, 109, [11, 52, 64, 65, 80, 81, 90, 106]
22/1 Polystichum setiferum 50, 51, 67, 74, 77, 82, 83, [57]
22/3 lonchitis [23, 36, 48, 68, 95, 103]

24/2 Thelypteris palustris 19, 51, [71]
24/3 phegopteris 84, [33, 61, H.15]
24/4 dryopteris 38, 45, 51, 52, 102, [11, 61]
24/5 robertiana 47, [37, 45]
26/1 Pilularia globulifera 26, 47, 64, 65, 76, [6, 24, 33]
28/1 Botrychium lunaria 20, 43, 79, 84, H.32
29/1 Ophioglossum vulgatum 74, 79
34/1 Juniperus communis H.8, [50, 52]
35/1 Taxus baccata 75, 76, 87
37/1 Trollius europaeus 79, 102, [H.32]
38/1 Helleborus foetidus (81, 82, 87, 89)
38/2 viridis (51, 102)
*39/1 Eranthis hyemalis 37, 42, Perthshire
43/1 Anemone nemorosa H.4, 7, 10, 11, 22
46/3 Ranunculus bulbosus H.4
46/5 arvensis 44, 48, 51, 68, 73, 75, 86, 109, [111]
46/7 sardous 47, 48, 49, 51, 56, 73, 79, 86, 87, 91–3, 95, 102, 104, [23, 32, 33, 82]
46/9 parviflorus 48, 65, 69, (H.39)
46/10 auricomus 74, 79, 85, 93, 98, H.9
46/11 lingua 35, 96, 109, [3, 4, 74, 105, 111]
46/16 hederaceus 86, 97, 99, H.22
46/17 lenormandii 74, 102, [7]
46/18 tripartitus [7, 15]
46/19 fluitans 90, 91, 103, [41, 44, 46, 48, 50, 63]
46/20 circinatus 5, 81, 86, 90, 92, [2, 10, 51, 65, 105]
46/21 trichophyllus 48, 50, 51, 74, 76, 79, 86, 97, 101, 102, 104–6, H.8, 13
46/23 baudotii 50, 51, 74, 81, 93, 96, 109, [30, 32, 38, 55, 57, 71, 80, 84, H.39]
*47/1 Adonis annua 5, 37, 50, 51, 54, 68, 75, 83, [1, 4]
48/1 Myosurus minimus 49, 51
49/1 Aquilegia vulgaris (79, 87, H.6, 33)
50/1 Thalictrum flavum 44, 51, 97, [45, 75, H.16]
50/3 minus 79, 96, [12, 30, 37, 61]
53/1 Berberis vulgaris 79, 92, 93, 101, 102, 110, [74]

*54/1 Mahonia aquifolium 9
55/1 Nymphaea alba 79, 86, 106, 111, H.4–7, 18, (35, H.12, 13)
56/1 Nuphar lutea 51, 79, 99, 102, H.4, 6, [1, H.20]
56/2 pumila [56, 58, 67, 73, 77, 91, 95]
57/1 Ceratophyllum demersum [44, 51, 84 ,11]
57/2 submersum [29, 83]
58/1 Papaver rhoeas 44, 76, 79, 86, 91, 93, 103, 109
58/2–3 dubium agg. 79, 87, 99–101, H.39
58/4 hybridum 59, (79), [H.4]
58/5 argemone 51, 78, 86–8, 101, 103, 106, 110
*58/6 somniferum 46
59/1 Meconopsis cambrica (9, 39, 86, 92)
61/1 Glaucium flavum 99
62/1 Chelidonium majus 78, 79, 91, 93, H.3–7, 15, 25, 26
65/3 Corydalis claviculata 79, 109, [71]
*67/2 Brassica napus 10, 28, 52
67/3 rapa 30, 51, 73, 78, 79, 86, 87, 89, 91, 93–7, 99, 100, 102, 103, 105–7, 109
67/4 nigra 65, 79, (81)
*68/1 Erucastrum gallicum 5, 45
69/1 Rhynchosinapis monensis 74, (82, 104), [4, 7, 11, 17, 21, 64]
*69/3 cheiranthos 56, 82
*70/2 Sinapis alba 78, 79, 86, 87, 89, 93, 96, 101, 106, 107, 110
*72/1 Diplotaxis muralis 40, 86, 109
72/2 tenuifolia 81, 86, [36]
74/1 Raphanus raphanistrum H.37
74/2 maritimus H.28, [46, 50, 52]
75/1 Crambe maritima 75, 107, [16, 60, H.10]
79/2 Lepidium campestre 42, 52, 67, 74, 79, 93, H.9
79/3 heterophyllum 65, 79
*79/4 Lepidium ruderale 26, 40, 42, 44, 45, 48, 73, 74, 80
79/6 latifolium (39, 44, 50, 51, 80, 107), [5, 60, H.1]
80/1 Coronopus squamatus 44, 86, 88–90, 92, 109, 110
*80/2 didymus 31, 79, 86, 106
*81/1 Cardaria draba 40, 76, 77
82/1 Isatis tinctoria (H.14), [30, 39]
84/1 Thlaspi arvense 109
84/3 perfoliatum [4, 30, 38, 53, 58, 64, 70]
84/4 alpestre 43, [3, 41, 50]
85/1 Teesdalia nudicaulis 79, 81, [6, 33, 34, 41, 42]
87/1 Hornungia petraea [3, 4]
88/1–3 Cochlearia officinalis agg. H.4, 37, [15, 16, 38]
88/4 scotica 93
88/5 danica 35, 109, [90, 93]
88/6 anglica 86, [48, 50, 57, 66, 77, 78, 83, 103, H.7]
89/1 Subularia aquatica 76, 94, 101, [71]
94/2 Draba norvegica 89, 104
94/3 incana [62, H.36]
94/4 muralis 62, 73
95/1–3 Erophila verna agg. H.3, 26
*96/1 Armoracia rusticana 73, 78–80, 84, 92, 96, H.40
97/2 Cardamine amara 35, 49, 79, [1, 2, 7, 8]
97/3 impatiens 49, [2, 4, 51, 53]
97/8 bulbifera (75)
98/1 Barbarea vulgaris 79, 102, 103, H.3
98/2 stricta [S, Ireland]
*98/3 intermedia 43, [9]
*98/4 verna 73, 74, 89, 101, [42]
99/1 Cardaminopsis petraea 100, 107
100/4 Arabis hirsuta 79, 87
100/5 brownii H.3
101/1 Turritis glabra 87, [14, 61]
102/3 Rorippa sylvestris 79, 104, 110
102/4 islandica 79, 87, 89, 92
102/5 amphibia 47, 60, 65, [3, 9, 69, 81, 83, 87, 92, 94, H.3]
103/1 Matthiola incana [45]
103/2 sinuata [14, 15, 51]
*104/1 Hesperis matronalis 46, 86, 100, H.30
105/1 Erysimum cheiranthoides 45, 60, 73, 81, 84, 96, 100, 109

*106/1 Cheiranthus cheiri 79, 86–9, 96, 101, 109, H.14, [H.18]
107/1 Alliaria petiolata 86, 100, 105, H.17, [H.3, 15]
108/1 Sisymbrium officinale 109
*108/2 irio [1, 2, 10, 14, 20, 22, 24, 26, 27, 53]
*108/4 orientale H.30, 34
*108/5 altissimum 12, 42–3, 45, 47–9, 68, 72–5, 77–81, 87, 89, 92–4, 111, H.3, 16, 30, 40
109/1 Arabidopsis thaliana 102
111/1 Descurainia sophia 44, 86, 88, 92, 107, 109
112/1 Reseda luteola 73, 79, 86, 93, 96, 107, H.1, 3, 4, 30
112/2 lutea 42, 48, 50, 84, 86, 92, 96, 106
113/1 Viola odorata H.20, (74, 86, 89, H.32)
113/2 hirta 59, 72
113/5 reichenbachiana 79, 102–4, 109, 112
113/6 canina 42, 76, 86, 107
113/7 lactea [6, 27, 29, 37, 48, 50, 63]
113/8 stagnina [11, 41, 53]
113/9 palustris 84, H.22, 23, 31, [26]
113/11 lutea [1, 4, 5, 26, 52, 60, 100, 102, 104, 110–12]
114/3 Polygala calcarea [1, 29, 108]
114/4 amara [63, S]
115/1 Hypericum androsaemum 106 (81)
*115/3 hircinum 51, Kent
115/5 perforatum 79, 106, H.27, 29
115/6 maculatum 77, 79, 82, 91, 94, 99, 102, 103, [21, 61, 85]
115/8 tetrapterum (111)
115/9 humifusum 109, 112
115/10 linarifolium (40)
115/12 hirsutum 45, 49, 74, 84, 86, 111, [H.9]
115/13 montanum 44, [7, 8, 14, 20, 30, 75, 106]
115/14 elodes 51, H.32, [7, 82]
118/1 Helianthemum chamaecistus 45, [46]
118/3 canum [42]
*120/1–2 Tamarix gallica agg. 16, [12]
121/1 Frankenia laevis (4, 66)
122/1 Elatine hexandra 102, 103, [37, 55, H.39]
122/2 hydropiper 50
123/1 Silene vulgaris 87, H.40
123/2 maritima 35, 66, 77
123/3 conica (32, 46, 66), [7]
123/6 gallica 75, 86, 93, 96
123/10 nutans [27, 32, 33]
123/11 italica 91, [14]
123/12 noctiflora (75, 76, 86, 92), [35]
123/14 alba 76, 79, H.23
124/2 Lychnis viscaria [48, 72, 106]
125/1 Agrostemma githago 37, 44, 63, 84, 86–93, 102, 104, 106, 109, H.37
127/1 Dianthus armeria 52, (65, 73, 74), [35, 87]
127/8 deltoides 44, 50, 75, (23, Cork), [3, 87, 101, 106, H.39]
*129/1 Saponaria officinalis 74, 93, H.16, 23, 27, 30, 36
130/1 Kohlrauschia prolifera [Cornwall, Perthshire]
131/1 Cerastium cerastoides 70, 107
131/2 arvense 49, 52, 79, 86,111
131/4 alpinum [94, 99]
131/5 arcticum [87, 90, 108, 110]
131/8 glomeratum 111
131/10 atrovirens 22, 76, 99, H.37, [55]
131/11 pumilum [1, 37, 41, 46]
131/12 semidecandrum 42, 43, 80, 86, 97, 107, [72]
132/1 Myosoton aquaticum 49, [48, 77, 82, 86, 90, 111]
133/1 Stellaria nemorum 42, 74, 79, 92, 109, [41]
133/3 pallida 40, 63, 102, H.14, 15, Cork
133/4 neglecta 28, 67, 89, 108, 109
133/5 holostea 110, 111, H.28
133/6 palustris 19, 58, 74, 79, [41, 49, 50]
135/1 Moenchia erecta 70, [35, 41]
136/1 Sagina apetala 102, [112]
136/2 ciliata 42, 44, 51, 52, 77–9, 87, 91
136/3 maritima 44, 50, 86

*207/5 Lathyrus tuberosus [35]
207/6   sylvestris 74, 91, 103, [68]
207/9   palustris [22, 41, 57, 59, 62, 69, 74]
207/11   montanus H.25, [26, 28]
*209/1 Spiraea salicifolia 21, 22, 73, 79
210/1 Filipendula vulgaris [72, 77, 91, 92, 109]
211/1 Rubus chamaemorus 45, 49, 73
211/2   saxatilis 59, 79, [1, 2, 51]
211/3   idaeus 111
211/9   caesius 86, 88, 91, [75, 77]
212/1 Potentilla fruticosa [77, H.4]
212/2   palustris (23), [7, 34]
212/3   sterilis [111, 112]
212/4   rupestris [48]
212/6   argentea 50, 81, [5, 7, 35, 49, 52]
*212/8   norvegica 39, 58, 70, 88, 92, 96, [7, 28]
212/11   tabernaemontani [3, 25, 39, 65, 68, 70, 81]
212/12   crantzii [46]
212/14   anglica 68, 78, 79, 81, 86, 87, 98–100, [90–6]
212/15   reptans 84, 86, 107, [105, 111, 112], 101
213/1 Sibbaldia procumbens 70, 73, 78, 91, 93
216/1 Geum urbanum 112
217/1 Dryas octopetala 87, [39, 57]
218/1 Agrimonia eupatoria 99, 109, H.40
218/2   odorata 79, 82
220/1 Alchemilla alpina [57, 62–4]
220/3   vulgaris H.12
222/1 Sanguisorba officinalis 6, 13, 45, 68, 79, [H.40]
223/1 Poterium sanguisorba 90, H.24
*223/2   polygamum 42–4, 46, 72, 77, 83, 90
225/1 Rosa arvensis [81, 84, 88, 89, 91, 106]
225/4   pimpinellifolia 39, 53, 86, 87, H.4, 5, 7, 37, [H.32]
225/14–17 rubiginosa agg. 76, 79, 107, H.16
*226/2 Prunus domestica 74, 78, 79, 89, 91, 98, 100, 107
*226/3   cerasifera 46
226/5   cerasus (59, 83, 90, 91, 94), [6, 30]
226/6   padus 74
*227/2 Cotoneaster simonsii 50, 79, 80
*227/4   microphyllus 39, 56, 65, 67, 70, 80, 88, 89
229/1 Crataegus oxyanthoides [62, 63, 66, 78, 90, 104, S, Ireland]
232/1 Sorbus aria 81, 96
232/7   torminalis 26, 44, 45
*233/1 Pyrus communis 5, 43, 88
233/2   cordata [34–6, 54]
234/1 Malus sylvestris 93, 106
235/1 Sedum rosea 60
235/2   telephium 51, 84, (105, H.13), [H.29]
*235/3   spurium 33
*235/4   dasyphyllum H.37, [H.39]
235/5   anglicum 65, 106, H.25, [19, 28, 38, 39, 92, 112]
235/6   album (79, H.37)
235/8   acre 79
235/10   forsteranum (37, North Yorkshire), [23]
*235/11   reflexum 67, 92, H.12, 32, [105]
235/12   villosum 98
238/1 Umbilicus rupestris 74
239/1 Saxifraga nivalis [83, 103, 111]
239/2   stellaris [42]
239/5   spathularis (H.34)
239/8   tridactylites 67, 68, 71, 79, 96, 104
239/9   granulata 86, 87, (H.37), [105, 111]
239/12   cespitosa 96
239/15   hypnoides 77, 81, 91, [111, H.17]
239/16   aizoides [41, 49, 71, 72]
239/17   oppositifolia 91, 109, [65]
242/2 Chrysosplenium alternifolium [24, 31, 32]
243/1 Parnassia palustris 50, 84, H.37, [46, 48, 110, H.38]
246/1 Ribes sylvestre 73, 79, 101, 109, Cork, [H.2]
246/2   spicatum 61

246/3 Ribes nigrum 46, 51, 52, 86, 99, [H.40]
246/5   alpinum (44, 76, 91, 99, 106), [41]
246/6   uva-crispa 109
247/1 Drosera rotundifolia 56
247/2   anglica 77, [13, 51, 65]
247/3   intermedia 5, 50, 51, 107, [30, 65]
249/1 Lythrum salicaria 77, 79
249/2   hyssopifolia [2, 70, H.12]
250/1 Peplis portula 51, 79, 86, 99, 108, 109
251/1 Daphne mezereum (63, 74, 75), [3, 5, 41, 59, 82]
251/2   laureola (86, 88, 91)
253/1 Ludwigia palustris [13]
254/1 Epilobium hirsutum 79, 87
254/4   lanceolatum [46, 48, 69–79]
254/5   roseum 52, 79, 89, [68, 70]
254/7   adnatum 72, 78, 95, 97, [88, 89, 111, H.21, 40]
254/8   lamyi Cork, [100]
254/9   obscurum 82, H.31
254/11   anagallidifolium 99
254/12   alsinifolium 91
*254/13   nerterioides 23
255/1 Chamaenerion angustifolium H.8, 31
*256/1 Oenothera biennis 77, 90, [35, 50, 51]
*256/3   stricta 18, 79, [32, 63]
258/2–3 Circaea alpina agg. 40, 50, 76, 93, 109, 110, [6, 11, 25, 45, 55, 67, 68, H.16]
259/1 Myriophyllum verticillatum 47, H.4, 9, [1, 41, 45, 50, 51, 95, 111]
259/2   spicatum 78, 79, 87, 93, 97, 99, 100, 107, 109
259/4   alterniflorum 66, 91, 99, [28]
261/1 Hippuris vulgaris 74, H.6, [47]
262/3 Callitriche obtusangula 21, 37, 51, [42]
262/4   intermedia H.5, 30
262/5   hermaphroditica 84, 87, 88, 94, [3, 9, 12, 15, 17, 19, 21, 25, 30, 36–8, 41, 46, 55, 56, 62, 65, 70
262/6   truncata 33,
263/1 Viscum album (67, 68, 70, 72, 74, 89)
264/1 Thesium humifusum [3, 30, 40]
265/1 Thelycrania sanguinea 1, (86, 92, 101, H.3, 32)
270/1 Sanicula europaea 79, 110
272/1 Eryngium maritimum 50, 77
272/2   campestre 1, [71]
273/1 Chaerophyllum temulentum 77, 99, 101
274/1 Anthriscus caucalis 74, 78, 79, 81, 86, 90, 93, [46, 111]
275/1 Scandix pecten-veneris 42, 43, 68, 74, 76, 78, 80, 86–8, 96, 98, 100–2, 106, 108, 109, 112
*276/1 Myrrhis odorata [1, 6, 19]
277/1 Torilis japonica 109
277/2   arvensis 42–4, 47–52, 59, 69
277/3   nodosa 43, 66, 67, 73, 92, 94, 102, 106, [H.22]
*278/1 Caucalis platycarpos 58, 59
*278/2   latifolia 44, 59, 92, 94, [14, 38]
*280/1 Smyrnium olusatrum 86, H.26
281/1 Physospermum cornubiense 1
282/1 Conium maculatum H.40
283/2 Bupleurum rotundifolium (48, 49, 74, 83, 92)
283/4   tenuissimum 52, [2, 3, 60]
284/1 Trinia glauca [7]
285/1 Apium graveolens 67, 89, 104
285/2   nodiflorum 77, 97, [72, 92–5, 112]
285/4   inundatum 88, 99, 107, H.12, 26
*286/1 Petroselinum crispum 12, 17, 40, 58, 75, [51]
286/2   segetum 50, [36, 49, 103]
287/1 Sison amomum 43, 44, [111]
288/1 Cicuta virosa 51, 66, 96–8, 101, [8, 9, 13–16, 35–7, 48, 49, 63, 64, 67, 70, 81, 92, 105, 111
291/1 Carum verticillatum 77, 96, 110
*291/2   carvi 4, 16, 33, 73, 74, 77, 91, H.3, 6, 26, 29
292/1 Bunium bulbocastanum (12, 33), [11]
294/1 Pimpinella saxifraga 97, 98, 109, H.37, 40, [112]

383/1 Blackstonia perfoliata 73, [67]
384/1 Gentiana pneumonanthe 15, [8, 44]
384/2   verna [62]
384/3   nivalis 108
385/1 Gentianella campestris 43, 44, 47, 56, 79, 84, [5, 6, 24, 30]
385/2   germanica [5, 7,   45, 57, 101]
385/3   amarella 47, 86, [H.1, 2, 6, 29, 32, 36–40]
385/4   anglica [14]
385/5   uliginosa [57, 65, 90, 96, 107]
387/1 Nymphoides peltata [54]
389/1 Cynoglossum officinale 66, 88
389/2   germanicum [90]
392/1 Symphytum officinale 79, (97, H.6, 18, 19, 24, 26, 29, 31)
392/6   tuberosum 79, (6, 25, 57, 61, Cork), [24, 29, 35]
*395/1 Pentaglottis sempervirens [42]
397/1 Lycopsis arvensis 79, 86, 87, 99
399/1 Pulmonaria longifolia [17, 51, 61, 63]
399/2   officinalis (66, 71, 73, 74)
400/2 Myosotis secunda 56, [21, 25, 26, H.14]
400/7   sylvatica 74, 86
400/10   ramosissima 73, 109, [75–7, 102]
401/2 Lithospermum officinale (76, 78, 82, 87, 91, 101, 111), [H.38]
401/3   arvense 44, 45, 48, 50, 51, 65, (73, 78, 80, 81, 86, 88, 96, 100, 101, 106)
402/1 Mertensia maritima [1, 4, 11, 54]
403/1 Echium vulgare 43, 73, 96, 98, 102, 109
403/2   lycopsis (10–12, 37)
405/1 Convolvulus arvensis 79, 87, 93, 107, H.18, 29
406/4 Calystegia soldanella [H.34]
407/1 Cuscuta europaea (39, 57, 69, 72, 77, 83, 92), [41, 53, 90]
407/2   epithymum 43, (72, 75–7, 85–7, 90, 92, 93), [42]
410/1 Atropa bella-donna 47, (42, 45, 76, 81, 86, 87, 98, 99)
411/1 Hyoscyamus niger (66, 72, 77, 101, 108)
413/1 Solanum dulcamara 97, [H.10]
413/3   nigrum (76, 86–9, 92)
*415/1 Datura stramonium H.18
416/1 Verbascum thapsus 102, 111, H.16
416/4   lychnitis (62, 64, 89, 99), [10, 14, 29]
416/5   pulverulentum (36)
416/7   nigrum (87–9, 95)
*416/9   blattaria 22, 48, [51]
416/10   virgatum (10, 39, 83), [22, 53]
417/1 Misopates orontium (65)
*418/1 Antirrhinum majus 33
*420/1 Linaria purpurea 74
420/3   repens (87, 93, 100, H.22), [H.40]
421/1 Chaenorhinum minus 49, [S]
422/1 Kickxia spuria [46]
422/2   elatine (75, 79, 86, 90)
424/2 Scrophularia aquatica 68, 78, [75–7, 87]
424/3   umbrosa 39, [1, 2, 13, 24]
424/4   scorodonia [20, H.1, 40]
*424/5   vernalis 18, 21, 73, 86, 93, 99, [27, 48, 55, 61, 75]
*425/1 Mimulus guttatus 32, H.19
*425/3   moschatus [H.37]
426/1 Limosella aquatica 44, [8, 49, 62, 90, 48]
430/2 Veronica anagallis-aquatica H.30
430/3   catenata 10, 90, 112, H.5, 12, 21, 24, 28, 33, 34, 36
430/4   scutellata H.31
430/6   montana 93, [31]
430/10   fruticans [107, 108]
430/12   alpina 72, 87
*430/14   peregrina Yorkshire
430/19   triphyllos [18, H.21]
430/20   hederifolia 74, 76, 86, 97, 99, 102, H.2–4, 16, 24, 29, 33, 40
430/22   polita 72, 76, 78, 79, 86, 91–3, 98, 102, 109, H.31, 33, 37
430/23   agrestis 52
*432/1 Pedicularis palustris 51
432/2   sylvatica H.22
433/1 Rhinanthus serotinus 60, 65, 69, 91, 96, 102, 106, [6, 20, 24, 30, 33 H.8]

434/1 Melampyrum cristatum [7, 8, 37, 40, 66, 70]
434/2   arvense [32, 33, 38, 58]
434/3   pratense 79, H.30, 31
434/4   sylvaticum [67, 68, 73, 75, 80, 94, 106, H.35, 38]
435/2 Euphrasia salisburgensis [3, H.3, 40]
437/1 Parentucellia viscosa 101, (49, 56), [85]
438/1 Bartsia alpina 97, 105
439/1 Lathraea squamaria 79, 97, H.25
440/2 Orobanche purpurea [1, 12, 13, 22, 41, 48, 56]
440/3   rapum-genistae 44, 45, [23, 53, 60, 85, 88, 110]
440/4   alba 62, H.29, [11, 41, 56]
440/5   caryophyllacea [26, 98, Devonshire]
440/6   elatior 10, 38, 39, 40, 42, 49
440/7   reticulata 27, [42]
440/8   minor 40, [48, 67]
440/9   picridis 27
440/10   hederae 11, 42
440/11   maritima [10, 14]
441/1 Pinguicula lusitanica 86, 96, H.30, [41, H.22]
441/3   vulgaris 84, H.37, [5, 8]
442/1-2 Utricularia vulgaris agg. 74, 106, 109, 111, [51, 77]
442/3   intermedia 26, 60, 85, 107, 111, H.4, [H.40]
442/4   minor 74, [8, 23, 99]
444/1 Verbena officinalis 65
445/2 Mentha pulegium 6, 51, [64, 65, 83, 85, 95, 109, H.38]
445/3   arvensis [111, 112]
*445/5   spicata 73, 84, 87, H.4, [74]
*445/6   longifolia 47, 58, 68, 70, 79, 82, 86, 96, 99, 107, 111, H.2, 5, 11, 20, 36, [10, 105, S]
445/7   rotundifolia (100, H.37, 40)
447/1 Origanum vulgare 74, 79, 93, 109, (102)
451/2 Calamintha ascendens 68, 85, 87
451/3   nepeta 33, 36, 38, [35, 43, 45, 49, 51, 54, H.2, 3]
452/1 Acinos arvensis 96, 107, (59, 74, 75, 77)
453/1 Clinopodium vulgare 74, 96, 98
*454/1 Melissa officinalis 25, 26, 28, 51, 63, 70, 80, 91, 93
455/2 Salvia pratensis [21, 25, 26, 34, 41, 52, 56, 62]
455/4   horminoides 44, 50, 51, 75, 86, 106, (48)
456/1 Melittis melissophyllum [34]
458/1 Betonica officinalis 76, [84, H.9, 10]
459/3 Stachys arvensis 84, 99, 107
459/1   germanica [3, 10, 11, 22, 24, 27, 57, 62]
459/5   alpina [14]
460/1 Ballota nigra 86, H.3, 6, 32, [76]
461/1 Galeobdolon luteum (72, 79, 86, 97)
462/1 Lamium amplexicaule 51, 78, 87
462/2   moluccellifolium 78, 92, [31, 50, 63, 84]
462/3   hybridum 67, 78, 80, 84, 86, 92–4, 99, 100, 106
462/5   album 109
465/1 Galeopsis angustifolia 56, 59, 65, 75, 86, 92, 99, [104, 111, H.5]
465/3   segetum 61, 62, [38, 59, 64, 71]
465/5   speciosa H.24, 30, (6, 10), [H.16]
466/1 Nepeta cataria 45, (74, 86), [44, H.3, 19]
468/1 Marrubium vulgare (73)
469/1 Scutellaria galericulata 51, 79, H.7, 10, 11
469/2   minor 19, 72, 74, 75, 77, [33]
*470/1 Teucrium chamaedrys 11, 44, 56, [28]
470/2   scordium [3, 26, H.5, 6]
470/4   scorodonia H.30, [112]
471/1 Ajuga chamaepitys [7, 10, 22, 32, 45, 47]
471/2   reptans [112]
472/2 Plantago media 79, 84, 86–8, 97, 99, (71)
472/4   maritima H.33, [H.13, 14]
472/5   coronopus 76, 92
473/1 Littorella uniflora 84, H.4, 8, 23, 34, [8]
474/1 Wahlenbergia hederacea 33, 37, 66
*475/3 Campanula rapunculoides 91
475/5   persicifolia (1, 84, 99)
475/7   rotundifolia H.37, [H.5]

588/1 Convallaria majalis 51, (68, 82, 86, 94, 109)
589/1 Polygonatum verticillatum 72, [21]
589/2 odoratum (9, 20, 25, 27, 28, 40, 47, 70, 92, 102), [8, 12, 16]
589/3 multiflorum 30, 50, (68), [1, 13]
591/1 Asparagus officinalis 51, (32), [46]
592/1 Ruscus aculeatus (73, 76, 100), [88, 109]
593/1 Lilium martagon (41)
594/1 Fritillaria meleagris 28, (53, 54, 75), [2, 62]
*595/1 Tulipa sylvestris 82
597/1 Gagea lutea 13, 92, 96, (58), [14, 41]
598/1 Ornithogalum umbellatum (42, 50, 92, 93, 102)
*598/2 nutans 50, Somerset, [9, 70]
598/3 pyrenaicum [11, 37]
599/1 Scilla verna [54]
600/1 Endymion non-scriptus H.22
602/1 Colchicum autumnale 24, 60, [10, 20, 51, 83, 85, 87, 94, 107, H.8, 21, 37]
603/1 Paris quadrifolia [H.2]
605/1 Juncus squarrosus H.23, [7]
*605/2 tenuis 51, 79, 83, 87
605/4 compressus 75, 100, 103, [41, 48, 72, 73, 80, 89, 102, 105, S]
605/5 gerardii 56, 86, 88, [7, 8, 40, 77]
605/6 trifidus 111
605/8 inflexus 75, 79, 86, 108, 109, [71]
605/12 filiformis 99, [95]
605/13 balticus [73]
605/14 maritimus 50, 76, 86, 96, H.4, [77]
605/15 acutus [1, 5, 6, 9, 18, 25, 33, 51, 54, 69, 73]
605/17 subnodulosus 43, 51, 75, 76, 102, [77, 83, 84, H.5]
605/18 acutiflorus 111
605/20 alpinoarticulatus [27, 41, 87, 90, 102]
605/24 castaneus 96, [99]
605/25 biglumis 87
605/26 triglumis 101
606/2 Luzula forsteri [6, 39, 41, 46, 70, 75, 109]
606/3 sylvatica 52, H.15
*606/4 luzuloides 38, 40, 60, 83
606/6 spicata 109, [49]
606/7 arcuata 89, 90
606/9 multiflora H.31
606/10 pallescens [85, H.2]
607/3 Allium scorodoprasum [1, 9, 37, 44, 48, 60, H.35]
607/5 vineale 79, 93
607/6 oleraceum 44, 45, 74, 81, 84, [30, 85, H.38]
*607/7 carinatum [H.21]
*607/10 triquetrum [13, 21, 27, 70]
611/2 Leucojum aestivum H.7, [29]
612/1 Galanthus nivalis (86, 89, 93)
614/1 Narcissus pseudonarcissus 43, (74, 76, 86, 87, 89, 92, Kerry)
615/1 Sisyrinchium bermudiana 1
616/3 Iris foetidissima (74, 100)
*618/1 Crocus nudiflorus [14, 34, 65]
*618/2 purpureus 5, 53, 54, [11]
619/1 Romulea columnae [41]
624/1 Cephalanthera damasonium [48, 55, 57, 62, 69, 70]
624/2 longifolia 15, 57, 67
625/1 Epipactis palustris 43, [21, 38, 46, 73, 104]
625/2 helleborine 74, 90, 100
625/3 purpurata [3, 4, 9, 10, 18, 27–9, 39, 46, 49, 54, 57, 61–5, 69, 70, 84, 89]
625/7 atrorubens 65, [3, 4, 6, 13, 17, 28, 34–6, 50, 102]
628/1 Listera ovata 106, H.3
628/2 cordata [3, H.17]
629/1 Neottia nidus avis 1, 44, 86, 91, 92, 94
631/1 Hammarbya paludosa 39, 57, 60, 89, 102
632/1 Liparis loeselii [54, H.11]
634/1 Herminium monorchis [21, 31, 32, 35]
635/1 Coeloglossum viride 18, 42, 79, 84, 87, 99, H.22, 30
636/1 Gymnadenia conopsea 31, 43, 79, 84, H.6

637/1 Leucorchis albida 51, 79, 99, [11, 15, 68, 112]
638/1 Platanthera chlorantha 79, 82, 92, 109
638/2 bifolia 43, [29, 57]
639/1 Neotinea intacta [H.27]
640/1 Ophrys apifera [72–7]
640/2 fuciflora [9, 16, 17, 26]
640/3 sphegodes [7, 30, 40]
640/4 insectifera [105]
641/1 Himantoglossum hircinum [34, 56, 57, 70, H.9]
642/1 Orchis purpurea [21, 24, 53, 54, Cork]
642/2 militaris [15, 16, 19]
642/4 ustulata [H.5, 21]
642/5 morio 43, [2]
643/2 Dactylorchis maculata H.19, 22, 24, 29
643/3 incarnata 38, [37]
643/4 praetermissa [47, 69, 70, 74, 79, 81, 89–92, 96, 104, 105, 110–12, H.3, 9, 16, 21, 34]
643/5 purpurella 44, H.5, 6, 17, 22, 29, 31, 32, 34, [41]

644/1 Aceras anthropophorum [9, 22, 24, 33, 54, 63]
645/1 Anacamptis pyramidalis 73
*646/1 Acorus calamus 65, 72, 86, 93, [8, 84]
649/1 Arum maculatum 79, 102
650/1 Lemna polyrhiza 44, 45, 49, 51, 52, 84
650/2 trisulca 45, 49, 75
650/4 gibba 75, 86, 109, [49, 52, 85, H.9]
652/2 Sparganium emersum 79, 99, H.6, 10, 12, 13, 15, 29
652/3 angustifolium [17, 27, 28, 34, 44, 45, 52, 54, 67, 71, 72, 74, 75, 79–86, 91, 93, 94]
652/4 minimum 79, 111, H.4, [33–5, H.40]
653/1 Typha latifolia 79, 107
653/2 angustifolia 68, [48]
654/2 Eriophorum gracile [37]
654/3 latifolium 51, 60, 81, 99, [7, 30, 53]
654/4 vaginatum 52, [7, 10, 15, 22, 23, 33]
655/2 Scirpus cespitosus [15]
655/3 maritimus 86
655/4 sylvaticus 44, 45, 79, 98
655/5 holoschoenus 11, S, [37]
655/6 triquetrus [14, 27, 55]
655/8 lacustris 79, 88, 99, H.4, [1]
655/9 tabernaemontani 70, 72, 74, 104, 106, H.5
655/10 setaceus 51. H.23
655/11 cernuus 51, 97, 106
655/12 fluitans 51, 99, 107, 109, H.29, [H.23]
656/1 Eleocharis parvula [8, H.29]
656/2 acicularis 26, 43, 51, 66, 74, 77, 78, 87, 96, [71, 108, S]
656/3 quinqueflora 45, 50, 79, 82, 87, H.37
656/4 multicaulis 44, 51, 65, 78, 79, 81, 86, 93, 109, H.6, [30, 31, 35, 47, 87]
656/6 uniglumis 45, 83, 88, 91, H.8, 21
657/1 Blysmus compressus [H.3]
657/2 rufus 50, 76, 77, 83, [25, 27, 46]
658/1 Cyperus longus [27, 39]
658/2 fuscus [62]
659/1 Schoenus nigricans 50, 57, 90, H.4, 6, 12, 29, 37, [63]
660/1 Rhynchospora alba 50–2, 77, 86, 106, H.4, 5, [23, 26, 61]
660/2 fusca [2, 4, 40, 42, 62, 98, H.38]
661/1 Cladium mariscus 46, 72, [51]
663/1 Carex laevigata 76, 78, [23, 33]
663/2 distans 86, [112]
663/3 punctata 52, [70]
663/4 hostiana 15, H.5, 6, 29, 32
663/5 binervis H.22
663/7 lepidocarpa 1, 2, 10, 18, 24, 31, 38, 39, 44, 46, 58, 74, 84–6, H.21, 32, 37
663/11 extensa 50, 99, [53, 83, 92]
663/12 sylvatica H.7
663/13 capillaris 104, 112, [87]
663/15 pseudocyperus 65, 66, Cornwall, [49, 102, H.21]

Appendix III

ACKNOWLEDGEMENTS

COUNTY LISTS

The following supplied more or less complete vice-county lists:

7–8	Wiltshire	J. D. Grose
9	Dorset	R. D'O. Good
15–16	Kent	F. Rose
17	Surrey	Flora Committee
20	Hertfordshire	J. G. Dony
21	Middlesex	D. H. Kent
25–26	Suffolk	Flora Committee
30	Bedfordshire	J. G. Dony
31	Huntingdonshire	J. L. Gilbert
36	Herefordshire	Flora Committee
38	Warwickshire	Flora Committee
39	Staffordshire	E. S. Edees
52	Anglesey	A. D. Q. Agnew and S. W. Greene
53–54	Lincolnshire	Miss J. Gibbons
56	Nottinghamshire	Mr and Mrs R. C. L. Howitt
57	Derbyshire	Flora Committee
61	East Yorkshire	R. D'O. Good and Miss E. Crackles
62 & 65	North Yorkshire	Miss C. M. Rob
71	Isle of Man	D. E. Allen
90	Angus	Miss U. K. Duncan
93	North Aberdeenshire	Flora Committee
95	Moray	Miss M. McCallum Webster
98	Argyll Main	K. N. G. MacLeay
102	South Ebudes	J. K. Morton
H.33	Fermanagh	R. D. Meikle

HELP WITH PARTICULAR SPECIES

The following provided additional data for particular species. Maps of those species marked with an asterisk have been published in the "Biological Flora" series in the *Journal of Ecology* and the year of publication is given in each case. Those marked with a dagger have been published elsewhere and the reference will be found in the Bibliography. A cross indicates species for which there are published papers giving localised distribution data. References to these will also be found in the Bibliography.

15/6	Asplenium viride	C. D. Pigott
21/2	†Dryopteris borreri	Miss P. J. Pugh (1953)
33/1	†Pinus sylvestris	H. M. Steven and A. Carlisle (1959)
45/1	*Clematis vitalba	E. Milne-Redhead
46/16	*Ranunculus hederaceus	C. D. K. Cook
46/23	baudotii	C. D. K. Cook
56/2	*Nuphar pumila	Mrs Y. Heslop Harrison (1955)
66.	Fumaria species	N. Y. Sandwith
87/1	*Hornungia petraea	Denis Ratcliffe (1959)
90/2	Bunias orientalis	B. M. G. Jones
94/4	*Draba muralis	Denis Ratcliffe (1960)
100/4	Arabis hirsuta	B. M. G. Jones
100/5	brownii	B. M. G. Jones
113/7	*Viola lactea	D. M. Moore (1958)
114.	Polygala species	D. R. Glendinning
117/1	*Tuberaria guttata	M. C. F. Proctor (1960)
118.	*Helianthemum species	M. C. F. Proctor (1956)
123/10	*Silene nutans	F. N. Hepper (1956)
127/1	Dianthus armeria	Miss S. S. Hooper
130/1	Kohlrauschia prolifera	Miss S. S. Hooper
131/11	Cerastium pumilum	E. Milne-Redhead

137.	Minuartia species	G. Halliday
142/1	*Spergula arvensis	J. K. New (1961)
146.	Herniaria species	D. E. Coombe and L. C. Frost
154.	Chenopodium species	J. P. M. Brenan
160.	Salicornia species	P. W. Ball
162/1	Tilia platyphyllos	C. D. Pigott
168/17	†Geranium purpureum	H. G. Baker (1955)
170/2	×Oxalis corniculata	D. P. Young (1958)
170/4	× europaea	D. P. Young (1958)
171/1	Impatiens noli-tangere	D. E. Coombe
171/2	capensis	D. E. Coombe
171/3	* parviflora	D. E. Coombe (1956)
180/1	Frangula alnus	P. W. Richards and others
187/2	Ulex gallii	M. C. F. Proctor
187/3	minor	M. C. F. Proctor
192/8	Trifolium molinerii	D. E. Coombe
192/12	bocconei	D. E. Coombe
192/14	strictum	D. E. Coombe
200/1	†Astragalus danicus	C. D. Pigott (1951)
217/1	†Dryas octopetala	C. D. Pigott (1956a)
229/1	Crataegus oxyacanthoides	A. D. Bradshaw
242/2	Chrysosplenium alternifolium	F. Rose
252/1	†Hippophae rhamnoides	E. W. Groves (1958)
254/13	*Epilobium nerterioides	Miss A. J. Davey (1961)
262.	Callitriche species	J. P. Savidge
285/3	Apium repens	R. D. Meikle
320/2	Polygonum raii	B. T. Styles
325.	Rumex species (rare)	J. E. Lousley
330.	Ulmus species	R. H. Richens
335/3	†Betula nana	Miss A. Conolly (1950)
343/21	†Salix herbacea	Miss A. Conolly (1950)
349/1	*Daboecia cantabrica	S. R. J. Woodell (1958)
353/1	*Arbutus unedo	J. R. Sealy and D. A. Webb (1950)
367/2	*Primula scotica	J. C. Ritchie (1954)
385.	Gentianella species	N. M. Pritchard
388/1	*Polemonium caeruleum	C. D. Pigott (1958)
392/2	Symphytum asperum	A. E. Wade
407/1	*Cuscuta europaea	B. Verdcourt (1948)
430/24	†Veronica filiformis	E. B. Bangerter and D. H. Kent (1957)
434/3	Melampyrum pratense	A. J. E. Smith
434/4	sylvaticum	A. J. E. Smith
445.	Mentha species	R. A. Graham and R. M. Harley
448.	*Thymus species	C. D. Pigott (1955)
462/5	Lamium album	Miss A. Conolly and Mrs D. Walker
485/6	Galium pumilum	K. M. Goodway
485/7	sterneri	K. M. Goodway
503.	†Galinsoga species	W. S. Lacey (1957)
506/4	×Senecio squalidus	D. H. Kent (1956, 1960)
513/2	Pulicaria vulgaris	C. D. Pigott
514/2	Filago apiculata	J. E. Lousley
514/3	spathulata	J. E. Lousley
540/6	Cirsium acaulon	C. D. Pigott (1956b)
544/5	×Centaurea jacea	E. M. Marsden-Jones and W. B. Turrill (1954)
559.	Crepis species	J. B. Marshall
562/1	Luronium natans	A. C. Jermy
564/1	Damasonium alisma	C. D. Pigott
573/1	†Scheuchzeria palustris	W. A. Sledge (1949)
577.	×Potamogeton species	J. E. Dandy and G. Taylor (1938–42)
602/1	*Colchicum autumnale	R. W. Butcher (1954)
625.	†Epipactis species	D. P. Young (1952)

641/1 †Himantoglossum hircinum	R. D'O. Good (1936)
643/6 ×Dactylorchis majalis	P. M. Hall (1937)
643/7 † traunsteineri	J. Heslop Harrison (1953)
643. Other Dactylorchis species	E. Milne-Redhead and V. S. Summerhayes
649/2 *Arum italicum	C. T. Prime (1954)
652. Sparganium species	C. D. K. Cook
657. Blysmus species	Mrs G. Crompton
663/10 †Carex serotina	Miss E. W. Davies (1953)
663/38 humilis	D. E. Coombe
663/56 diandra	G. Halliday
675/1 Nardurus maritimus	C. A. Stace
683/3 Bromus benekenii	A. Melderis
692/2 †Avena ludoviciana	Miss J. M. Thurston (1954)
701/1 *Agrostis setacea	R. B. Ivimey-Cook (1959)
715/1 *Nardus stricta	M. J. Chadwick (1960)
716. †Spartina species	Miss J. M. Lambert (1959)

COUNTY REFEREES

The following checked the field cards for the vice-counties listed:

1 O. V. Polunin	61 R. D'O. Good
1a (Scillies) J. E. Lousley	62 Miss C. M. Rob
2 R. W. David	63 & 64 W. A. Sledge
3 & 4 W. Keble Martin	65 Miss C. M. Rob
5 C. C. Townsend	66–68 D. H. Valentine
6 N. Y. Sandwith	69 G. Wilson
7 & 8 J. D. Grose	70 Derek Ratcliffe
9 R. D'O. Good	71 D. E. Allen
10 A. W. Westrup	72–74 H. Milne-Redhead
11 N. D. Simpson	75–77 R. Mackechnie
12 E. C. Wallace	78 P. S. Green
13 O. Buckle	79 & 80 B. L. Burtt
14 E. C. Wallace	81 P. S. Green
15 & 16 F. Rose	82–84 B. L. Burtt
17 D. P. Young	85 Miss C. W. Muirhead
18 & 19 S. Jermyn and B. T. Ward	86 & 87 B. W. Ribbons
20 J. G. Dony	88 & 89 J. Grant Roger
21 D. H. Kent	90 Miss U. K. Duncan
22 & 23 E. F. Warburg	91 J. Grant Roger
24 R. A. Graham and R. M. Harley	92–95 Miss M. McCallum Webster
25 & 26 F. W. Simpson	96 A. Slack
27 & 28 E. L. Swann	96b (Nairn) Miss M. McCallum Webster
30 J. G. Dony	97 E. C. Wallace
31 J. Gilbert	98 K. N. G. MacLeay
32 I. Hepburn	99 A. McG. Stirling
33 C. C. Townsend	100 D. Patton
34 C. C. Townsend and N. Y. Sandwith	101 Miss M. H. Cunningham
35 A. E. Wade	102 J. K. Morton
36 F. M. Day	103 Miss C. W. Muirhead
37 R. C. L. Burges	104 G. Halliday
38 J. G. Hawkes	105–108 E. C. Wallace
39 E. S. Edees	109 Miss M. McCallum Webster
40 R. C. L. Burges	110 J. W. Heslop Harrison and Miss M. S. Campbell
41–43 A. E. Wade	111 I. Hedge
44 Mrs I. M. Vaughan	112 D. H. N. Spence
45–47 A. E. Wade	S D. McClintock
48 P. Benoit	H.1–31 D. A. Webb
49–52 A. E. Wade	H.32 J. Heslop Harrison and Miss M. P. H. Kertland
53 & 54 Miss J. Gibbons	H.33 R. D. Meikle
55 T. G. Tutin	H.34 J. Heslop Harrison and Miss M. P. H. Kertland
56 R. C. L. Howitt	
57 A. R. Clapham and C. D. Pigott	H.35 D. A. Webb
58 W. D. Graddon	H.36–40 J. Heslop Harrison and Miss M. P. H. Kertland
59 & 60 Miss V. Gordon	

GENERAL ACKNOWLEDGEMENTS

The following organisations contributed records:

Natural History Societies
Aberdeen Natural History Society
Altrincham Natural History Society
Andersonian Naturalists of Glasgow
Barnsley Naturalist and Scientific Society
Belfast Naturalists' Field Club
Birmingham Natural History and Philosophical Society
Bishop's Stortford and District Natural History Society
Blackburn Naturalists' Field Club
Bournemouth Natural Science Society
Bradford Naturalists' Society
Bristol Naturalists' Society
Bury Field Naturalists' Society
Cambridge Natural History Society
Caradoc and Severn Valley Field Club
Carlisle Natural History Society
Castleford and District Naturalists' Society
Chepstow Society
Cotteswold Naturalists' Field Club
Craven Naturalists' and Scientific Association
Croydon Natural History and Scientific Society
The Devonshire Association for the Advancement of Science, Literature and Art
Dublin Naturalists' Field Club
Edinburgh Botanical Society
Essex Field Club
Exeter Natural History Society
Hampshire Field Club
Harrogate and District Naturalist and Scientific Society
Haslemere Natural History Society
Herefordshire Botanical Society
Horsham Natural History Society
Isle of Wight Natural History and Archaeological Society
Kettering and District Naturalists' Society and Field Club
Leicester Literary and Philosophical Society
Liverpool Botanical Society
Liverpool Naturalists' Field Club
London Natural History Society
London Region Youth Hostels Association (Field Group)
Loughborough Naturalists' Club
Mid-Somerset Naturalist Society
Montgomeryshire Field Society
Norfolk and Norwich Naturalists' Society
Northamptonshire Natural History Society and Field Club
North Gloucestershire Naturalists' Society
Oswestry Branch, B.E.N.A.
Otley Naturalists' Society
Perthshire Society of Natural Science
Reading and District Natural History Society
Rye Natural History Society
Société Jersiaise
Somerset Archaeological and Natural History Society (Natural History Section)
South Essex Natural History Society
Southampton Natural History Society
South-west Ross Field Study Group
Suffolk Naturalists' Society
Torridge and District Branch, B.E.N.A.
Wakefield Naturalists' Society
West Wales Field Society
Weybridge Natural History Society
Wharfedale Naturalists' Society
Whitby Naturalists' Club, Botanical Section
Wigan and District Field Club
Wild Flower Society
Worcestershire Naturalists' Club
Yorkshire Naturalists' Union

Schools and Colleges
Ackworth School, Yorkshire
Askrigg C. E. School, N. Yorkshire
Babington House School, London, S.E.9
Bartley County Secondary School, Southampton
Batley Primary School, W. Yorkshire
Bedales School, Petersfield, Hampshire
Bede Grammar School, Sunderland
Bedstone School, Shropshire
Beverley Park Camp School, Pateley Bridge, Yorkshire
Bishop's Stortford College, Hertfordshire
Blundell's School, Tiverton, Devon
Blyth School, Norwich
Bootham School, York
Brighton College, Sussex
Bridlington High School for Girls
Brinkley School, Cambridgeshire
Brockhill School, Hythe, Kent
Buckhaven High School, Fife
Cedars School, Leighton Buzzard, Bedford
Clunbury C.E. School, Shropshire
Collyer's School, Horsham, Sussex
County Secondary Grammar School, Newport, Isle of Wight
Dauntsey's School, Devizes, Wiltshire
Doncaster Grammar School, Yorkshire
Dulwich College, London, S.E.21
East End Boys' School, Pembroke
East Suffolk Primary Schools
Essex Primary Schools
Eton College, Buckinghamshire
Fakenham Grammar School, Norfolk
Felsted School, Essex
Friends' School, Saffron Waldon, Essex
Furzedown College, London, S.W.17
Glendale County Secondary School, Wooler, Northumberland
Greek Street High School, Stockport, Cheshire
Gresham's School, Holt, Norfolk
Halstead Grammar School, Essex
Hammond's Grammar School, Swaffham, Norfolk
Hanbury Secondary Modern School, Church Langton, Leicestershire
Harrogate Grammar School, Yorkshire
Harrow School, Middlesex
Hill Brow School, Brent Knoll, Somerset
Hull Grammar School, E. Yorkshire
Hutton Grammar School, Preston, Lancashire
Ilfracombe Grammar School, N. Devon
Ipswich High School, Suffolk
Kennington County Secondary School, London, S.W.9
Kichen Grammar School, Bitterne, Hampshire
Kimbolton School, Huntingdonshire
King Edward's High School for Girls, Birmingham, 15
Kingham Hill School, Kingham, Oxfordshire
King's School, Worcester
King's Warren School, London, S.E.18
Kirkby Lonsdale National School, Westmorland
Lady Eleanor Holles School, Hampton, Middlesex
Lancing College, Sussex
Leighton Park School, Reading, Berkshire
Liskeard Grammar School, Cornwall
Llandysul Grammar School, Cardiganshire
Lord Wandsworth College, Long Sutton, Hampshire
Magdalen College School, Oxford
Manchester High School, Manchester, 14
Marlborough College, Wiltshire
Merthyr Tydfil County Grammar School, Glamorgan
Midhurst Grammar School, Sussex
Minehead Grammar School, Somerset
Neath Grammar School, Glamorgan
Newquay County School, Cornwall

Norfolk Primary Schools
Otterburn County Primary School, Northumberland
Oundle School, Northamptonshire
Palmer's School, Grays, Essex
Penistone Grammar School, W. Yorkshire
Pontygof Girls' School, Ebbw Vale, Monmouthshire
Queen's University Natural History Society, Belfast
Richmond High School, Yorkshire
St Hilda's School, Whitby, N. Yorkshire
St Margaret's P.N.E.U. School, Ludlow, Shropshire
Scalby Modern School, Newby, Scarborough, N. Yorkshire
Scarborough Girls' High School, N. Yorkshire
Sexey's School, Bruton, Somerset
Southport High School for Girls, Lancashire
Stoke Row Primary School, Henley, Oxford
Stranmillis Training College Field Study Society
Thornbury County Secondary School, Gloucestershire
Tiffin Boys' School, Kingston-on-Thames, Surrey
Uppingham School, Rutland
Upton County School, Wiveliscombe, Somerset
Ursuline Convent School, Co. Sligo
Wellingborough School, Northamptonshire
Wellingborough Grammar School, Northamptonshire
Wellingborough High School, Northamptonshire
Whittingham C.E. School, Alnwick, Northumberland
Winchester College, Hampshire
Withycombe C.E. School, Somerset

Other organisations
Field Study Centres
    Dale Fort, Pembrokeshire
    Flatford Mill, Suffolk
    Juniper Hall, Surrey
    Malham Tarn, Yorkshire
    Preston Montford, Shropshire
    Slapton, Devon
Officers of the National Agricultural Advisory Service

INDIVIDUAL CONTRIBUTORS

W. L. Abbott, Mr F. W. Adams, The Revd Canon J. H. Adams, Mr. K. J. Adams, Dr E. M. Adcock, Mr A. D. Q. Agnew, Dr J. K. Aiken, Mr J. Ainsworth, Mr J. R. Aitken, Miss P. Alexander, Mr D. E. Allen, Miss G. E. Allen, Mr K. G. Allenby, Miss R. Allinson, Mrs E. L. Almond, Mr A. H. G. Alston, Dr K. L. Alvin, Mr F. Ambrose, Miss. S. R. Amner, Mr D. J. Anderson, Mrs E. Anderson, Miss M. C. Anderson, Mr C. E. A. Andrews, Mr M. V. Angel, Dr T. H. Angel, Mrs A. B. Angus, Mr J. Anthony, Mrs J. Appleyard, Miss K. M. Archard, Miss M. F. Archer, Mrs B. Archibald, Miss P. R. K. Armitage, Mr M. A. Arnold, Mr G. M. Ash, Miss M. Ash, Mrs G. Ashton, Mr D. L. Ashworth, Mr J. Ashworth, Mrs D. Aston, Mr R. S. Atkinson, Mrs B. Auld, Mr R. A. Avery, Mr P. G. Awcock.

Capt. D. H. L. Back, Miss H. Baelz, Mrs G. D. Baigent, The Revd Dr D. S. Bailey, Mr G. S. Bailey, Miss E. C. Bain, Miss M. Baker, Mr A. Ball, Mrs M. D. Ball, Dr P. W. Ball, Miss R. M. Ball, Mr J. O. Ballard, Mr E. B. Bangerter, Miss S. M. Banks, Mr S. D. Bannister, Mr P. J. T. Barbary, Mrs Barker, Miss J. Barker, Mrs J. E. Barker, Miss L. Barker, Mr P. A. Barker, Mr R. W. Barker, Mr W. G. Barker, Mr F. C. Barnes, Miss R. M. Barnes, Miss M. E. Barnsdale, Mr W. M. M. Baron, Miss E. M. Barraud, Mr C. P. Barrett, Mrs J. H. M. Bartlett, Mr D. D. Bartley, Miss F. M. Barton, Col. H. R. Barton, Mr T. H. C. Bartrop, Mr E. B. Basden, Miss M. E. Bastow, Lady Olivia Bates, Mr S. Batey, Mr D. J. Bauer, Miss D. Baylis, Miss J. M. Beadon, Mrs. L. W. Beasley, Miss E. P. Beattie, Mr K. A. Beckett, The Revd A. J. C. Beddow, Miss G. Bell, Mr T. M. Bell, Mr M. Bendix, Mr H. Bendorffe, Mrs M. W. Benn, Mr P. M. Benoit, Lieut.-Col. C. J. F. Bensley, Mr B. Bentley, Mr F. Bentley, Miss P. M. Bevan, Dr Biggar, Miss E. I. Biggar, Miss M. Biller, Miss M. B. Bing, Mr F. J. Bingley, Mr J. D. Birchall, Mr A. J. Bird,

Mr E. L. Birse, Mr O. N. Bishop, Miss D. B. Blackburn, Dr K. B. Blackburn, Miss P. Blackhall, Miss F. M. Blackhurst, Dr H. Blackler, Miss E. M. Blackwell, Mrs K. M. Blades, Miss N. M. Blaikley, Mrs P. Bland, Mr D. W. Bloodworth, Mrs M. L. Bolitho, Mr W. A. Bollard, Mrs A. V. Bolton, Mrs G. F. Bolton, Miss K. Bolton, Mr C. J. Bond, Mr R. A. Boniface, Mr H. W. Boon, Miss B. Booth, Miss E. M. Booth, Mr B. N. Boothby, Mr S. T. Bormond, Mr T. A. Bowbeer, Dr H. J. M. Bowen, Miss S. Bower, Miss J. Bowman, Mr R. P. Bowman, Miss M. P. Boycott, Mrs A. K. Boyd, Mr H. J. D. Boyd, Mr J. Boyd, Mr R. K. Braddon, Miss V. Bradley, Mr A. D. Bradshaw, Mr E. Bradshaw, Dr M. E. Bradshaw, Mr J. P. M. Brenan, Mrs M. A. Brewins, Lady Anne Brewis, Mr J. N. Brierly, Dr D. Briggs, Mrs M. Briggs, Miss M. A. Briggs, Mr D. Brightmore, Mr J. Bromwich, Mr B. S. Brookes, Mr H. A. Brookman, Miss B. Brown, Mrs E. D. Brown, Mr G. M. Brown, Mr J. J. Brown, Miss M. I. Brown, Dr N. Brownbridge, Mr R. K. Brummitt, Mr J. P. Brunker, Mr C. J. Bruxner, Miss J. Buchanan, Mr C. Bucke, Mr O. Buckle, Mr K. Budd, Mr A. L. Bull, Mr A. T. Bull, Miss F. M. Bull, Mr J. G. Bull, Mr K. E. Bull, Miss E. R. Bullard, Mrs D. E. Bunce, Mr T. F. Buntin, Mr B. T. Bunting, Mr W. Bunting, Mr D. G. Burch, Miss C. Burchardt, Miss P. H. Burford, Dr R. C. L. Burges, Mr I. H. Burkill, Mr K. F. P. Burkitt, Sir David Burnett, Mr J. Burns, Jnr., Mr J. S. Burns, Miss B. A. Burrough, Mr E. A. Burrows, Mr J. S. Burton, Mr R. F. Burton, Mr B. L. Burtt, Miss E. A. Bush, Dr R. W. Butcher, Mrs E. Butler, Miss G. Butler, Miss J. D. Butler, Mrs Y. M. Butler, Mr A. Butterfield.

Miss D. A. Cadbury, Mr J. Cadbury, Mrs C. M. A. Cadell, Mr J. R. Cadman, Mr A. G. Cadogan, Dr H. L. Caldwell, Miss M. W. Caldwell, Dr E. O. Callen, Miss K. Calverley, Mr W. D. Calvert, Mr W. V. Calvert, Dr A. J. Campbell, Mrs G. Campbell, Miss M. Campbell, Miss M. S. Campbell, Mr J. F. M. Cannon, Mr K. Scott Cansdale, Mrs E. N. Cardo, Mr J. Carlyle, Miss D. Carr, Mr J. W. Carr, Mr E. N. Carrothers, Miss V. M. Caswall, Mr R. G. Cave, Mr M. J. Chadwick, Mr T. Chadwick, Mr D. Chamberlain, Mr E. Chambers, Miss M. Chambers, Mrs A. H. Chandler, Mr J. H. Chandler, Mr G. M. Chapman, Professor V. J. Chapman, Mr S. G. Charles, Mr A. O. Chater, Dr E. H. Chater, Mr D. M. Cheason, Mr E. Chicken, Miss L. M. Child, Mr D. M. Chillingworth, Mr L. J. Churchill, Professor A. R. Clapham, Mr C. Clapham, Mr M. C. Clark, Dr W. A. Clark, Mrs D. A. Clarke, Mr J. H. Clarke, Mr J. W. Clarke, Mr R. Clarke, Mr R. A. R. Clarke, Mrs R. R. Clarke, Mr S. J. Clarke, Mr D. V. Clish, Dr R. S. Clymo, Miss L. E. Cobb, Miss A. W. Cochrane, Mrs O. Cochrane, Mr M. H. Cocke, Lieut-Col. J. Codrington, Miss C. I. Coe, Mrs M. C. Cohen, Mr J. A. Cole, Dr M. J. Cole, Miss I. Coles, Mr M. G. Collett, Mr T. G. Collett, Miss P. Collier, Mr R. E. Collins, Miss E. Collyer, Miss E. R. T. Conacher, Mr W. Condry, Miss A. Conolly, Dr C. D. K. Cook, Mr F. Cooke, Dr D. E. Coombe, Mrs J. I. Coomber, Mrs H. C. Copestake, Dr R. E. C. Copithorne, Mr W. O. Copland, Mrs M. Cordiner, Mrs E. A. Cormack, Mr R. Corner, Mrs W. Corry, Mr V. Cory, Mr C. P. J. Coulcher, Mr D. A. Coult, Mrs J. M. Courtenay, Mr J. B. Coutts, Mr L. A. Cowcill, Mrs K. Coxhead, Mr J. A. Crabbe, Miss E. Crackles, Mr E. Crapper, Mr W. S. Craster, Mr S. A. Craven, Dr A. Crawford, Mr G. I. Crawford, Miss M. Crocker, Mrs H. E. Crockett, Mrs G. Crompton, Mrs M. Crookston, Mr A. C. Crundwell, Mr H. Crute, Mr J. Cullen, Mrs M. Cullen, The Revd J. C. Culshaw, Mr B. M. Cunliffe, Miss M. H. Cunningham, Miss C. Curle, Mr A. Currie, Mrs G. C. Curtis, Mrs R. Cusack, Mr J. Cusden.

Miss A. Daisley, Dr D. H. Dalby, Mrs H. Dales, Miss D. G. D'Alton, Mr M. D'Alton, Miss E. Dampier-Child, Mr J. E. Dandy, Miss D. Darlow, Mrs A. Daughty, Dr A. J. Davey, Mrs M. Davey, Mr F. David, Mr R. W. David, Dr J. H. Davie, Mr H. Q. Davies, Mr R. G. Davies, Mr S. J. J. Davies, Mr W. J. Davies, Dr P. H. Davis, Mr T. A. W. Davis, Mr R. Davison, Miss V. Davison, Miss N. Dawson, Mr F. M. Day, Mr G. Day, Mr R. O. J. Day, Mr B. de Graff, Mrs E. Deighton, Miss M. P. Denne, Mr T. E. Dennis, Mrs B. Denny, Miss K. Dent, Mr B. J. Deverall, Miss D. E. de Vesian, Dr A. F. Devonshire, Mrs G. E. Dickinson, Mr J. Dickson, Mr R. Dix, Mr A. J. Dodd, Mr J. D. Dodge, Mr C. J. Dolloway, Mr H. P. Donald, Mrs W. E. Donaldson, Dr J. G. Dony,

Mrs E. F. Doran, Miss N. J. Drake, Mr R. Drane, Mr M. Dransfield, Miss G. Drennan, Miss J. M. Drewitt, Mr B. F. T. Ducker, Miss B. Duckers, Mrs D. L. Duckworth, Dr E. Duffey, Mrs J. E. Duncan, Mrs M. J. Duncan, Miss U. K. Duncan, Mrs E. Dunn, Dr M. D. Dunn, Mrs K. Dunstan, Mr T. W. J. D. Dupree, Mr R. C. Dyason, Mr W. J. P. Dyce.

Dr N. B. Eales, Mr L. H. Easson, Miss J. G. Eastwood, Dr N. B. Eastwood, Mrs G. M. Ede, Mr E. S. Edees, Mr G. A. Edgill, Mr M. Edmunds, Mr R. Edwards, Mrs R. S. Edwards, Dr W. J. Eggeling, Mrs F. M. Elder, Lady Alethea Eliot, Mr T. T. Elkington, Mrs E. I. A. Ellershaw, The Revd E. A. Elliott, Mrs G. Elliott, Dr R. J. Elliott, Mr E. A. Ellis, Miss G. Elwell, Mr R. E. Emms, Miss M. Etherington, Mrs S. W. Eunson, Mr A. S. Evans, Dr E. M. Evans, Mrs F. Evans, Mr G. Evans, Mr G. L. Evans, Mr I. M. Evans, Mr J. B. Evans, Mr M. Evans, Mr T. Evans, Mrs B. Everard, Mr G. H. Ewing, Mr A. W. Exell.

Mr J. Fairweather, Miss M. W. Fallows, Mr D. B. Fanshawe, Mrs E. Farnol, Mr D. Farren, Mr J. R. Faulkner, Dr A. F. Fenton, Mr R. E. C. Ferreira, Miss J. M. Ferrier, Mr W. E. H. Fiddian, Mr J. A. Field, Mr J. L. Fielding, Mr F. Fincher, Mr I. C. Fitch, Mr G. W. Gill, Mr R. S. R. Fitter, Mrs V. H. Fitzgerald, Mrs M. A. Fixsen, Mr B. Flannigan, Mr G. J. Fleming, Mr J. R. Flenley, Mr W. W. Fletcher, Miss D. Fonge, Miss C. Forrest, Mr J. D. Forrest, Mr J. L. Forrest, Mr J. T. Forrest, Miss C. Forsyth, Mrs E. Foster, The Revd A. F. Fountain, Dr B. W. Fox, Mrs W. Frame, Mrs S. Francis, Mr K. A. Franey, Mr B. Frankland, Mr J. N. Frankland, Mr J. Frankton, Miss A. D. French, Lieut-Col. F. H. R. French, Dr L. C. Frost, Miss L. W. Frost, Miss D. M. Frowde, Mr F. G. Fuller, Mr L. Fullerton, Mr J. Fulton, Mrs S. Furse.

Mr J. C. Gardiner, Mr R. J. Garland, Mr G. W. Garlick, Mrs B. E. M. Garratt, Mr D. Garrett, Miss M. Garvie, Mrs E. Gass, Mr A. Gaunt, Dr J. Gay, Dr P. A. Gay, Mr C. J. Gent, Miss M. B. Gerrans, Miss J. Gibbons, Miss A. R. Gibbs, Mrs A. N. Gibby, Mrs G. M. Gibson, Mr J. L. Gilbert, Mr O. L. Gilbert, Mr H. Gilbert-Carter, Mr B. W. Gill, Mr J. A. Gilleghan, Mrs S. Gillett, Mr J. B. Gillies, The Revd R. A. Gilman, Mr J. S. L. Gilmour, Mrs M. R. Gilson, Dr C. H. Gimingham, Mr D. R. Glassford, Mr D. R. Glendinning, Mrs J. Glennie, Miss G. Glover, Professor H. Godwin, Professor R. D'O. Good, Mr A. A. Goodhead, Mr F. D. Goodliffe, Miss C. M. Goodman, Mr G. T. Goodman, Dr K. M. Goodway, Dr A. K. Gordon, Mrs S. Gordon, Miss V. Gordon, Mr M. A. T. Gove, Mr W. D. Graddon, Mrs E. Graham, Mrs E. I. Graham, The Revd G. G. Graham, Mr R. A. Graham, Mr G. G. Grahame, Mr D. R. Grant, Mr T. E. C. Graty, Miss I. F. Gravestock, Mrs D. H. Grayson, Mrs M. Green, Mrs M. R. Green, Mr P. S. Green, Mrs B. M. Greene, Mr S. W. Greene, Miss A. Greenwood, Mrs M. Greenwood, Dr J. Greer, Mr J. B. Gregory, Miss M. Gregory, Capt. O. Greig, Mrs J. Grieve, Miss G. A. Griffiths, Mr J. D. Grose, Mrs R. G. B. Groundwater, Mr E. W. Groves, Miss J. M. Guest, Mrs A. H. Gurney, Miss C. Gurney.

Mr E. C. M. Haes, Mr F. T. Hall, Mr & Mrs P. C. Hall, Mr R. H. Hall, Mr W. A. Hall, Mr & Mrs A. D. Hallam, Dr G. Halliday, Mrs J. Halliday, Mr M. P. Halstead, Mrs A. P. S. Hamilton, Miss M. N. Hamilton, Mr T. S. Hamilton, Miss S. Hamilton-Meikle, Miss M. F. Hancock, Miss M. Handley, Mr W. Handyside, Mr M. K. Hanson, Mrs M. Harber, Mr W. H. Hardaker, Mr R. E. Hardy, Mrs E. N. Harford, Mrs A. Hargreaves, Mr R. M. Harley, Mr C. E. J. Harper, Mr J. Harper, Miss A. S. Harris, Mrs J. A. Harris, Mrs S. Harris, Dr B. D. Harrison, Professor & Mrs J. Heslop Harrison, Professor J. W. Heslop Harrison, Mrs R. N. Harrison, Mrs H. G. Harvey, Mr M. J. Harvey, Dr J. K. Hasler, Mrs A. W. Haslett, Mr J. P. Hatton, Professor J. G. Hawkes, Mr J. H. Hawkins, Mrs N. M. Hawkins, Mr J. A. Hay, Miss M. G. Hay, Mrs D. E. Haythornthwaite, Dr L. A. F. Heath, Mr I. C. Hedge, Mr G. E. C. Hemingway, Miss C. Hemsley, Mr J. H. Hemsley, Mr D. M. Henderson, Mrs P. Henderson, Mrs R. Henning, Miss P. Hensman, Mr I. Hepburn, Mrs M. I. Hepburn, Mr F. N. Hepper, Mr T. F. Hering, The Revd Canon G. A. K. Hervey, Miss R. T. Heward, Mr D. G. Hewett, Dr V. H. Heywood, Mr R. H. Higgins, Mrs R. Hild, Mrs E. M. Hill, Miss P. M. Hill, Mr S. T. Hill,

Mrs M. P. Hill-Cottingham, Miss E. M. Hillman, Mrs J. C. Hilton, Mr G. Hind, Mr D. J. Hinson, Mrs M. C. Hockaday, Miss B. Hodgson, Mr J. Hodgson, Mrs L. Hoggett, Miss S. Holbourn, Miss K. M. Hollick, Mr H. C. Holme, Mr P. H. Holway, Miss B. Hooper, Miss S. S. Hooper, Dr J. F. Hope-Simpson, Mr B. Hopkins, Miss B. A. Hopkins, Mr W. J. Hopkins, Mr J. M. Hopkinson, Mrs S. E. Hopton, Dr D. A. Hopwood, Mrs M. H. Hornby, Mr W. B. Hornby, Mr J. Horsman, Mrs L. M. Hort, Mr E. K. Horwood, Mrs G. I. Hoskins, Mrs F. Houseman, Miss M. A. Howard, Mr C. A. Howe, Miss J. S. Howes, Mr & Mrs R. C. L. Howitt, Dr C. E. Hubbard, Miss J. M. Hubbard, Miss G. M. Hughes, Dr M. G. Hughes, Mr R. Hull, Mr D. C. Hulme, Miss S. A. Humphreys, Mr D. A. J. Hunford, Mrs A. K. Hunt, Mrs D. E. Hunt, Mr J. B. Hunt, Mr P. F. Hunt, Mrs S. Hurd, Miss B. Hurst, Mrs C. C. Hurst, Mr G. R. Hutson, Mrs L. T. Hyde.

Mr B. Ing, Capt. E. R. Ingles, Mrs W. Inglesfield, Miss H. Inglis, Mr H. A. P. Ingram, Miss S. M. Ingram, Mr W. F. Irvine, Miss M. Irwin, Miss E. M. C. Isherwood, Dr R. B. Ivimey-Cook.

Mr A. L. Jackson, Mr K. Jackson, Mr P. G. Jackson, Mrs J. T. Jacobs, Miss P. Jagoe, Mr H. I. James, Miss R. D. Jaques, Miss S. Jaques, Mr N. Jardine, Dr F. M. Jarrett, Mr N. Jee, Mr R. L. Jefferies, Mr C. Jeffrey, Miss R. Jellis, Mr A. C. Jermy, Mr S. T. Jermyn, Miss H. Jerrold, Mrs S. Jewell, Mrs R. V. Johns, Miss M. S. Johnson, Miss F. K. Johnston, Mr T. O. Johnston, Mr M. F. Johnstone, Mr A. S. Jones, Mr A. Vaughan Jones, Mr B. D. Jones, Mr B. M. G. Jones, Dr D. G. Jones, Dr E. W. Jones, Mr H. S. Jones, Miss K. Jones, Dr M. D. G. Jones, Mr T. W. W. Jones, Mr K. Joy, Mr B. E. Juniper.

Mr J. M. Kahn, Mr S. L. M. Karley, Mr R. Kaye, Mrs G. D. Keech, Mrs A. Kelham, Mrs C. Kendon, Mr F. M. Kendrick, Miss A. Kennedy, Mr A. Kenneth, Mr D. H. Kent, Miss M. B. Kent, Miss M. P. H. Kertland, Mr D. Keys, Mr L. N. Kidd, Mr J. A. Kiernan, Dr B. A. Kilby, Miss R. Kilby, Mr & Mrs D. E. Kimmins, Mrs E. L. King, Miss J. Kirk, Mr F. W. Knaggs, Miss B. A. Kneller, Dr J. T. H. Knight, Mr M. E. Knight, Mr R. Knowles, Mr R. B. Knox.

Dr W. S. Lacey, Miss J. Lamb, Dr J. M. Lambert, Mr R. A. Langdale-Smith, Miss N. E. Langley, Mr C. Langridge, Mr D. A. Lannon, Mr L. Larsen, Mr J. R. Laundon, Mr G. F. Lawrence, Mr I. C. Lawrence, Mr G. E. Laycock, Mr F. Leach, Miss V. M. Leather, Mr C. H. Lee, Mr A. W. Leftwich, Lady Edith Legard, Miss E. Lemon, Lady Lennard, Mrs F. Le Sueur, Dr B. J. Levy, Mr H. R. Lewis, Miss I. D. Lewis, Mrs L. Lewis, Mr R. Lewis, Mr W. D. Linton, Mrs M. Little, Mr K. G. Littlejohn, Mrs E. R. Littlewood, Miss A. Livesey, Miss S. Livingstone, Mr P. S. Lloyd, Miss S. F. Lloyd, Mrs N. Lockhart-Mure, Mrs Long, Mr A. G. Long, Miss D. A. C. Long, Mr R. A. Long, Miss C. E. Longfield, Mr R. E. Longton, Mr J. W. Longworth, Mr J. E. Lousley, Mr P. F. Lumley, Miss F. E. Lutener, Dr A. G. Lyon, Dr A. M. Lysaght.

Miss C. McAlister, Mr D. McClintock, Miss J. McClure, Miss A. M. McCosh, Mr D. J. McCosh, Miss M. M. Macdonald, Mr W. A. B. Macdonald, Miss B. L. McFarlane, Mr J. McGrath, Mrs M. MacInnes, Mrs F. M. McIver, Mr J. R. Mackay, Miss M. McKechnie, Mr R. Mackechnie, Mr G. C. Mackenzie, Mrs G. Mackie, Mrs M. P. Mackie, Professor K. N. G. MacLeay, Mr G. MacLennan, Mr A. MacLeod, Dr A. M. Macleod, Mr R. MacLeod, Mr R. D. Macleod, Miss C. Macmahon, Mrs N. McMillan, Mrs V. J. Macnair, Mr I. H. McNaughton, Dr J. McNeill, Dr P. Macpherson, Dr D. N. McVean, Mr I. M'Whan, Mr H. Mace, Mr L. Magee, Mrs M. B. Mallinson, The Revd T. Maloney, Miss A. M. Maltby, Mr S. A. Manning, Miss A. L. Marchant, Mr L. J. Margetts, Mr M. J. Marriott, Miss P. Marsh, Mr J. B. Marshall, Miss A. Martin, Mr J. Martin, The Revd W. Keble Martin, Mr W. R. Masefield, Mrs K. M. Mason, Mr G. A. Matthews, Mr R. F. May, Mr R. Maycock, Mr H. T. Mayo, Mr T. F. Medd, Miss H. D. Megaw, Miss S. Megaw, Mr R. D. Meikle, Professor M. F. M. Meiklejohn, Dr A. Melderis, Dr R. Melville, Miss M. J. Mence, Mr G. Messenger, Mr H. Meyer, Miss B. Miall, Lady Miles,

Mr B. Miles, Mr G. G. H. Miller, Mr G. R. Miller, Mr H. Miller, Miss H. M. Miller, Dr J. N. Mills, Mr A. T. Milne, Mr G. G. Milne, Dr J. Milne, Mr E. Milne-Redhead, Dr H. Milne-Redhead, Miss M. E. Milward, Mr R. Minor, Mrs N. Mitchinson, Mr H. Molgaard, Mrs E. J. Montgomery, Mr J. McK. Moon, Mr D. M. Moore, Miss F. A. Moore, Mrs L. Moore, Miss B. M. C. Morgan, Mr G. H. Morgan, Mrs M. J. Morgan, Mr L. E. Morris, Mr M. Morris, Mr C. M. Morrison, Miss E. M. Morrison, Dr M. E. S. Morrison, Mr N. R. Morrison, Mrs R. H. Mortis, Mr J. D. Morton, Dr J. K. Morton, Miss M. Morton, Miss A. Mowat, Miss C. W. Muirhead, Miss V. Muller, Mrs I. S. Mulloy, Mrs D. Munro-Cape, Mrs M. V. Murdoch, Mr F. Murgatroyd, Miss V. W. Mugatroyd, Miss R. J. Murphy, Mrs C. W. Murray, Mrs D. Murray, Mr M. Murray, Mrs M. A. L. Murray, Mrs M. S. Myers.

Miss B. Nash, Mrs C. D. Needham, Miss M. Needham, Mr E. C. Neighbour, Dr G. A. Nelson, Mr P. J. M. Nethercott, Mrs M. Neville, Miss B. Nevinson, Dr J. Newbould, Dr P. J. Newbould, Miss S. Newlands, Miss S. E. Newton, Miss I. G. Nicholson, Mr E. Nimmo, Miss E. R. Noble, Mr C. R. Nodder, Mrs J. M. Nolder, Miss G. Norman, Dr L. G. Norman, Mrs M. Norman, Miss M. M. Norman, Mr P. R. Norman, Mr R. F. Norris, Dr V. Norris, Miss D. E. North, Mr M. Northway, Mr A. P. Norwood.

Mr T. G. & Miss S. Odell, Mrs J. E. Ogston, The Revd Father P. H. O'Kelly, Miss H. Öpik, Mr E. E. Orchard, Mr P. D. Orton, Mr J. T. Osborne, Mr P. H. Oswald, Miss F. E. Other, Mr J. Ounsted, Miss S. E. Outhwaite, Mr G. Owen, Mr R. E. Owens, Mr J. E. C. Oxenham.

Mr J. E. Page, Mr R. C. Palmer, Mr W. H. Palmer, Mr J. S. R. Pankhurst, Mrs C. B. Parker, Mr H. Parker, Mr P. F. Parker, Mr R. E. Parker, Mrs M. Parkinson, Mrs P. A. Parr, Mrs A. Parris, Miss G. I. Parry, Mrs D. Parsons, Miss N. Parsons, Mrs F. Partridge, Dr J. S. Pate, Mr A. Paterson, Mrs E. F. Paterson, Mrs D. Paton, Mrs P. C. Paton, Dr D. Patton, Mrs V. N. Paul, Miss E. M. Payne, Mrs J. Payne, Mr R. M. Payne, Mr B. W. Pearson, Mr E. L. P. Pearson, Dr P. C. Pecket, Miss H. A. Peden, Mrs M. H. Peebles, Mrs J. H. Pennington, Mr T. D. Pennington, Dr M. S. Percival, Mr C. Perraton, Mr S. Perring, Dr C. P. Petch, Mr J. H. G. Peterken, Mrs J. D. Peterson, Mrs M. C. Peterson, Mrs J. Petford, Mr A. Pettet, Miss A. K. Pettit, Mr D. Philcox, The Hon. Gwenllian Philipps, Miss L. G. Phillipps, Miss V. I. Phillips, Mr G. G. Pierce, Dr C. D. Pigott, Mr F. M. Pilkington, Dr M. G. Pitman, Mr P. B. Pitman, Mr H. J. Platt, Mr G. Platten, Miss D. Pocock, Mr R. J. Pollitt, Mr O. V. Polunin, Professor & Mrs M. E. D. Poore, Mrs H. M. Porteous, Mr I. T. Prance, Mr H. M. Pratt, Mr T. F. Preece, Mrs R. Prendergast, Mrs A. E. Pressly, Mrs J. M. R. Price, Mr J. W. Price, The Revd A. L. Primavesi, Dr C. T. Prime, Dr N. M. Pritchard, Mr G. A. Probert, The Revd H. G. Proctor, Dr M. C. F. Proctor, Dr D. C. Prowse, Mr R. D. Pucknell, Mr D. Punter, Miss M. I. Pye, Mrs T. H. Pyke, Mrs J. Pym, Mrs C. Pyrah.

Mr O. Rackham, Miss S. Ramage, Dr D. S. Ranwell, Dr Denis Ratcliffe, Dr Derek Ratcliffe, Mr J. E. Raven, Mrs E. M. Rawson, Miss M. N. Read, Mr R. C. Readett, Mr D. E. Redfearn, Dr A. N. Redfern, Mr E. J. Redshaw, Miss M. Rees, Mr B. G. Reeves, Miss P. Reidy, Mrs M. E. Reis, Mr G. C. Rhodes, Mr B. W. Ribbons, Mr D. J. Rice, Miss A. N. Richards, Miss C. M. Richards, Mrs M. Richards, Professor P. W. Richards, Mr F. D. S. Richardson, Dr J. A. Richardson, Mr W. E. Richardson, Mr R. H. Richens, Miss V. I. Ricketts, Mrs V. J. Ricketts, Mr H. Riley, Miss C. M. Rob, Miss W. Roberts, Miss B. F. Robertson, Miss M. A. Robertson, Mr A. E. Robinson, Mr C. A. Robinson, Mr F. A. Robinson, Mr A. W. Robson, Mrs C. I. Robson, Mr N. K. B. Robson, Dr J. Roche, Capt. & Mrs R. G. B. Roe, Mr J. Grant Roger, Mrs B. C. Rogers, Miss C. M. Rogers, Miss S. Rogers, Mr J. Rogerson, Miss M. E. Roper, Mr R. Roscoe, Dr F. Rose, Mr R. Ross, Dr E. M. Rosser, Mr J. Rossiter, Mr G. L. Rowe, Dr C. H. Fraser Rowell, Mr P. Rowling, Mr D. Royle, Mr A. Runnalls, Mrs B. H. S. Russell, Mr A. Russell-Smith, Mr E. M. Rutter, Mr M. G. Rutterford, Mr J. S. Ryland.

Dr H. A. Salzen, Mr R. Sandell, Mrs C. Sandeman, Mr N. Y. Sandwith, Mr J. Sankey, Mr B. P. Sargeant, Mr H. B. Sargent, Mrs G. Satow, Miss A. Saunders, Mrs N. Saunders, Mr J. P. Savidge, Mr M. M. Sayer, Miss M. Scannell, Miss C. Schelwald, Miss J. Schofield, Miss S. M. Schofield, Mr T. Schofield, Miss M. A. R. S. Scholey, Miss I. Scholley, Miss D. Schuftan, Mr D. I. G. Scott, Dr E. Scott, Dr G. A. M. Scott, Mr R. Scott, Mr Walter Scott, Mr Wilfred Scott, Mr W. A. Scott, Mr W. K. Scudamore, Mrs M. M. Seabroke, Miss A. Seabrook, Mr A. Searle, Miss A. V. Seaton, Mr M. R. D. Seaward, Mrs P. Seaward, Dr B. Seddon, Mrs J. Sell, Mr P. D. Sell, Mr B. F. C. Sennitt, Dr E. R. Seppings, Mr C. Seymour, Miss C. Shaddick, Mr J. E. M. Shakeshaft, Miss M. R. Shanahan, Mrs D. Sharrock, Mr C. Shaw, The Revd C. E. Shaw, Mr G. A. Shaw, Miss M. S. Shaw, Mr P. G. Sheasby, Mr R. Sheldrick, Mrs I. L. Shelton, Mrs K. Sidaway, Mr & Mrs A. G. Side, Mrs A. M. Simmonds, Mr P. H. Simon, Mr B. Simpson, Mr F. W. Simpson, Mrs M. H. Simpson, Mr N. D. Simpson, Mr M. Sinclair, Mr R. Sinclair, Mr C. Sinker, Mrs A. Skimming, Miss K. M. Skinner, Mr A. Slack, Mrs A. Slansky, Mrs F. M. Slater, Mr R. J. Slatter, Dr W. A. Sledge, Mrs P. Sleigh, Mr P. Sleight-Holme, Mrs L. M. Small, Mr A. Smith, Mr A. E. Smith, Mr A. J. E. Smith, Mr A. Malins Smith, Mr A. R. Smith, Mr D. L. Smith, Mr E. R. Smith, Mr E. S. Smith, Mr G. L. Smith, Mrs G. W. Smith, Miss H. D. Smith, Mrs J. E. Smith, Mrs K. Pickard Smith, Mr L. R. Smith, Mrs M. Smith, Miss M. Smith, Mrs M. M. Smith, Miss N. Smith, Mr P. Bevington Smith, Mr R. Smith, Miss U. K. Smith, Mr R. W. Snaydon, Mrs M. L. Sneyd, Mrs A. H. Sommerville, Miss M. Songhurst, Mr R. Soper, Mr J. E. S. Souster, Mr A. J. Souter, Mrs M. Southwell, Mr W. H. B. Sowerby, Miss I. Spalding, Dr D. H. N. Spence, Mr A. G. Spencer, Mr P. Spicer, Miss M. R. Spiller, Dr T. A. Sprague, Miss A. V. Spratt, Mrs B. Spurgon, Mr B. Spurgon, Mr C. A. Stace, Mr J. Langford Stacey, Dr P. C. Stafford, Miss F. E. A. Stafforth, Dr M. W. Stanier, Miss E. Starr, Dr W. T. Stearn, Mr A. W. Stelfox, Mr J. D. Stephen, Mrs E. K. Stephenson, Mrs G. M. Steuart, Miss A. B. Stevens, Miss E. H. Stevenson, Mrs D. L. Stewart, Miss J. C. Stewart, Mr A. McG. Stirling, Miss M. Stirling, Mr B. Storer, Miss E. E. Storey, Miss A. Straker, Mr J. P. A. Strange, Mr D. T. Streeter, Dr J. M. Stuart, Miss B. M. Sturdy, Mrs M. A. Stutchbury, Dr B. T. Styles, Mr V. S. Summerhayes, Mr J. Summerton, Miss C. A. Swain, Dr G. A. Swan, Mr E. L. Swann, Dr T. D. V. Swinscow, Mrs E. K. Swinton, Miss R. M. Sworder.

Mr J. Taggart, Mr J. M. Taverner, Lady Taylor, Mr F. J. Taylor, Sir George Taylor, Miss I. Taylor, Mr J. Taylor, Mr P. Taylor, Mr H. F. Tebbs, Miss M. I. Tetley, Mr T. F. Teversham, Mr A. Tewnion, Mr W. L. Theobald, Mrs E. Thesigner, Dr E. Thomas, Dr G. E. Thomas, Mr J. D. Thomas, Mr J. F. Thomas, Mrs A. Thompson, Mrs J. Thompson, Mr M. G. V. Thompson, Mr R. Thompson, Mrs A. Thomson, Mr A. L. Thorpe, Miss S. C. Thurman, Miss J. M. Thurston, Mr J. Timson, Mr A. R. Tindall, Mrs J. Tinsley, Miss J. M. Tobias, Mr C. Toft, Mr M. R. Tomlinson, Mr C. C. Townsend, Major E. W.

Townsend, Mr P. J. O. Trist, Mr L. C. Tromans, Miss P. A. Tromans, Mrs M. Trost, Mr T. Trought, Mr J. E. Trowell, Miss J. Tucker, Mr W. E. Tucker, Mr W. H. Tucker, Miss E. M. Tully, Miss J. M. Tunstall, Mrs A. R. Turnbull, Mr C. Turner, Mrs M. Turner, Miss M. A. Turner, Professor T. G. Tutin, Dr F. H. Tyrer, Mr P. Tyson.

Mrs E. Underwood, Mr J. G. Urquhart.

Professor D. H. Valentine, Mr F. R. Vane, Mr A. H. Vaughan, Mrs I. M. Vaughan, Miss K. A. Veitch, Mr L. S. V. Venables, Brig. F. E. W. Venning, Mr B. Verdcourt, Mr H. S. Vere-Hodge, Miss N. G. Vernon, Mrs P. G. Vicars-Miles, Miss M. E. Vince, Mr C. J. Vyle.

Mr N. M. Wace, Mr A. E. Wade, Mr G. B. Wakefield, Mr R. W. Wakeley, The Hon. Mrs Waldy, Mr A. D. Walker, Dr & Mrs D. Walker, Miss D. R. Walker, Dr S. Walker, Mr D. W. Wallace, Mr E. C. Wallace, Mr R. Wallace, Mr. T. J. Wallace, Miss L. J. Walley, Mr O. H. Wallis, Mr R. C. Walls, Mr M. L. Walsh, Mr G. Walton, Professor J. R. Walton, Mr T. W. Wanless, Mr P. J. Wanstall, Dr E. F. Warburg, Mr B. T. Ward, Mr C. W. Ward, Mr H. G. Ward, Mr J. F. C. Ward, Professor P. F. Wareing, Mrs S. Warren, Mr W. E. Warren, Mr R. L. Wastie, Mrs D. M. Watson, Dr E. V. Watson, Miss M. Watson, Dr A. S. Watt, Dr G. Watt, Mr J. S. Watt, Miss O. E. Watt, Mr W. Boyd Watt, Mr W. Watts, Dr W. A. Watts, Professor D. A. Webb, Mrs I. Webster, Miss M. McCallum Webster, Mr A. Wegener, Mr J. Weir, Mrs B. Welch, Mr D. Welch, Mr H. Weller, Mr R. V. Wells, Mr E. J. Wenham, Dr C. West, Mr J. B. Weston, Mr A. W. Westrup, Mr J. A. Whellan, Mr D. White, Miss K. White, Mrs K. M. W. White, Mr P. H. F. White, Miss T. White, Dr F. H. Whitehead, Mrs L. E. Whitehead, Dr H. L. K. Whitehouse, Miss M. M. Whiting, Dr T. C. Whitmore, Miss J. M. Whyte, Mr E. D. Wiggins, Miss A. M. Wigham, Mrs C. A. Wilkins, Mr D. A. Wilkins, Mrs B. M. Wilkinson, Miss G. M. Wilkinson, Mr P. R. Wilkinson, Mrs V. M. Wilkinson, Mrs D. Will, Mr & Mrs J. E. Willé, Mr J. T. Williams, Mr J. W. Williams, Mr L. M. Williams, Mrs M. R. Williams, Dr W. B. Williams, Miss J. D. Williamson, Dr A. J. Willis, Miss J. C. N. Willis, Mr J. H. Willis, Mr P. W. Wilmot-Dear, Mrs D. P. Wilson, Mrs E. Wilson, Mr G. Wilson, Mrs J. Y. Wilson, Miss M. C. E. Wise, Mr J. R. Ironside Wood, Mrs M. Woodall, Dr S. R. J. Woodell, Mr M. S. Woodgates, Mrs F. L. Woodman, Mrs D. V. G. Woods, Lady Woodward, Mr E. V. Wray, Dr F. R. Elliston Wright, Mr J. D. Wright, Miss P. M. Wright, Mr W. S. Wright, Mr T. C. Wrigley, Miss M. M. Wynn.

Mrs A. Yendell, Dr P. F. Yeo, Miss B. E. Young, Miss B. J. Young, Dr D. P. Young, Mrs H. J. Younger, Mr G. Youngs.

Mrs R. Ziman.

Bibliography

BAKER, H. G. (1955). *Geranium purpureum* Vill. and *G. robertianum* L. in the British flora. I. *Geranium purpureum. Watsonia*, **3**, 160–7.

BANGERTER, E. B. & KENT, D. H. (1957). *Veronica filiformis* Sm. in the British Isles. *Proc. B.S.B.I.*, **2**, 197–217.

BENNETT, A., SALMON, C. E. & MATTHEWS, J. R. (1929–30). Second supplement to Watson's Topographical Botany. *J. Bot. Lond.*, **67 & 68** (suppl.).

BENTHAM, G. & HOOKER, J. D. (1924). *Handbook of the British Flora.* Ed. vii revised by A. B. Rendle. Ashford, Kent.

CHRISTY, M. (1884). On the species of the genus *Primula* in Essex. *Trans. Essex Field Club*, **3**, 148–211.

CLAPHAM, A. R., TUTIN, T. G. & WARBURG, E. F. (1952). *Flora of the British Isles.* Cambridge.

CONOLLY, A. P. *et al.* (1950). Studies in post-glacial history of British vegetation. XI. Late-glacial deposits in Cornwall. *Phil. Trans. Roy. Soc.* **234**, B, 397 –469 [Maps of *Betula nana* L. and *Salix herbacea* L.].

DANDY, J. E. (1958). *List of British Vascular Plants.* London.

DANDY, J. E. & TAYLOR, G. (1938–42). Studies of British Potamogetons. I–XVIII. *J. Bot. Lond.*, 76–80.

DAVIES, E. W. (1953). Notes on *Carex flava* and its allies. I. A Sedge new to the British Isles. *Watsonia*, **3**, 66–9.

DRUCE, G. C. (1932). *Comital Flora of the British Isles.* Arbroath.

GODWIN, H. (1956). *History of the British Flora.* Cambridge.

GOOD, R. D'O. (1936). On the distribution of the Lizard Orchid (*Himanto-glossum hircinum* Koch). *New Phyt.*, **35**, 142–70.

GROVES, E. W. (1958). *Hippophae rhamnoides* in the British Isles. *Proc. B.S.B.I.*, **3**, 1–21.

HALL, P. M. (1937). The Irish Marsh Orchids. *Rept Bot. Soc. & Exch. Club*, **11**, 330–52.

HARRISON, J. HESLOP (1953). Studies in *Orchis* L. II. *Orchis traunsteineri* Saut. in the British Isles. *Watsonia*, **3**, 1–6.

HOFFMANN, H. (1860). Vergleichende Studien zur Lehre von der Bodenstetig-keit der Pflanzen. *Ber. Oberhess. Ges. Natur. & Heilk.*, **8**, 1–12.

HOFFMANN, H. (1867–9). Pflanzenarealstudien in den Mittelrheingegenden. *Ber. Oberhess. Ges. Natur. & Heilk.*, **12**, 51–60; **13**, 1–63.

HOFFMANN, H. (1879). Nachträge zur Flora des Mittelrheingebietes. *Ber. Oberhess. Ges. Natur. & Heilk.*, **18**, 1–48.

HULTÉN, E. (1950). *Atlas över Kärlväxterna i Norden.* Stockholm.

IHNE, E. (1879). Studien zur Pflanzengeographie. *Ber. Oberhess. Ges. Natur. & Heilk.*, **18**, 49–82.

KENT, D. H. (1956). *Senecio squalidus* L. in the British Isles. I. Early records (to 1877). *Proc. B.S.B.I.*, **2**, 115–18.

KENT, D. H. (1960). *Senecio squalidus* L. in the British Isles. II. The spread from Oxford (1879–1939). *Proc. B.S.B.I.*, **3**, 375–9.

LACEY, W. S. (1957). A comparison of the spread of *Galinsoga parviflora* and *G. ciliata* in Britain. In LOUSLEY (1957), 109.

LAMBERT, J. M. in GOODMAN, P. J. *et al.* (1959). Investigations into 'Die-back' in *Spartina townsendii* agg. *J. Ecol.*, **47**, 651–77.

LINTON, D. L. (Ed.) (1956). *Sheffield and its Region.* Brit. Assn. Sheffield.

LOUSLEY, J. E. (Ed.) (1951). *The Study of the Distribution of British Plants.* Arbroath.

LOUSLEY, J. E. (Ed.) (1957). *Progress in the Study of the British Flora.* Arbroath.

MARSDEN-JONES, E. M. & TURRILL, W. B. (1954). *British Knapweeds.* London.

MATTHEWS, J. R. (1937). The geographical relationships of the British Flora. *J. Ecol.*, **25**, 1–90.

MATTHEWS, J. R. (1942). The germination of *Trientalis europaea. J. Bot. Lond.*, **80**, 12–16.

METEOROLOGICAL OFFICE (1938). Air Ministry Publication no. 421.

MEYER, A. (1926). Über einige Zusammenhänge zwischen Klima und Boden in Europa. *Chemie der Erde*, **2**, 209–347.

MOSS, C. E. (1914). *The Cambridge British Flora.* Vol. 2. Cambridge.

PIGOTT, C. D. (1951). In LOUSLEY (1951), 19. [Map of *Astragalus danicus* Retz.]

PIGOTT, C. D. (1956a). The vegetation of Upper Teesdale in the North Pennines. *J. Ecol.*, **44**, 545–86. [Map of *Dryas octopetala* L.].

PIGOTT, C. D. (1956b). In LINTON (1956), 86. [Map of *Cirsium acaulon* (L.) Scop.].

PRAEGER, R. Ll. (1901). *Irish Topographical Botany.* Dublin.

PRAEGER, R. Ll. (1902). On types of distribution in the Irish flora. *Proc. Roy. Irish Acad.*, **24**, B, 1–60.

PRAEGER, R. Ll. (1909). *Tourist's Flora of the West of Ireland.* Dublin.

PRAEGER, R. Ll. (1934). *The Botanist in Ireland.* Dublin.

PUGH, P. J. (1953). The distribution of *Dryopteris borreri* Newm. in the British Isles. *Watsonia*, **3**, 57–65.

SALISBURY, E. J. (1932). The East Anglian flora. *Trans. Norf. & Norw. Nat. Soc.*, **13**, 191–263.

SKENE, M. (1924). *The Biology of the Flowering Plants.* London.

SLEDGE, W. A. (1949). The distribution and ecology of *Scheuchzeria palustris* L. *Watsonia*, **1**, 24–35.

STAPF, O. (1917). A cartographic study of the southern element in the British flora. *Proc. Linn. Soc. Lond.*, sess. **129**, 81–92.

STEVEN, H. M. & CARLISLE, A. (1959). *The Native Pinewoods of Scotland.* Edinburgh.

THURSTON, J. M. (1954). A survey of wild oats in England and Wales in 1951. *Ann. Appl. Biol.*, **41**, 619–36.

WATSON, H. C. [1832]. *Outlines of the Geographical Distribution of British Plants.* Edinburgh.

WATSON, H. C. (1835). *Remarks on the Geographical Distribution of British Plants.* London.

WATSON, H. C. (1836). Observations on the construction of maps for illus-trating the distribution of plants. *Loud. Mag. Nat. Hist.*, **9**, 17–21.

WATSON, H. C. (1843). *Geographical Distribution of British Plants.* 'Ed. iii'. London.

WATSON, H. C. (1847). *Cybele Britannica.* London.

WATSON, H. C. (1873). *Topographical Botany.* London.

WEBB, D. A. (1952). Irish plant records. *Watsonia*, **2**, 217–36.

WEBB, D. A. (1955). The Distribution Maps Scheme: a provisional extension to Ireland of the British National Grid. *Proc. B.S.B.I.*, **1**, 316–19.

WEST, G. (1904). A comparative study of the dominant phanerogamic and higher cryptogamic flora of aquatic habit in three lake areas of Scotland [Flora of Scottish Lakes]. *Proc. Roy. Soc. Edin.*, **25**, 967–1023.

WEST, G. (1910). A further contribution to a comparative study of the dominant phanerogamic and higher cryptogamic flora of aquatic habit in Scottish lakes [Flora of Scottish Lakes]. *Proc. Roy. Soc. Edin.*, **30**, 65–181.

YOUNG, D. P. (1952). Studies in the British *Epipactis. Watsonia*, **2**, 253–76.

YOUNG, D. P. (1958). *Oxalis* in the British Isles. *Watsonia*, **4**, 51–69.

An index and bibliography

TO DISTRIBUTION MAPS PUBLISHED
BETWEEN 1962 AND 1989

C.D. PRESTON

Biological Records Centre, Monks Wood Experimental Station, Abbots Ripton, Huntingdon PE17 2LS

INTRODUCTION

This index lists distribution maps of vascular plants produced between the publication of the first edition of the *Atlas of the British Flora* (1962) and the end of 1989. Maps are included if they satisfy the following criteria.

1 They cover the whole of the British Isles. I have also included a few maps which only cover Great Britain, or which only map occurrences in specific habitats: a note after the reference indicates the works containing such maps.

2 They plot records in 10-km grid squares.

3 They are revised or updated versions of maps originally published in the *Atlas*, new maps of species which were mapped in the *Atlas* or maps of taxa which were not included in the *Atlas*. I have not included the revised maps in the second and third editions of the *Atlas*, as these are published in this volume, nor have I cited references to works which simply reprint the map originally published in the first edition.

4 They are published. I have excluded maps in unpublished theses and maps which, although technically published (e.g. on microfiche), are not available through normal channels.

LIST OF TAXA MAPPED

Acaena novae-zelandiae: Gynn & Richards, 1985.
Acer pseudoplatanus: Scott, 1975.
Achillea millefolium: Scott, 1975.
Acorus calamus: Greenwood, 1974.
Actaea spicata: Usher, 1986a.
Adiantum capillus-veneris: Jermy et al., 1978.
× Agropogon littoralis (Agrostis stolonifera × Polypogon monspeliensis): Perring & Sell, 1968.
Agropyron repens: see Elymus repens.
Agrostemma githago: Nature Conservancy Council, 1977; Firbank, 1988.
Agrostis curtisii: Proctor, 1969.
A. setacea: see A. curtisii.
Alchemilla acutiloba: Perring & Sell, 1968; Bradshaw, 1962; 1970.
A. conjuncta: Perring & Sell, 1968.
A. filicaulis subsp. filicaulis: Bradshaw, 1962; Perring & Sell, 1968.
A. filicaulis subsp. vestita: Perring & Sell, 1968.
A. glabra: Perring & Sell, 1968.
A. glaucescens: Perring & Sell, 1968.
A. glomerulans: Bradshaw, 1962; Perring & Sell, 1968.
A. minima: Perring & Sell, 1968.

A few genera or subgenera (e.g. *Carex, Rubus* subgenus *Rubus*) are covered by works which provide a comprehensive set of maps. In these cases (which are noted in the species list below) I have not cited earlier maps.

All maps are listed under the genus or species name. Nomenclature follows the *Excursion Flora of the British Isles* (Clapham et al., 1981) as far as possible. I have used the generally accepted name of taxa which are not included in the *Excursion Flora*. If the name used in the *Atlas* differs from that in the *Excursion Flora*, the *Atlas* name is included in the list with an appropriate cross-reference. If the name in the reference cited differs both from that used in the *Atlas* and from that used in the *Excursion Flora*, it is also included.

Some published maps have become very misleading because of subsequent taxonomic change. Where there is a danger that the reader might be confused or misled by the bland citation of such maps I have indicated the change in an explanatory note.

Although I have included all the relevant maps that I know, I cannot claim that the list is comprehensive. I would be grateful for details of any omissions.

A. mollis: Perring & Sell, 1968.
A. monticola: Bradshaw, 1962; Perring & Sell, 1968.
A. subcrenata: Perring & Sell, 1968; Bradshaw, 1962; 1970.
A. tytthantha: Perring & Sell, 1968.
A. wichurae: Perring & Sell, 1968; Bradshaw, 1962; 1970.
A. xanthochlora: Perring & Sell, 1968.
Alopecurus myosuroides: Naylor, 1972.
× Ammocalamagrostis baltica (Ammophila arenaria × Calamagrostis epigejos): Perring & Sell, 1968.
Ammophila arenaria: Huiskes, 1979.
Anagallis arvensis subsp. arvensis forma azurea: Perring & Sell, 1968.
A. arvensis subsp. foemina: Perring & Sell, 1968.
Andromeda polifolia: Scannell, 1973.
Anemone nemorosa: Shirreffs, 1985.
Angelica sylvestris: Scott, 1975.
Anogramma leptophylla: Jermy et al., 1978.
Anthemis arvensis: Kay, 1971b.
A. cotula: Kay, 1971a.
Anthoxanthum odoratum: Scott, 1975.
Anthyllis vulneraria, all subspecies: Perring & Sell, 1968.
Aphanes arvensis: Perring & Sell, 1968.
A. microcarpa: Perring & Sell, 1968.
Apium × moorei (A. inundatum × nodiflorum): Perring & Sell, 1968.

Daucus carota subsp. gummifer: Perring & Sell, 1968.
Deschampsia cespitosa: Scott, 1975; Davy, 1980.
Desmazeria rigida: Clark, 1974.
D. rigida subsp. majus: Perring & Sell, 1968.
Dianthus armeria: Wells, 1967.
Diphasiastrum alpinum: Jermy et al., 1978.
D. issleri: Jermy et al., 1978.
Draba incana: Conolly & Dahl, 1970; Godwin, 1975; Anderson & Shimwell, 1981.
Drosera × obovata (D. anglica × rotundifolia): Perring & Sell, 1968.
Dryas octopetala: Bradshaw, 1970; Conolly & Dahl, 1970; Elkington, 1971; Godwin, 1975.
Dryopteris, all species and hybrids then recognized: Jermy et al., 1978. Earlier maps are not listed here.
Eleocharis austriaca: Perring & Sell, 1968.
E. palustris subsp. palustris: Perring & Sell, 1968, as E. palustris subsp. microcarpa.
Elodea canadensis: Simpson, 1984; 1986.
E. ernstiae: Simpson, 1984; 1986, as E. callitrichoides.
E. nuttallii: Natural Environment Research Council, 1976; Perring, 1976; Simpson, 1984; 1986 (see note 13).
Elymus repens: Palmer & Sagar, 1963.
Empetrum nigrum: Bell & Tallis, 1973; Pigott, 1974.
Epilobium adenocaulon: see E. ciliatum.
E. angustifolium: see Chamerion angustifolium.
E. brunnescens: Usher, 1973.
E. ciliatum: Crawley, 1987.
E. ciliatum: Preston, 1989.
E. lanceolatum: Perring, 1974.
E. nerterioides: see E. brunnescens.
Equisetum, all species and hybrids then recognized: Jermy et al., 1978. Earlier maps are not cited here.
Erica cinerea: Bannister, 1965.
E. × stuartii (E. mackaiana × tetralix): Perring & Sell, 1968, as E. × praegeri.
E. tetralix: Bannister, 1966.
E. × watsonii (E. ciliaris × tetralix): Perring & Sell, 1968.
Eriophorum vaginatum: Wein, 1973; Teagle, 1978.
Erodium cicutarium subsp. dunense: Perring & Sell, 1968 (see note 2).
E. glutinosum: Perring & Sell, 1968 (see note 2).
Erophila verna subsp. spathulata: Perring & Sell, 1968.
Euphrasia, all species except E. salisburgensis: Perring & Sell, 1968.
Euphorbia amygdaloides: Gent, 1989.
Festuca altissima: Conolly & Dahl, 1970.
F. brevipila: Wilkinson & Stace, 1989 (see note 3).
F. juncifolia × Vulpia fasciculata: Ainscough et al., 1986.
F. lemanii: Wilkinson & Stace, 1989 (see note 3).
F. longifolia: see note 3.
F. rubra × Vulpia bromoides: Ainscough et al., 1986.
F. rubra × V. fasciculata: Ainscough et al., 1986.
F. rubra × V. myuros: Ainscough et al., 1986.
F. tenuifolia: Perring & Sell, 1968.
× Festulolium loliaceum (Festuca pratensis × Lolium perenne): Perring & Sell, 1968.
Filipendula ulmaria: Scott, 1975.
Frankenia laevis: Brightmore, 1979.
Fraxinus excelsior: Scott, 1975.
Fritillaria meleagris: Perring & Walters, 1971; Perring, 1974; Bradshaw, 1977.
Fumaria muralis subsp. muralis: Perring & Sell, 1968.
F. muralis subsp. neglecta: Perring & Sell, 1968.
F. officinalis subsp. wirtgenii: Perring & Sell, 1968.
F. × painteri (F. muralis × officinalis): Perring & Sell, 1968.
Galeobdolon luteum: see Lamiastrum galeobdolon.
Galium aparine: Scott, 1975.
G. palustre: Scott, 1975.
G. × pomeranicum (G. mollugo × verum): Perring & Sell, 1968.
Gastridium ventricosum: Trist, 1986.
Genista tinctoria subsp. littoralis, and other prostrate forms: Perring & Sell, 1968.

Gentiana pneumonanthe: Nature Conservancy Council, 1977.
G. verna: Elkington, 1963; Bradshaw, 1970, as G. alpina.
Gentianella amarella subsp. druceana: Perring & Sell, 1968.
G. amarella subsp. septentrionalis: Perring & Sell, 1968.
Geranium purpureum subsp. forsteri: Perring & Sell, 1968.
G. purpureum subsp. purpureum: Perring & Sell, 1968.
G. robertianum: Scott, 1975.
G. robertianum subsp. celticum: Perring & Sell, 1968.
G. robertianum subsp. maritimum: Perring & Sell, 1968.
Geum × intermedium (G. rivale × urbanum): Perring & Sell, 1968.
Glaucium flavum: Scott, 1963b.
Glyceria × pedicellata (G. fluitans × plicata): Perring & Sell, 1968.
Gymnadenia conopsea: Smith, 1980.
G. conopsea subsp. densiflora: Perring & Sell, 1968.
Gymnocarpium dryopteris: Jermy et al., 1978.
G. robertianum: Jermy et al., 1978.
Halimione portulacoides: Akeroyd & Preston, 1984.
Hedera helix: Scott, 1975.
H. helix cultivar Hibernica: Rutherford, 1976; 1979.
Helianthemum canum: Bradshaw, 1970; Godwin, 1975.
H. canum subsp. canum: Perring & Sell, 1968.
H. canum subsp. levigatum: Perring & Sell, 1968.
H. chamaecistus: see H. nummularium.
H. nummularium: Ratcliffe & Conolly, 1972.
Helictotrichon pratense: see Avenula pratensis.
Heracleum mantegazzianum: Drever & Hunter, 1970; Clegg & Grace, 1974; Williamson & Forbes, 1972; Anon., 1983.
H. sphondylium: Scott, 1975.
Hieracium (including Pilosella), all species, subspecies and hybrids then recognized except H. pilosella: Perring & Sell, 1968.
Himantoglossum hircinum: Perring, 1974.
Hippocrepis comosa: Proctor, 1969; Bradshaw, 1970; Fearn, 1973.
Hippophaë rhamnoides: Ward, 1972; Godwin, 1975; Crawford, 1976.
Holcus lanatus: Scott, 1975.
Hornungia petraea: Clapham, 1969.
Huperzia selago: Proctor, 1969; Jermy et al., 1978.
Hymenophyllum tunbrigense: Richards & Evans, 1972; Jermy et al., 1978.
H. wilsonii: Richards & Evans, 1972; Jermy et al., 1978.
Hypericum maculatum subsp. maculatum: Perring & Sell, 1968.
H. maculatum subsp. obtusiusculum: Perring & Sell, 1968.
Hypochaeris maculata: Wells, 1976.
H. radicata: Scott, 1975; Turkington & Aarssen, 1983.
Hypochoeris: see Hypochaeris.
Ilex aquifolium: Peterken & Lloyd, 1968; Pollard et al., 1974; Peterken, 1975.
Impatiens glandulifera: Usher, 1986b.
Isoetes echinospora: Jermy et al., 1978.
I. histrix: Jermy et al., 1978.
I. lacustris: Jermy et al., 1978.
Jasione montana: Parnell, 1985 (see note 4).
Juncus alpinoarticulatus: see J. alpinus.
J. alpinus: Bradshaw, 1970.
J. castaneus: Conolly & Dahl, 1970.
J. × diffusus (J. effusus × inflexus): Perring & Sell, 1968.
J. effusus: Scott, 1975.
J. filiformis: Blackstock, 1981.
J. foliosus: Cope & Stace, 1978.
J. ranarius: Cope & Stace, 1978, as J. ambiguus.
J. squarrosus: Welch, 1966.
J. trifidus: Conolly & Dahl, 1970.
Juniperus communis: Jeffers, 1972; Bowen, 1974; Duffey et al., 1974.
J. communis subsp. communis: Perring & Sell, 1968; Ward, 1973; 1977 Smith, 1980.
J. communis subsp. nana: Perring & Sell, 1968; Ward, 1977.
Kobresia simpliciuscula: Bradshaw, 1970; Conolly & Dahl, 1970.
Koenigia islandica: Godwin, 1975.
Kohlrauschia prolifera: see Petrorhagia nanteuilii.
Lamiastrum galeobdolon: Wells, 1982; Packham, 1983.
L. galeobdolon subsp. galeobdolon: Wegmüller, 1971; Packham, 1983.
L. galeobdolon subsp. montanum: Wegmüller, 1971; Packham, 1983.

Polystichum, all species and hybrids: Jermy et al., 1978.
Populus nigra: Milne-Redhead, 1975; Beckett & Beckett, 1979.
Potamogeton alpinus: Godwin, 1975 (see note 9).
P. × cooperi (P. crispus × perfoliatus): Perring & Sell, 1968.
P. filiformis: Godwin, 1975.
P. × fluitans (P. lucens × natans): Perring & Sell, 1968.
P. × lintonii (P. crispus × friesii): Perring & Sell, 1968.
P. × nitens (P. gramineus × perfoliatus): Perring & Sell, 1968.
P. × olivaceus (P. alpinus × crispus): Perring & Sell, 1968.
P. praelongus: Godwin, 1975 (see note 9).
P. × salicifolius (P. lucens × perfoliatus): Perring & Sell, 1968.
P. × sparganifolius (P. gramineus × natans): Perring & Sell, 1968.
P. × suecicus (P. filiformis × pectinatus): Perring & Sell, 1968; Godwin, 1975.
P. × zizii (P. gramineus × lucens): Perring & Sell, 1968.
Potentilla anserina: Ockendon & Walters, 1970; Scott, 1975.
P. crantzii: Smith, 1963b; 1971; Godwin, 1975.
P. erecta: Richards, 1973.
P. erecta subsp. strictissima: Richards, 1973.
P. fruticosa: Elkington & Woodell, 1963; Bradshaw, 1970.
P. tabernaemontani: Smith, 1963a; 1971.
Primula elatior: Wells, 1982; Richards, 1989.
P. farinosa: Bradshaw, 1970; Wells & Oswald, 1988; Richards, 1989.
P. scotica: Richards, 1989.
P. × tommasinii (P. veris × vulgaris): Perring & Sell, 1968.
P. veris: Richards, 1989.
P. vulgaris: Perring & Sell, 1968; Richards, 1989.
Prunella × intermedia (P. laciniata × vulgaris): Perring & Sell, 1968.
P. vulgaris: Scott, 1975.
Prunus padus: Anderson & Shimwell, 1981.
Pseudotsuga menziesii: Jeffers, 1972.
Pteridium aquilinum: Jermy et al., 1978.
Puccinellia distans: Scott & Davison, 1982; Scott, 1985.
P. maritima: Gray & Scott, 1977.
Pulicaria vulgaris: Nature Conservancy Council, 1977.
Pulsatilla vulgaris: Wells & Barling, 1971; Nature Conservancy Council, 1977; Gent, 1989.
Pyrola rotundifolia subsp. maritima: Perring & Sell, 1968; Perring, 1974.
Quercus petraea: Gardiner, 1974.
Q. robur: Gardiner, 1974.
Ranunculus acris: Scott, 1975.
R. ficaria subsp. bulbilifer: Perring & Sell, 1968; Taylor & Markham, 1978; both as R. ficaria subsp. bulbifer.
R. ficaria subsp. ficaria: Perring & Sell, 1968; Taylor & Markham, 1978.
R. penicillatus subsp. penicillatus: Webster, 1988.
R. penicillatus subsp. pseudofluitans var. pseudofluitans: Webster, 1988.
R. penicillatus subsp. pseudofluitans var. vertumnus: Webster, 1988.
R. repens: Scott, 1975.
Raphanus raphanistrum, white-flowered and yellow-flowered plants: Perring & Sell, 1968.
Reynoutria japonica: Conolly, 1977.
R. sachalinensis: Conolly, 1977.
Rhamnus catharticus: Smith, 1980.
Rhinanthus minor, all subspecies: Perring & Sell, 1968.
Rhododendron ponticum: Cross, 1975; Usher, 1986b.
Rorippa islandica: Perring, 1974 (see note 10).
R. microphylla: see Nasturtium microphyllum.
R. nasturtium-aquaticum: see Nasturtium officinale.
R. × sterilis: see Nasturtium × sterile.
Rubus chamaemorus: Clapham, 1969; Taylor, 1971.
Rubus subgenus Rubus (the segregates of R. fruticosus), all species: Edees & Newton, 1988. Earlier maps are not cited here.
Rumex acetosa: Scott, 1975.
R. crispus: Cavers & Harper, 1964.
R. obtusifolius: Cavers & Harper, 1964; Scott, 1975.
R. tenuifolius: Perring & Sell, 1968.
Sagittaria sagittifolia: Greenwood, 1974.
Salix herbacea: Conolly & Dahl, 1970; Morris, 1970.

S. nigricans: Donald, 1981.
S. phylicifolia: Godwin, 1975.
S. reticulata: Conolly & Dahl, 1970; Godwin, 1975.
Sambucus nigra: Scott, 1975.
Samolus valerandi: Ranwell, 1972.
Sarothamnus scoparius: see Cytisus scoparius.
Saussurea alpina: Godwin, 1975.
Saxifraga × polita (S. hirsuta × spathularis): Perring & Sell, 1968.
S. spathularis: Scannell, 1973.
S. stellaris: Conolly & Dahl, 1970; Pigott, 1978.
Schoenoplectus × carinatus (S. lacustris × triqueter): Perring & Sell, 1968, as Scirpus × carinatus.
S. × kuekenthalianus (S. tabernaemontani × triqueter): Perring & Sell, 1968, as Scirpus × kuekenthalianus.
Schoenus nigricans: Sparling, 1968.
Scleranthus perennis subsp. perennis: Perring & Sell, 1968.
S. perennis subsp. prostratus: Perring & Sell, 1968.
Scutellaria × hybrida (S. galericulata × minor): Perring & Sell, 1968.
Sedum anglicum: Proctor, 1969.
Selaginella kraussiana: Jermy et al., 1978.
S. selaginoides: Jermy et al., 1978.
Senecio integrifolius: Smith, 1979.
S. jacobaea: Scott, 1975.
S. squalidus: Usher, 1973; Gent, 1989.
S. vulgaris: Scott, 1975.
S. vulgaris subsp. denticulatus: Perring & Sell, 1968.
S. vulgaris subsp. vulgaris var. hibernicus: Perring & Sell, 1968 (see note 11).
Sesleria albicans: Bellamy, 1970; Bradshaw, 1970; Dixon, 1982; Usher, 1986a.
S. caerulea: see S. albicans.
Sibbaldia procumbens: Coker, 1966.
Sibthorpia europaea: Proctor, 1969.
Silene acaulis: Jones & Richards, 1962.
S. alba × dioica: Perring & Sell, 1968.
S. maritima: Baker & Dalby, 1980.
S. nutans: Clapham, 1969.
S. otites: Walters, 1978.
Sorbus, all native species except S. aucuparia and S. torminalis: Perring & Sell, 1968.
S. intermedia: Perring & Sell, 1968.
S. torminalis: Beckett & Beckett, 1979; Wells, 1982.
Sparganium angustifolium: Godwin, 1975.
S. erectum, all subspecies: Perring & Sell, 1968.
Spartina glabra: Perring & Sell, 1968.
S. maritima: Goodman et al., 1969.
S. × townsendii sens. lat.: Goodman et al., 1969.
S. × townsendii sens. str.: Perring & Sell, 1968; Goodman et al., 1969.
Spiranthes romanzoffiana: Perring, 1974.
Stachys × ambigua (S. palustris × sylvatica): Perring & Sell, 1968.
Stellaria media: Sobey, 1981.
S. nemorum subsp. glochidisperma: Perring & Sell, 1968.
Stratiotes aloides: Greenwood, 1974.
Symphytum × uplandicum (S. asperum × officinale): Perring & Sell, 1968.
Taraxacum laevigatum: Perring & Sell, 1968 (see note 12).
T. palustre: Perring & Sell, 1968 (see note 12).
T. spectabile: Perring & Sell, 1968 (see note 12).
Taxus baccata: Jeffers, 1972; Tittensor, 1980.
Teucrium scorodonia: Hutchinson, 1968.
Thalictrum alpinum: Bradshaw, 1970; Conolly & Dahl, 1970; Godwin, 1975.
T. minus: Usher, 1986a.
Thelycrania sanguinea: see Cornus sanguinea.
Thelypteris dryopteris: see Gymnocarpium dryopteris.
T. oreopteris: see Oreopteris limbosperma.
T. palustris: see T. thelypteroides.
T. phegopteris: see Phegopteris connectilis.
T. robertiana: see Gymnocarpium robertianum.
T. thelypteroides: Jermy et al., 1978.

Thlaspi perfoliatum: Rich et al., 1989.
Tilia platyphyllos: Pigott, 1981.
Tofieldia pusilla: Bradshaw, 1970.
Trichomanes speciosum: Jermy et al., 1978.
Trifolium occidentale: Perring & Sell, 1968; Milton, 1984.
T. ochroleucon: Wells, 1982; Wells & Oswald, 1988.
T. pratense: Scott, 1975.
T. repens: Scott, 1975; Burdon, 1983.
T. suffocatum: Proctor, 1969.
Tussilago farfara: Scott, 1975.
Ulex gallii: Proctor, 1969; Anderson & Shimwell, 1981.
Umbilicus rupestris: Proctor, 1969; Ratcliffe & Conolly, 1972; Anderson & Shimwell, 1981.
Urtica dioica: Scannell, 1973; Scott, 1975.
Utricularia australis: Perring & Sell, 1968, as U. neglecta.
U. vulgaris: Perring & Sell, 1968.
Vaccinium × intermedium (V. myrtillus × vitis-idaea): Perring & Sell, 1968.
V. myrtillus: Proctor, 1969.

V. vitis-idaea: Clapham, 1969; Ratcliffe & Conolly, 1972; Anderson & Shimwell, 1981.
Valerianella locusta subsp. dunensis: Perring & Sell, 1968.
Veronica chamaedrys: Scott, 1975.
V. filiformis: Crawley, 1987.
V. peregrina: Bangerter, 1966.
V. scutellata var. villosa: Perring & Sell, 1968.
V. serpyllifolia subsp. humifusa: Perring & Sell, 1968.
V. spicata subsp. hybrida: Perring & Sell, 1968.
V. spicata subsp. spicata: Perring & Sell, 1968.
Viburnum lantana: Ratcliffe & Conolly, 1972; Pigott, 1974; Smith, 1980.
Vicia orobus: Wells & Oswald, 1988.
Viola palustris subsp. juressi: Perring & Sell, 1968.
V. riviniana: Scott, 1975.
Viscum album: Perring, 1973.
Vulpia fasciculata: Stace & Cotton, 1976; Watkinson, 1978.
V. membranacea: see V. fasciculata.
Wahlenbergia hederacea: Anderson & Shimwell, 1981.
Woodsia alpina: Jermy et al., 1978.
W. ilvensis: Jermy et al., 1978.

NOTES

1 The taxonomy of *Bromus hordeaceus* subsp. *thominii* has been reassessed by Smith (1968). The true subsp. *thominii* is a coastal plant; inland records mapped by Perring & Sell (1968) are probably *B.* × *pseudothominii (B. hordeaceus × lepidus).*

2 I have departed here from the nomenclature of Clapham et al. (1981), who treat *E. glutinosum* as *E. cicutarium* subsp. *bipinnatum*. Under this treatment *E. cicutarium* subsp. *dunense* is regarded as an intermediate between subsp. *cicutarium* and subsp. *bipinnatum*.

3 The plant mapped as *Festuca longifolia* in successive editions of the *Atlas* consists of two species, *F. brevipila* and *F. lemanii*, both of which are mapped by Wilkinson & Stace (1989).

4 The caption to the map of *Jasione montana* published by Parnell (1985) is erroneous. Circles denote records made before 1950 (not 1930), dots records made from 1950 (not 1930) onwards.

5 *Mentha scotica* is probably best included in *M. spicata* (Harley, 1975) but I have listed it separately here as its identity has not yet been satisfactorily resolved.

6 *Mentha x webberi* is an enigmatic mint, perhaps referable to *M. x rotundifolia (M. longifolia x suaveolens)* but requiring cytological study (Harley, 1975).

7 An alternative taxonomic treatment of the plants mapped by Perring & Sell (1968) as *Odontites verna* subsp. *pumila* is suggested by Snogerup (1982). Her paper includes a distribution map of *O. litoralis* subsp. *litoralis*.

8 The plant mapped as *Kohlrauschia prolifera* in successive editions of the *Atlas* consists of two species, now placed in the genus *Petrorhagia. P. nanteuilii*, the native British species, is mapped by Perring & Sell (1968). The introduced *P. prolifera* has not been mapped.

9 The maps of *Potamogeton alpinus* and *P. praelongus* in Godwin (1975) show records determined by J.E. Dandy & G. Taylor as closed circles, and other records as open circles. The caption stating that these symbols indicate records made before 1930 and from 1930 onwards is erroneous.

10 The plant mapped as *Rorippa islandica* in successive editions of the *Atlas* is now treated as two species. The rare northern diploid *R. islandica* sens. str. is the plant mapped by Perring (1974). Almost all the records on the *Atlas* map are of the commoner tetraploid, *R. palustris*.

11 The name *Senecio vulgaris* subsp. *vulgaris* forma *ligulatus* D.E. Allen, applied to *S. vulgaris* subsp. *vulgaris* var. *hibernicus* by Perring & Sell (1968), is a nomen nudum.

12 The taxonomy of the British *Taraxacum* species has been radically revised in recent years. The plants mapped as *T. laevigatum, T. palustre* and *T. spectabile* by Perring & Sell (1968) do not correspond to any species now recognized.

13 The plant mapped as *Elodea nuttallii* in the first edition of the *Atlas* is actually the rare native *Hydrilla verticillata*. The true *E. nuttallii*, an alien species, is mapped by the authors cited here.

REFERENCES

AINSCOUGH, M.M., BARKER, C.M. & STACE, C.A. (1986). Natural hybrids between *Festuca* and species of *Vulpia* section *Vulpia*. Watsonia, 16 : 143–151.

AKEROYD, J.R. & PRESTON, C.D. (1984). *Halimione portulacoides* (L.) Aellen on coastal rocks and cliffs. Watsonia, 15 : 95–103.

ANDERSON, P. & SHIMWELL, D. (1981). *Wild flowers and other plants of the Peak District*. Ashbourne.

ANONYMOUS (1983). *Giant Hogweed. The problem and its control*. (North of Scotland College of Agriculture Leaflet no. 49). Aberdeen.

BAKER, A.J.M. & DALBY, D.H. (1980). Morphological variation between some isolated populations of *Silene maritima* With. in the British Isles with particular reference to inland populations on metalliferous soils. New Phytol., 84: 123–138.

BANGERTER, E.B. (1966). Further notes on *Veronica peregrina* L. Proc. bot. Soc. Br. Isl., 6. 215–220.

BANNISTER, P. (1965). *Erica cinerea* L. (Biological Flora of the British Isles no. 100). J. Ecol., 53 : 537–542.

BANNISTER, P. (1966). *Erica tetralix* L. (Biological Flora of the British Isles no. 102). J. Ecol., 54: 795–813.

BARRY, R. & WADE, P.M. (1986). *Callitriche truncata* Guss. (Biological Flora of the British Isles no. 162). J. Ecol., 74: 289–294.

BECKETT, K. & BECKETT, G. (1979). *Planting native trees and shrubs*. Norwich.

BEDDOWS, A.R. (1967). *Lolium perenne* L. (Biological Flora of the British Isles no. 107). J. Ecol., 55: 567–587.

BEDDOWS, A.R. (1973). *Lolium multiflorum* Lam. (Biological Flora of the British Isles no. 131). J. Ecol., 61: 587–600.

BELL, J.N.B. & TALLIS, J.H. (1973). *Empetrum nigrum* L. (Biological Flora of the British Isles no. 130). J. Ecol., 61: 289–305.

BELLAMY, D.J. (1970). The Vegetation, in DEWDNEY, J.C., ed. Durham County and City with Teesside, pp. 133–141. Durham.

BLACKSTOCK, T.H. (1981). The distribution of *Juncus filiformis* L. in Britain. Watsonia, 13: 209–214.

BOORMAN, L.A. (1967). *Limonium vulgare* Mill. and *Limonium humile* Mill. (Biological Flora of the British Isles no. 166). J. Ecol., 55: 221–232.

BOWEN, H.J.M. (1974). *Air pollution and its effects on plants*, in PERRING, F. H., ed. The flora of a changing Britain, pp. 119–127. Faringdon.

BRADSHAW, A.D. (1977). Conservation problems in the future. Proc. R. Soc., ser. B, 197: 77–96.

BRADSHAW, M.E. (1962). The distribution and status of five species of the *Alchemilla vulgaris* L. aggregate in Upper Teesdale. *J. Ecol.*, 50: 681–706.

BRADSHAW, M.E. (1970). *The Teesdale Flora*, in DEWDNEY, J.C., ed. *Durham County and City with Teesside*, pp. 141–152. Durham.

BRIGHTMORE, D. (1979). *Frankenia laevis* L. (Biological Flora of the British Isles no. 146). *J. Ecol.*, 67: 1097–1107.

BRIGHTMORE, D. & WHITE, P.H.F. (1963). *Lathyrus japonicus* Willd. (Biological Flora of the British Isles no. 94). *J. Ecol.*, 51: 795–801.

BURDON, J.J. (1983). *Trifolium repens* L. (Biological Flora of the British Isles no. 154). *J. Ecol.*, 71: 307–330.

CARLISLE, A. & BROWN, A.H.F. (1968). *Pinus sylvestris* L. (Biological Flora of the British Isles no. 109). *J. Ecol.*, 56: 269–307.

CAVERS, P.B. & HARPER, J.L. (1964). *Rumex obtusifolius* L. and *Rumex crispus* L. (Biological Flora of the British Isles no. 98). *J. Ecol.*, 52: 737–766.

CLAPHAM, A.R., ed. (1969). *Flora of Derbyshire*. Derby.

CLAPHAM, A.R., TUTIN, T.G. & WARBURG, E.F. (1981). *Excursion Flora of the British Isles*, edn 3. Cambridge.

CLARK, S.C. (1974). *Catapodium rigidum* (L.) C.E. Hubbard. (Biological Flora of the British Isles no. 136). *J. Ecol.*, 62: 325–339.

CLEGG, L.M. & GRACE, J. (1974). The distribution of *Heracleum mantegazzianum* (Somm. & Levier) near Edinburgh. *Trans. Proc. bot. Soc. Edinb.*, 42: 223–229.

COKER, P.D. (1966). *Sibbaldia procumbens* L. (Biological Flora of the British Isles no. 104). *J. Ecol.*, 54: 823–831.

COKER, P.D. & COKER, A.M. (1973). *Phyllodoce caerulea* (L.) Bab. (Biological Flora of the British Isles no. 133). *J. Ecol.*, 61: 901–913.

CONOLLY, A.P. (1977). The distribution and history in the British Isles of some alien species of *Polygonum* and *Reynoutria*. *Watsonia*, 11: 291–311.

CONOLLY, A.P. & DAHL, E. (1970). *Maximum summer temperature in relation to the modern and quaternary distributions of certain arctic-montane species in the British Isles*, in WALKER, D. & WEST, R.G. eds. *Studies in the vegetational history of the British Isles*, pp. 159–223. Cambridge.

COPE, T.A. & STACE, C.A. (1978). The *Juncus bufonius* L. aggregate in western Europe. *Watsonia*, 12: 113–128.

CRAWFORD, R.M.M. (1976). *Mineral nutrition*, in HALL, M.A., ed. *Plant Structure, Function and Adaptation*, pp. 197–253. London.

CRAWFORD, R.M.M. & PALIN, M.A. (1981). Root respiration and temperature limits to the north-south distribution of four perennial maritime plants. *Flora*, 171: 338–354.

CRAWLEY, M.J. (1987). *What makes a community invasible?*, in GRAY, A.J., CRAWLEY, M.J. & EDWARDS, P.J., eds *Colonization, Succession and Stability*, pp. 429–453. Oxford.

CROSS, J.R. (1975). *Rhododendron ponticum* L. (Biological Flora of the British Isles no. 137). *J. Ecol.*, 63: 345–364.

DAVID, R.W. & KELCEY, J.G. (1985). *Carex muricata* L. aggregate (Biological Flora of the British Isles no. 159). *J. Ecol.*, 73: 1021–1039.

DAVY, A.J. (1980). *Deschampsia caespitosa* (L.) Beauv. (Biological Flora of the British Isles no. 149). *J. Ecol.*, 68: 1075–1096.

DAWSON, F.H. & WARMAN, E.A. (1987). *Crassula helmsii* (T.Kirk) Cockayne: is it an aggressive alien aquatic plant in Britain? *Biol. Conserv.*, 42: 247–272.

*DIXON, J.M. (1982). *Sesleria albicans* Kit. ex Schultes (Biological Flora of the British Isles no. 151). *J. Ecol.*, 70: 667–684.

DONALD, D.R. (1981). Dark-leaved willow *Salix nigricans* in Cambridgeshire. *Nature Cambs.*, no. 24: 50–52.

DONY, J.G. (1968). The B.S.B.I. in a changing Britain. *Proc. bot. Soc. Br. Isl.*, 7: 315–323

DREVER, J.C. & HUNTER, J.A.A. (1970). Giant Hogweed dermatitis. *Scott. med. J.*, 15: 315–319.

DUFFEY, E., MORRIS, M.G., SHEAIL, J., WARD, L.K., WELLS, D.A. & WELLS, T.C.E. (1974). *Grassland Ecology and Wildlife Management*. London.

EDEES, E.S. & NEWTON, A. (1988). *Brambles of the British Isles*. London.

ELKINGTON, T.T. (1963). *Gentiana verna* L. (Biological Flora of the British Isles no. 91). *J. Ecol.*, 51: 755–767.

ELKINGTON, T.T. (1964). *Myosotis alpestris* E.W. Schmidt (Biological Flora of the British Isles no. 96). *J. Ecol.*, 52: 709–722.

ELKINGTON, T.T. (1971). *Dryas octopetala* L. (Biological Flora of the British Isles no. 124). *J. Ecol.*, 59: 887–905.

ELKINGTON, T.T. & WOODELL, S.K.J. (1963). *Potentilla fruticosa* L. (Biological Flora of the British Isles no. 92). *J. Ecol.*, 51: 769–781.

FARMER, A.M. (1989). *Lobelia dortmanna* L.(Biological Flora of the British Isles no. 165). *J. Ecol.*, 77: 1161–1173.

FEARN, G.M. (1973). *Hippocrepis comosa* L. (Biological Flora of the British Isles no. 134). *J. Ecol.*, 61: 915–926.

FEARN, G.M. (1975). Variation of *Polygala amarella* Crantz in Britain. *Watsonia*, 10: 371–383.

FIRBANK, L.G. (1988). *Agrostemma githago* L. (*Lychnis githago* (L.) Scop.). (Biological Flora of the British Isles no. 165). *J. Ecol.*, 76: 1232–1246.

GARDINER, A.S. (1974). *A history of the taxonomy and distribution of the native oak species*, in MORRIS, M.G. & PERRING, F.H., eds. *The British Oak*, pp. 13–26. Faringdon.

GENT, G.M. (1989). *Wild Flowers*, in COLSTON, A. & PERRING, F., eds. *The Nature of Northamptonshire*, pp. 108–113. Buckingham.

GODWIN, H. (1975). *The History of the British Flora*, edn 2. Cambridge.

GOODMAN, P.J., BRAYBROOKS, E.M., LAMBERT, J.M. & MARCHANT, C.J. (1969). *Spartina* Schreb. (Biological Flora of the British Isles no. 116). *J. Ecol.*, 57: 285–313.

GORNALL, R.J. (1988). The coastal ecodeme of *Parnassia palustris* L. *Watsonia*, 17: 139–143.

GRAY, A.J. & SCOTT, R. (1977). *Puccinellia maritima* (Huds.) Parl. (Biological Flora of the British Isles no. 140). *J. Ecol.*, 65: 699–716.

GREENWOOD, E.F. (1974). Herbicide treatments on the Lancaster canal. *Nature Lancashire*, 4: 24–36.

GYNN, E.G. & RICHARDS, A.J. (1985). *Acaena novae-zelandiae* T. Kirk (Biological Flora of the British Isles no. 161). *J. Ecol.*, 73: 1055–1063.

HARLEY, R.M. (1975). *Mentha* L., in STACE, C.A., ed. *Hybridization and the Flora of the British Isles*, pp. 383–390. London.

HASLAM, S.M. (1972). *Phragmites communis* Trin. (Biological Flora of the British Isles no. 128). *J. Ecol.*, 60: 585–610.

HORRILL, A.D. (1972). *Melampyrum cristatum* L. (Biological Flora of the British Isles no. 125). *J. Ecol.*, 60: 235–244.

HOWARTH, S.E. & WILLIAMS, J.T. (1968). *Chrysanthemum leucanthemum* L. (Biological Flora of the British Isles no. 110). *J. Ecol.*, 56: 585–595.

HOWARTH, S.E. & WILLIAMS, J.T. (1972). *Chrysanthemum segetum* L. (Biological Flora of the British Isles no. 127). *J. Ecol.*, 60: 573–584.

HUISKES, A.H.L. (1979). *Ammophila arenaria* (L.) Link (Biological Flora of the British Isles no. 144). *J. Ecol.*, 67: 363–382.

HUTCHINGS, M.J. (1987). The population biology of the Early Spider Orchid, *Ophrys sphegodes* Mill. I. A demographic study from 1975 to 1984. *J. Ecol.*, 75: 711–727.

HUTCHINGS, M.J. (1989). *Population biology and conservation of Ophrys sphegodes*, in PRITCHARD, H.W., *Modern methods in orchid conservation*, pp. 101–115. Cambridge.

HUTCHINGS, C.S. & SEYMOUR, G.B. (1982). *Poa annua* L. (Biological Flora of the British Isles no. 153). *J. Ecol.*, 70: 887–901.

HUTCHINSON, T.C. (1966). The occurrence of living and sub-fossil remains of *Betula nana* L. in Upper Teesdale. *New Phytol.*, 65: 351–357.

HUTCHINSON, T.C. (1968). *Teucrium scorodonia* L. (Biological Flora of the British Isles no. 113). *J. Ecol.*, 56: 901–911.

*IDLE, E.T. (1986). *Evaluation at the local scale: a region in Scotland*, in USHER, M.B., ed. *Wildlife Conservation Evaluation*, pp. 181–198. London.

JEFFERS, J.N.R. (1972). *Conifers and nature conservation*, in NAPIER, E., ed. *Conifers in the British Isles*, pp. 59–73. London. [Some maps cover Great Britain only].

*JERMY, A.C., ARNOLD, H.R., FARRELL, L. & PERRING, F.H. eds. (1978). *Atlas of Ferns of the British Isles*. London.

JERMY, A.C., CHATER, A.O. & DAVID, R.W. (1982). *Sedges of the British Isles*. (B.S.B.I. Handbook no. 1, edn. 2). London.

JONES, D.A. & TURKINGTON, R. (1986). *Lotus corniculatus* L. (Biological Flora of the British Isles no. 163). *J. Ecol.*, 74: 1185–1212.

JONES, V. & RICHARDS, P.W. (1962). *Silene acaulis* (L.) Jacq. (Biological Flora of the British Isles no.83). *J. Ecol.*, 50: 475–487.

KAY, Q.O.N. (1971a). *Anthemis cotula* L. (Biological Flora of the British Isles no. 122). *J. Ecol.*, 59: 623–636.

KAY, Q.O.N. (1971b). *Anthemis arvensis* L. (Biological Flora of the British Isles no.123). *J. Ecol.*, 59: 637–648.

LEACH, S. & RICH, T. (1989). Scurvy-grasses take to the road. *B.S.B.I. News*, no. 52: 15–16. [Only roadside records mapped].

LESLIE, A.C. & WALTERS, S.M. (1983). The occurrence of *Lemna minuscula* in the British Isles. *Watsonia*, 14: 243–248.

MARSHALL, J.K. (1967). *Corynephorus canescens* (L.) Beauv. (Biological Flora of the British Isles no. 105) *J. Ecol.*, 55: 207–220.

MCNAUGHTON, I.H. & HARPER, J.L. (1964). *Papaver* L. (Biological Flora of the British Isles no. 99). *J. Ecol.*, 52: 767–793.

MILNE-REDHEAD, E. (1975). Black Poplar Survey. *B.S.B.I. News*, no. 9: 10–12.

MILTON, J.N.B. (1984). *Trifolium occidentale* D.E. Coombe (Atlantic Clover) new to North Devon. *Watsonia*, 15: 120–121.

MITCHELL, N.D. & RICHARDS, A.J. (1979). *Brassica oleracea* L. subsp. *oleracea* (Biological Flora of the British Isles no. 145). *J. Ecol.*, 67: 1087–1096.

MORRIS, M.G. (1966). *Ceuthorhynchus unguicularis* C.G. Thomson (Col., Curculionidae) new to the British Isles, from the Suffolk Breckland and the Burren, Co. Clare. *Entomologist's mon. Mag.*, 101: 279–286.

MORRIS, M.G. (1970). *Phyllodecta polaris* Schneider (Col., Chrysomelidae) new to the British Isles from Wester Ross and Inverness-shire, Scotland. *Entomologist's mon. Mag.*, 106: 48–53.

MORRIS, M.G. (1976). *Apion sicardi* Desbrochers, a species of weevil (Col., Apionidae) new to Britain. *Entomologist's mon. Mag.*, 111: 165–171.

MYERSCOUGH, P.J. (1980). *Epilobium angustifolium* L. (Biological Flora of the British Isles no. 148). *J. Ecol.*, 68: 1047–1074.

NATURAL ENVIRONMENT RESEARCH COUNCIL (1976). *Biological Surveillance*. (The N.E.R.C. Publications series B, no. 18). London.

NATURE CONSERVANCY COUNCIL (1977). *Nature conservation and agriculture*. London.

NAYLOR, R.E.L. (1972). *Alopecurus myosuroides* Huds. (Biological Flora of the British Isles no. 129). *J. Ecol.*, 60: 611–622.

NOBLE, J.C. (1982). *Carex arenaria* L. (Biological Flora of the British Isles no. 152). *J. Ecol.*, 70: 867–886.

OCKENDON, D.J. & WALTERS, S.M. (1970). Studies in *Potentilla anserina* L. *Watsonia*, 8: 135–144.

PACKHAM, J.R. (1978). *Oxalis acetosella* L. (Biological Flora of the British Isles no. 141). *J. Ecol.*, 66: 669–693.

PACKHAM, J.R. (1983). *Lamiastrum galeobdolon* (L.) Ehrend. & Polatschek (Biological Flora of the British Isles no. 155). *J. Ecol.*, 71: 975–997.

PALIN, M.A. (1988). *Ligusticum scoticum* L. (*Haloscias scoticum* (L.) Fr.). (Biological Flora of the British Isles no. 164). *J. Ecol.*, 76: 889–902.

PALMER, J.H. & SAGAR, G.R. (1963). *Agropyron repens* (L.) Beauv. (Biological Flora of the British Isles no. 93). *J. Ecol.*, 51: 783–794.

PARNELL, J.A.N. (1985). *Jasione montana* L. (Biological Flora of the British Isles no. 157). *J. Ecol.*, 73: 341–358.

PERRING, F. [H.] (1973). Mistletoe, in GREEN, P.S., ed. *Plants: wild and cultivated*, pp. 139–145. Hampton.

PERRING, F.H. (1974). Changes in our native vascular plant flora, in HAWKSWORTH, D.L., ed. *The Changing Flora and Fauna of Britain* (Systematics Association Special Volume no. 6), pp. 7–25. London.

PERRING, F.H. (1976). Records for leisure and profit. *New Scient.*, 72: 725–727.

PERRING, F.H. & SELL, P.D. eds. (1968). *Critical Supplement to the Atlas of the British Flora*. London.

PERRING, F.H. & WALTERS, S.M. (1971). Conserving rare plants in Britain. *Nature*, 229: 375–377.

PETERKEN, G.F. (1975). Holly Survey. *Watsonia*, 10: 297–299. [Only records in hedges mapped].

PETERKEN, G.F. & LLOYD, P.S. (1968). *Ilex aquifolium* L. (Biological Flora of the British Isles no. 108). *J. Ecol.*, 55: 841–858.

PIGOTT, C.D. (1968). *Cirsium acaulon* (L.) Scop. (Biological Flora of the British Isles no. 111). *J. Ecol.*, 56: 597–612.

PIGOTT, C.D. (1974). *The response of plants to climate and climatic change*, in PERRING, F. [H.], ed. *The flora of a changing Britain*, pp. 32–44. Faringdon.

PIGOTT, C.D. (1978). *Climate and vegetation*, in CLAPHAM, A.R., ed. *Upper Teesdale, the area and its natural history*, pp. 102–121. London.

PIGOTT, C.D. (1981). *The status, ecology and conservation of* Tilia platyphyllos *in Britain*, in SYNGE, H., ed. *The Biological Aspects of Rare Plant Conservation*, pp. 305–317. Chichester.

POLLARD, E., HOOPER, M.D. & MOORE, N.W. (1974). *Hedges*. London. [Only records in hedges mapped].

PRESTON, C.D. (1989). The spread of *Epilobium ciliatum* Raf. in the British Isles. *Watsonia*, 17: 279–288.

PRESTON, C.D. & SELL, P.D. (1989). The Aizoaceae naturalized in the British Isles. *Watsonia*, 17: 217–245.

PRESTON, C.D. & WHITEHOUSE, H.L.K. (1986). The habitat of *Lythrum hyssopifolia* L. in Cambridgeshire, its only surviving English locality. *Biol. Conserv.*, 35: 41–62.

PRIME, C.T. (1973). *Arums – Lords-and-Ladies*, in GREEN, P.S., ed. *Plants: wild and cultivated*, pp. 167–171. Hampton.

PROCTOR, M.C.F. (1969). *Flora and vegetation*, in BARLOW, F., ed. *Exeter and its region*, pp. 97–116. Exeter.

RANDALL, R.E. (1977). The past and present status and distribution of Sea Pea, *Lathyrus japonicus* Willd., in the British Isles. *Watsonia*, 11: 247–251.

RANWELL, D.S. (1972). *Ecology of salt marshes and sand dunes*. London.

RATCLIFFE, D. & CONOLLY, A.P. (1972). *Flora and vegetation*, in PYE, N., ed. *Leicester and its region*, pp. 133–159. Leicester.

RATCLIFFE, D.A. & OSWALD, P.H. eds. (1988). *The Flow Country*. Peterborough.

RAVEN, P.H. (1963). *Circaea* in the British Isles. *Watsonia*, 5: 262–272.

RICH, T.C.G., KITCHEN, M.A.R. & KITCHEN, C. (1989). *Thlaspi perfoliatum* L. (Cruciferae) in the British Isles : distribution. *Watsonia*, 17: 401–407.

RICHARDS, A.J. (1973). An upland race of *Potentilla erecta* (L.) Räusch. in the British Isles. *Watsonia*, 9: 301–317.

*RICHARDS, [A.]J. (1989). *Primulas of the British Isles*. Aylesbury.

RICHARDS, P.W. & EVANS, G.B. (1972). *Hymenophyllum* (Biological Flora of the British Isles no. 126). *J. Ecol.*, 60: 245–268.

RUTHERFORD, A. (1976). B.S.B.I. Irish Ivy Survey. *B.S.B.I. News* no. 13: 17–19.

RUTHERFORD, A. (1979). The BSBI Irish Ivy Survey. *B.S.B.I. News* no. 22: 8–9

SAGAR, G.R. & HARPER, J.L. (1964). *Plantago major* L., *P. media* L. and *P. lanceolata* L. (Biological Flora of the British Isles no. 95). *J. Ecol.*, 52: 189–221.

SARGENT, C., MOUNTFORD, O. & GREEN, D. (1986). The distribution of *Poa angustifolia* L. in Britain. *Watsonia*, 16: 31–36.

SCANNELL, M.J.P. (1973). The Ireland we live in. The Plants. *Capuchin Annual*, 1973: 146–167.

SCOTT, D.W. ed. (1975). *BSBI mapping scheme: distribution maps of 50 common plants*. [Abbots Ripton].

SCOTT, G.A.M. (1963a). *Mertensia maritima* (L.) S.F. Gray (Biological Flora of the British Isles no. 89). *J. Ecol.*, 51: 733–742.

SCOTT, G.A.M. (1963b). *Glaucium flavum* Crantz. (Biological Flora of the British Isles no. 90). *J. Ecol.*, 51: 743–754.

SCOTT, G.A.M. & RANDALL, R.E. (1976). *Crambe maritima* L. (Biological Flora of the British Isles no. 139). *J. Ecol.*, 64: 1077–1091.

SCOTT, N.E. (1985). The updated distribution of maritime species on British roadsides. *Watsonia*, 15: 381–386. [Only roadside records mapped].

SCOTT, N.E. & DAVISON, A.W. (1982). De-icing salt and the invasion of road verges by maritime plants. *Watsonia*, 14: 41–52. [Only roadside records mapped].

SCURFIELD, G. (1962). *Cardaria draba* (L.) Desv. (Biological Flora of the British Isles no. 84). *J. Ecol.*, 50: 489–499.

SELL, P.D. (1986). The genus *Cicerbita* Wallr. in the British Isles. *Watsonia*, 16: 121–129.

SHIRREFFS, D.A. (1985). *Anemone nemorosa* L. (Biological Flora of the British Isles no. 158). *J. Ecol.*, 73: 1005–1020.

SIMPSON, D.A. (1984). A short history of the introduction and spread of *Elodea* Michx in the British Isles. *Watsonia*, 15: 1–9. [Only first vice-county records mapped].

SIMPSON, D.A. (1986). Taxonomy of *Elodea* Michx in the British Isles. *Watsonia*, 16: 1–14.

SMITH, C.J. (1980). *Ecology of the English Chalk*. London. [Maps cover Great Britain only].

SMITH, G.L. (1963a). Studies in *Potentilla* L. I. Embryological investigations into the mechanism of agamospermy in British *P. tabernaemontani* Aschers. *New Phytol.*, 62: 264–282.

SMITH, G.L. (1963b). Studies in *Potentilla* L. II. Cytological aspects of apomixis in *P. crantzii* (Cr.) Beck ex Fritsch. *New Phytol.*, 62: 283–300.

SMITH, G.L. (1971). Studies in *Potentilla* L. III. Variation in British *P. tabernaemontani* Aschers. and *P. crantzii* (Cr.) Beck ex Fritsch. *New Phytol.*, 70: 607–618.

SMITH, P. [M.] (1968). The *Bromus mollis* aggregate in Britain. *Watsonia*, 6: 327–344.

SMITH, P.M. (1973). Observations on some critical bromegrasses. *Watsonia*, 9: 319–332.

SMITH, U.K. (1979). *Senecio integrifolius* (L.) Clairv. (Biological Flora of the British Isles no. 147). *J. Ecol.*, 67: 1109–1124.

SNOGERUP, B. (1982). *Odontites litoralis* Fries subsp. *litoralis* in the British Isles. *Watsonia*, 14: 35–39.

SOBEY, D.G. (1981). *Stellaria media* (L.) Vill. (Biological Flora of the British Isles no. 150). *J. Ecol.*, 69: 311–335.

SPARLING, J.H. (1968). *Schoenus nigricans* L. (Biological Flora of the British Isles no. 115). *J. Ecol.*, 56: 883–899.

STACE, C.A. & COTTON, R. (1976). Nomenclature, comparison and distribution of *Vulpia membranacea* (L.) Dumort. and *V. fasciculata* (Forskål) Samp. *Watsonia*, 11: 117–123.

SUMMERFIELD, R.J. (1974). *Narthecium ossifragum* (L.) Huds. (Biological Flora of the British Isles no. 135). *J. Ecol.*, 62: 325–339.

TASCHEREAU, P.M. (1985). Taxonomy of *Atriplex* species indigenous to the British Isles. *Watsonia*, 15: 183–209.

TASCHEREAU, P.M. (1989). Taxonomy, morphology and distribution of *Atriplex* hybrids in the British Isles. *Watsonia*, 17: 247–264.

TAYLOR, K. (1971). *Rubus chamaemorus* L. (Biological Flora of the British Isles no. 121). *J. Ecol.*, 59: 293–306.

TAYLOR, K. & MARKHAM, B. (1978). *Ranunculus ficaria* L. (Biological Flora of the British Isles no. 142). *J. Ecol.*, 66: 1011–1031.

TEAGLE, W.G. (1978). *The Endless Village*. Shrewsbury.

TIMSON, J. (1966). *Polygonum hydropiper* L. (Biological Flora of the British Isles no. 103). *J. Ecol.*, 54: 815–821.

TITTENSOR, R.M. (1980). Ecological history of yew *Taxus baccata* L. in southern England. *Biol. Conserv.*, 17: 243–265.

TRIST, P.J.O. (1986). The distribution, ecology, history and status of *Gastridium ventricosum* (Gouan) Schinz & Thell. in the British Isles. *Watsonia*, 16: 43–54.

TURKINGTON, R. & AARSSEN, L.W. (1983). *Hypochoeris radicata* L. (Biological Flora of the British Isles no. 156). *J. Ecol.*, 71: 999–1022.

USHER, M.B. (1973). *Biological Management and Conservation*. London.

*USHER, M.B. (1986a). *Wildlife conservation evaluation: attributes, criteria and values*, in USHER, M.B., ed. *Wildlife Conservation Evaluation*, pp. 3–44. London.

USHER, M.B. (1986b). Invasibility and wildlife conservation: invasive species on nature reserves. *Phil. Trans. R. Soc. Lond.*, ser. B, 314: 695–710.

WALTERS, S.M. (1978). *The Eastern England Rare Plant Project in the University Botanic Garden, Cambridge*, in SYNGE, H. & TOWNSEND, H., eds. *Survival or Extinction*, pp. 37–46. Kew.

WARD, L.K. (1972). *The status of* Hippophaë *as part of the British flora*, in RANWELL, D.S., ed. *The management of sea buckthorn* Hippophaë rhamnoides L. *on selected sites in Great Britain*. Norwich.

WARD, L.K. (1973). The conservation of juniper. I. Present status of juniper in southern England. *J. appl. Ecol.*, 10: 165–188.

WARD, L.K. (1977). The conservation of juniper: the associated fauna with special reference to southern England. *J. appl. Ecol.*, 14: 81–120.

WATKINSON, A.R. (1978). *Vulpia fasciculata* (Forskål) Samp. (Biological Flora of the British Isles no. 143). *J. Ecol.*, 66: 1033–1049.

WEBSTER, S.D. (1988). *Ranunculus penicillatus* (Dumort.) Bab. in Great Britain and Ireland. *Watsonia*, 17: 1–22.

WEGMÜLLER, S. (1971). A cytotaxonomic study of *Lamiastrum galeobdolon* (L.) Ehrend. & Polatschek in Britain. *Watsonia*, 8: 277–288.

WEIN, R.W. (1973). *Eriophorum vaginatum* L. (Biological Flora of the British Isles no. 132). *J. Ecol.*, 61: 601–615.

WELCH, D. (1966). *Juncus squarrosus* L. (Biological Flora of the British Isles no. 101). *J. Ecol.*, 54: 535–548.

WELLS, D.A. & OSWALD, P.H. (1988). *The conservation of meadows and pastures*. Peterborough. [Maps cover Great Britain only].

WELLS, T.C.E. (1967). *Dianthus armeria* L. at Woodwalton Fen, Hunts. *Proc. bot. Soc. Br. Isl.*, 6: 337–342.

WELLS, T.C.E. (1976). *Hypochoeris maculata* L. (Biological Flora of the British Isles no. 138). *J. Ecol.*, 64: 757–774.

WELLS, T.C.E. (1982). Some wild flowers of the Trust's Reserves in Huntingdonshire. *Watsonia*, 17: 1–22.

WELLS, T.C.E. & BARLING, D.M. (1971). *Pulsatilla vulgaris* Mill. (Biological Flora of the British Isles no. 120). *J. Ecol.*, 59: 275–292.

WELLS, T.C.E. & SHEAIL, J. (1988). *The effects of agricultural change on the wildlife interest of lowland grasslands*, in PARK, J.R., ed. *Environmental Management in Agriculture European Perspectives*, pp. 186–210. London & New York.

WILKINSON, M.J. & STACE, C.A. (1989). The taxonomic relationship and typification of *Festuca brevipila* Tracey and *F. lemanii* Bastard (Poaceae). *Watsonia*, 17: 289–299.

WILLIAMS, J.T. (1963). *Chenopodium album* L. (Biological Flora of the British Isles no. 87). *J. Ecol.*, 51: 711–725.

WILLIAMS, J.T. (1969). *Chenopodium rubrum* L. (Biological Flora of the British Isles no. 117). *J. Ecol.*, 57: 831–841.

WILLIAMSON, J.A. & FORBES, J.C. (1982). Giant Hogweed (*Heracleum mantegazzianum*): its spread and control with glyphosate in amenity areas. *Proc. Br. Crop Prot. Conf., Weeds, Brighton, 1982*, 3: 967–972.

WOODELL, S.R.J. & KOOTIN-SANWU, M. (1971). Intraspecific variation in *Caltha palustris*. *New Phytol.*, 70: 173–186.

* References marked with an asterisk map species with two symbols, a dot and a circle, but do not explain the meaning of the symbols in the text. The dot denotes records made from 1950 onwards, the circle earlier records. In the case of Richards (1989), this only applies to the maps of *Primula elatior*, *P. farinosa* and *P. scotica*; all records of *P. veris* and *P. vulgaris* are shown by dots.

Index

The roman numerals refer to the Introduction; the letters *a* and *b* which follow them refer to the left and right hand column of the page respectively. The arabic numerals give the page on which the map is to be found.

437

e–counties

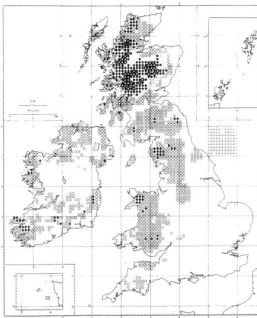

Altitude
○ Squares with some land over 1,000 ft
● Squares with some land over 2,500 ft
◉ Squares with some land over 3,500 ft

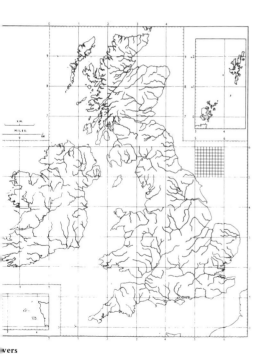

vers

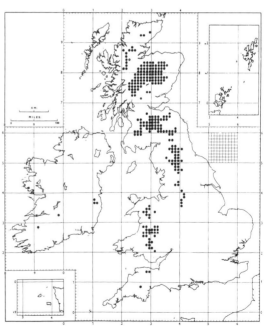

Altitude
● Squares with no land below 500 ft

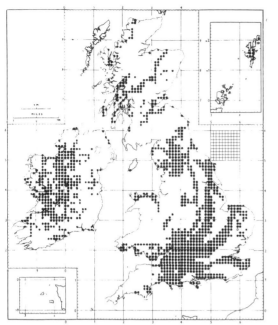

Geology – Chalks & Limestones
• Squares with outcrop greater than 5% of area

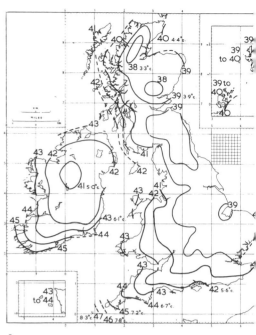

January mean
Average means of daily mean temperature
1901–1930 January

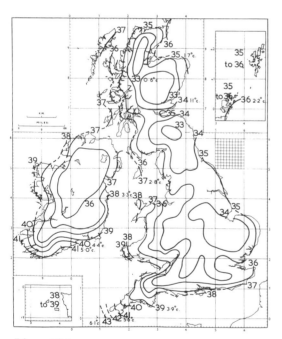

February Minimum
Average means of daily minimum temperature
1901–1930 February

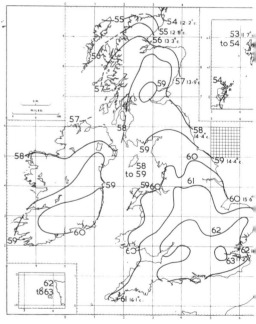

July Mean
Average means of daily mean temperature
1901–1930 July

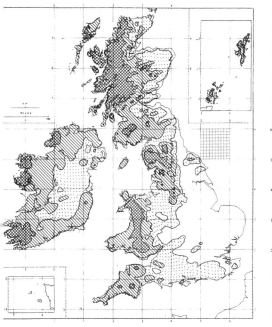

ainfall
verage annual rainfall 1901–1930
☐ under 30 inches ▨ 40–60 inches
▦ 30–40 inches ▨ over 60 inches

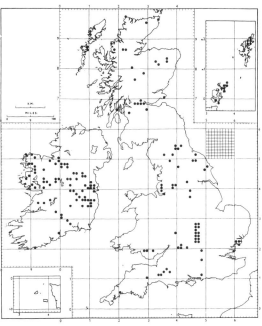

Under-recorded squares
● Squares with relatively fewer species recorded

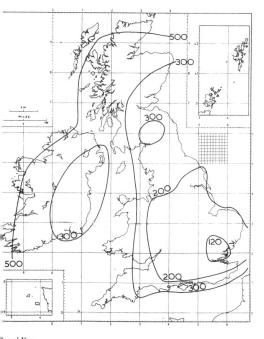

Humidity
Meyer's Precipitation – Saturation Deficit Quotient

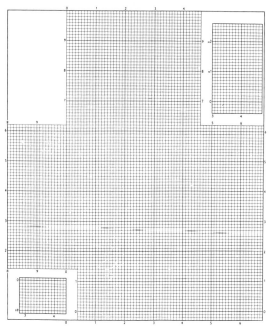

Grid Squares

B.S.B.I. PUBLICATIONS
24 GLAPTHORN ROAD
OUNDLE
PETERBOROUGH PE8 4JQ
0832 - 273388